Sterilized by the State

Eugenics, Race, and the Population Scare in Twentieth-Century North America

This book is the first comprehensive analysis of eugenics in North America focused on the second half of the twentieth century. Based on new research, Randall Hansen and Desmond King show why eugenic sterilization policies persisted after the 1940s in the United States and Canada. Through extensive archival research, King and Hansen show how both superintendents at homes for the "feebleminded" and pro-sterilization advocates repositioned themselves after 1945 to avoid the taint of Nazi eugenics. Drawing on interviews with victims of sterilization and primary documents, this book traces the post-1940s development of eugenic policy and shows that both eugenic arguments and committed eugenicists informed population, welfare, and birth control policy in postwar America. In providing revisionist histories of the choice movement, the anti–population growth movement, and the Great Society programs, this book contributes to public policy and political and intellectual history.

Randall Hansen is Director of the Centre for European Russian and Eurasian Studies and Canada Research Chair in political science at the University of Toronto. Hansen has served as Fellow and tutor in politics at Merton College, Oxford, and has held a research fellowship at Christ Church, Oxford. He is author of *Citizenship and Immigration in Postwar Britain*, *Fire and Fury: The Allied Bombing of Germany 1942–44* (shortlisted for the Governor General's Literary Award), and *Disobeying Hitler: German Resistance after July 20, 1944*.

Desmond King holds the Andrew W. Mellon Chair of American Government at the University of Oxford and is a Fellow of Nuffield College, Oxford. His research on race, welfare, and liberalism includes *In the Name of Liberalism: Illiberal Social Policy in the United States and Britain*; *Making Americans: Immigration, Race, and the Origins of the Diverse Democracy*; *The Liberty of Strangers: Making the American Nation*; and *Still a House Divided: Race and Politics in Obama's America* (with Rogers M. Smith), selected as a Choice Outstanding Academic Title. He is a Fellow of the British Academy.

Sterilized by the State

Eugenics, Race, and the Population Scare
in Twentieth-Century North America

RANDALL HANSEN
University of Toronto

DESMOND KING
Nuffield College, University of Oxford

CAMBRIDGE
UNIVERSITY PRESS

CAMBRIDGE
UNIVERSITY PRESS

32 Avenue of the Americas, New York, NY 10013-2473, USA

Cambridge University Press is part of the University of Cambridge.

It furthers the University's mission by disseminating knowledge in the pursuit of education, learning, and research at the highest international levels of excellence.

www.cambridge.org
Information on this title: www.cambridge.org/9781107659704

First published 2013

Printed in the United States of America

A catalog record for this publication is available from the British Library.

Library of Congress Cataloging in Publication Data
Hansen, Randall.
Sterilized by the state : eugenics, race, and the population scare in twentieth-century North America / Randall Hansen, University of Toronto, Desmond King, Nuffield College, University of Oxford.
    pages   cm
ISBN 978-1-107-03292-7 – ISBN 978-1-107-65970-4 (paperback)
1. Eugenics – United States – History – 20th century.   2. Eugenics – Canada – History – 20th century.   3. Sterilization (Birth control) – United States – History – 20th century.   4. Sterilization (Birth control) – Canada – History – 20th century.
I. King, Desmond S.   II. Title.
HQ755.5.U5H36   2013
363.9'20973–dc23   2013015860

ISBN 978-1-107-03292-7 Hardback
ISBN 978-1-107-65970-4 Paperback

Contents

Acknowledgments

We have benefited from much support over several years on both sides of the Atlantic during the course of this project. We are grateful to the Wellcome Foundation and the Social Sciences and Humanities Research Council of Canada for major research grants that made research for this book possible and to Nuffield College, University of Oxford, for supporting research undertaken by the second author as part of the project.

We owe an immense debt to a group of colleagues who participated in a workshop in Toronto in December 2011 and offered trenchant, constructive commentaries on a draft of the manuscript. We are pleased to thank Ian Dowbiggin, Gary Gerstle, Mark Largent, Véronique Mottier, Alexandra Minna Stern, and Annette Timm. These colleagues' comments and suggestions for revision both saved us from unintended errors and helped us think through key aspects of the manuscript. We also thank Professor Sharon M. Leon for responding to queries about sterilization initiatives in Ohio.

At Cambridge University Press, Lewis Bateman has been an outstanding editor who has always provided encouragement and sage advice about various aspects of the manuscript. We are also grateful to the Press's two anonymous readers for their valuable comments on the draft manuscript.

We are immensely grateful to the victims of sterilization for agreeing to meet and allowing us to record their deeply personal stories. For their participation, our sincere thanks belong to Velma Hayes, Dennis Heath, Judy Lytton, Ted McNeil, Ruth Morris, Leilani Muir, Ken Nelson, Ken Newman, and Laverne Throckhorn. We also thank those individuals interviewed whose experience in a professional capacity with sterilization and victims of coerced sterilization provided great insight into our research: Dr. Ernest Brown, Gordon Bullivant, Jon Faulds, Dr. Warren Hern, Dr. Robert Lampard, Dr. Everett W. Lovrien, Bill Lynch, and John Murphy. Special thanks also go out to Bruce Uditsky, CEO of the Alberta Association of Community Living in Edmonton and to William

West, Adult and Family Case Coordinator, the Association for Retarded Citizens, Portland, Oregon, for going to great lengths to help us secure interviews with sterilization victims in Alberta and Oregon. Thanks also go to Julie Sullivan, then of the *Oregonian*, for agreeing to an interview and for putting us in touch with Velma Hayes.

For research assistance, we are grateful to Joseph Hawker, Paul Lawrie, Isabella Matauschek, and Guy Tourlamain.

PART A

I

Coerced Sterilization
Outcomes, Theories, Methods

For almost a century, politicians, lawmakers, doctors, bureaucrats, scientists, and citizens embarked on an ambitious social engineering project: coerced sterilization.

In North America, it began in the 1880s, with one prison doctor's desire to prevent masturbation among his inmates. In the following decades, hundreds of thousands of people – above all, those deemed to be "feebleminded" and therefore likely to reproduce that trait – were sterilized in dozens of states, provinces, and countries around the world. The United States forcibly sterilized at least 60,000 feebleminded[1] patients from the 1910s to the 1970s; Nazi Germany (the most widely known instance) sterilized approximately 360,000 such individuals in the 1930s; Canada eugenically sterilized approximately 3,000 people (more than 90 percent of such sterilizations occurred in the province of Alberta); and the countries of Scandinavia coercively sterilized 35,500, with tens of thousands more sterilized under quasi-voluntary conditions thereafter.[2] In the United States,

[1] For obvious reasons, there is much debate on how to refer to those with learning disabilities. The correct term today is "people with developmental disabilities." Rather than using this description anachronistically, we have opted to use the term "feebleminded" when we refer to such individuals in historical context, as that is how people with mental disabilities were understood and constructed in the pre–Second World War period. This construction, in turn, had an influence on how state agencies treated them. When we are speaking generally or discussing a contemporary context, we use the term "people with developmental disabilities." The distinction between historical and contemporary usage is not always clear-cut, but we strive to maintain it and, more important, the correct use of the terms throughout the manuscript.

[2] See Mattias Tydén, "The Scandinavian States: Reformed Eugenics Applied," in *The Oxford Handbook of the History of Eugenics*, ed. Alison Bashford and Philippa Levine (Oxford: Oxford University Press, 2010), figure from page 370. For a comparative perspective, see Alberto Spektorowski, "The Eugenic Temptation in Socialism: Sweden, Germany and the Soviet Union," *Comparative Studies in Society and History* 46, no. 1 (2004): 84–106.

the majority of coerced sterilizations occurred within state institutions: chiefly homes for the feebleminded but also in state hospitals and prisons.

Eugenics provided the main justification for sterilizing the mentally handicapped in the first half of the twentieth century. To oversimplify somewhat, eugenics is the doctrine that states that the fostering of good genes and the elimination of bad ones will serve the cause of national "racial" health by permitting better breeding of a nation's "stock" of people. Early social science research purported to show that there were large numbers of feebleminded and, on top of that, that they were producing children at a disproportionately high rate. Therefore, doctors, mental health superintendents, psychologists, and other professionals concluded that the inevitable result would be a gradual decline in overall national intelligence. Prevailing theories of heredity and their influential advocates maintained that inferior traits were necessarily transmitted without modification from generation to generation. As a eugenic report published in 1918 on the "Care of the Insane" in California put it: "the whole stream of human life is being constantly polluted by the admixture of the tainted blood of the extremely defective."[3]

The story of the people who arranged and carried out sterilizations in North America, how and why they did it, and the story of those who were sterilized are the subject of this book. Our aim is twofold: (1) to understand why these eugenic sterilizations occurred, and (2) why they continued to occur after 1945. In answering the latter question, we seek an understanding of why, despite the revelations of the German National Socialists' mass sterilization program and their mass murder of the mentally handicapped, sterilization in North America continued and, in some states, increased in the following decades.

Understanding coerced sterilization requires sifting the copious sets of primary documents on the topic. The book relies on archives from about twenty collections in four countries. In addition, there is a rich historical literature on eugenics and sterilization. Dozens of scholars have written meticulously researched and carefully argued books on eugenic ideas and eugenic policy in the United Kingdom,[4] the United States,[5] and

---

[3] Report by John R. Haynes, an influential eugenicist advocate and administrator on the California State Board of Charities Corrections, cited in Alexandra Minna Stern, *Eugenic Nation: Faults and Frontiers of Better Breeding in Modern America* (Berkeley: University of California Press, 2005), 103 and 253.

[4] G. R. Searle, *Eugenics and Politics in Britain, 1900–1914* (Leyden: Noordhoff International Publishing, 1976); David Barker, "The Biology of Stupidity: Genetics, Eugenics and Mental Deficiency in the Inter-War Years," *British Journal of the History of Science* 22, no. 3 (1989): 347–75; Richard A. Soloway, *Demography and Degeneration: Eugenics and the Declining Birthrate in Twentieth Century Britain* (Chapel Hill: University of North Carolina Press, 1990); Mathew Thomson, *The Problem of Mental Deficiency: Eugenics, Democracy, and Social Policy in Britain, c. 1870–1959* (Oxford: Oxford University Press, 1998).

[5] Mark H. Haller, *Eugenics: Hereditarian Attitudes in American Thought.* (New Brunswick, NJ: Rutgers University Press, 1963); Philip R. Reilly, *The Surgical Solution: A History of Involuntary*

Germany,[6] as well as comparative studies of these and other countries.[7] We cannot and we do not intend to replicate the exhaustive detail of existing case studies, and, in part for this reason, we devote as much space to the understudied postwar period as we do to the prewar one.[8] Our aim in the prewar chapters is rather to use this literature to draw out further comparisons (including in particular the neglected and important case of Canada)[9] and to reflect on the general factors determining whether states adopted coerced sterilization policies.

The following portions of the book can be read in several ways: as organizational histories of some of the chief pro-sterilization lobby groups, as oral histories of the victims of coerced sterilization, and as what might be called "alternate histories." In the last, we offer in particular revisionist histories of two great social movements and one transformative social program: the choice movement, the anti-population growth movement, and the Great Society programs. All three had important eugenic undercurrents, and many of the

Sterilization in the United States (Baltimore: Johns Hopkins University Press, 1991); Elazar Barkan, *The Retreat of Scientific Racism: Changing Concepts of Race in Britain and the United States between the Two World Wars* (Cambridge: Cambridge University Press, 1992); Ian R. Dowbiggin, *Keeping America Sane: Psychiatry and Eugenics in the United States and Canada, 1880–1940* (Ithaca: Cornell University Press, 1997); Wendy Kline, *Building a Better Race: Gender, Sexuality, and Eugenics from the Turn of the Century to the Baby Boom* (Berkeley: University of California Press, 2001); Stern, *Eugenic Nation*; Mark A. Largent, *Breeding Contempt: The History of Coerced Sterilization in the United States* (New Brunswick: Rutgers University Press, 2008); Paul A. Lombardo, ed., *A Century of Eugenics in America: From the Indiana Experiment to the Human Genome Era* (Bloomington: Indiana University Press, 2011).

[6] Gisela Bock, *Zwangssterilisation im Nationalsozialismus* (Opladen: Westdeutscher Verlag, 1986); Peter Weingart, "German Eugenics between Science and Politics," *Osiris* 2nd series, vol. 5 (1989): 260–82; Sheila F. Weiss, "Race and Class in Fritz Lenz's Eugenics," *Medizinhistorisches Journal: Internationale Vierteljahresschrift für Wissenschaftsgeschichte* 27, nos. 1–2 (1992): 5–25; Paul Weindling, *Health, Race and German Politics between National Unification and Nazism* (Cambridge: Cambridge University Press, 1989); Weindling, "The Survival of Eugenics in 20th Century Germany," *American Journal of Human Genetics* 52, no. 3 (1993): 643–9; Michael Burleigh, *Death and Deliverance: Euthanasia in Germany c. 1900–1945* (Cambridge: Cambridge University Press, 1994); Benno Müller-Hill, *Murderous Science: Elimination by Scientific Selection of Jews, Gypsies, and Others, Germany, 1933–1945* (Oxford: Oxford University Press, 1988).

[7] Daniel J. Kevles, *In the Name of Eugenics: Genetics and the Uses of Human Heredity* (Cambridge: Harvard University Press, 1995); Dowbiggin, *Keeping America Sane*; Diane B. Paul, "Eugenics and the Left," in *Politics of Heredity: Essays on Eugenics, Biometrics, and the Nature-Nature Debate* (Albany: State University of New York Press, 1998), chapter 2; Nancy Leys Stepan, *The Hour of Eugenics* (Ithaca, NY: Cornell University Press, 1991); and Ian Dowbiggin, *The Sterilization Movement and Global Fertility in the Twentieth Century* (Oxford: Oxford University Press, 2008).

[8] See Johanna Schoen, *Choice and Coercion: Birth Control, Sterilization and Abortion in Public Health and Welfare* (Chapel Hill: University of North Carolina Press, 2005); Matthew Connelly, *Fatal Misconception: The Struggle to Control World Population* (Cambridge: Harvard University Press, 2008).

[9] For major exceptions, see Dowbiggin, *Keeping America Sane*; Angus McLaren, *Our Own Master Race: Eugenics in Canada, 1885–1945* (Toronto: McClelland & Stewart, 1990); Carolyn Strange and Jennifer A. Stephen, "Eugenics in Canada: A Checkered History 1850s–1990s," in *Oxford Handbook of the History of Eugenics*, ed. Bashford and Levine, 523–38.

individuals involved in all three had previously enthusiastically supported eugenics and coerced eugenic sterilization.

TRANSLATING IDEAS INTO POLICY: COERCED EUGENIC STERILIZATION

Political scientists have devoted sparse attention to the study of eugenics and coerced sterilization. This oversight is curious. Sterilization policy across the United States was a significant plank of public policy, one that directly affected tens of thousands of people and indirectly affected hundreds of thousands more. The study of policy is basic to the discipline, and work on sterilization should be mainstream, not marginal, in policy studies. Furthermore, a core interest of political scientists is power, and the sterilization of the mentally handicapped was a case par excellence of the exercise and abuse of power. Over the past century, the mentally ill and developmentally disabled have been stigmatized, isolated, and institutionalized; their bodies have been poked, prodded, restrained, electrically shocked, beaten, and mutilated. They have been robbed of their dignity, their reproductive power, their citizenship, and, at times, their lives. The mistreatment of these individuals is one of the great human rights abuses of the twentieth century, and political scientists have all but ignored it.

Two factors likely account for political scientists' neglect of coerced sterilization. First, coerced sterilization is a quintessential example of illiberal social policy, and scholars have paid little attention to illiberal policy as public policy. Although recognized as a category,[10] there has been a marked hesitancy on the part of scholars to apply established social science techniques to the study of policies such as genocide, mass expulsion, slavery, sterilization, or even the more mundane illiberal policies such as conditional welfare benefits, workfare, and other punitive social policies. Scholars have accorded extensive attention to anti-liberal political parties and, to some degree, to the movements supporting them, but, even here, they are viewed as distinct spheres of inquiry rather than as mainstream social science.[11]

Second, coerced sterilization continues to be viewed as prewar history. It is in fact also part of postwar politics. The practice of eugenic sterilization persisted in major liberal democracies until the 1970s, and, in some cases, it is still with us today.[12] Eugenic sterilization did not end with the Second World War. It also did not end when the world began to learn that doctors in National Socialist Germany had sterilized hundreds of thousands of mental patients, murdered

---

[10] Desmond King, *In the Name of Liberalism: Illiberal Social Policy in the United States and Britain* (Oxford: Oxford University Press, 1999).

[11] Hence there is the hesitation to view right-wing, anti-immigrant, Islamophobic movements in Europe as "new social movements."

[12] Alison Bashford, "Epilogue: Where Did Eugenics Go?" in *Oxford Handbook to the History of Eugenics*, ed. Bashford and Levine, 539–58.

80,000 more, and used them to conduct experiments in the very gassing techniques used to liquidate the Jews of Europe. On the contrary, the coerced sterilization of the mentally ill in North America continued and, in other states, increased. It did not end in most states until the 1970s. In some places, it did not end at all.[13]

The question of why policies adopted in the 1910s, 1920s, and 1930s could last so long and at such human cost does not admit of easy answers, and it is dealt with in the second part of the book. The explanation rests on the examination of archival records and through conversations with dozens of individuals who were sterilized unknowingly, against their will, and sometimes after an impossible choice between a barren freedom and a fertile imprisonment. We have gathered, corroborated, and weaved into the narrative the personal stories of some of those who were forcibly sterilized.

In offering an account of why coerced sterilizations continued after the war – and therefore after the Holocaust and the human rights revolution – we focus on the interaction among institutions, ideas, and agents. The decision to sterilize was made by officials (Eugenics Boards populated with doctors, asylum superintendents, and social workers) holding power over the mentally ill and developmentally disabled. The key figure in our account is the superintendent, and the key institution is the home for the feebleminded.[14] The vast majority of official sterilizations occurred within institutions (North Carolina is the exception), and the rules and ethos of the mental health institution are central to understanding how sterilization could continue, especially in the postwar years.[15]

In the institutions, the superintendent had near-complete legal power over the patients, and he (it was almost invariably a "he") was the central figure in deciding (a) whether the institution would sterilize its patients at all and (b) whether an individual's sterilization would be authorized. Institutional practice

---

[13] In the late 1990s, a semi-voluntary sterilization program in Peru resulted in hundreds of thousands of coercive sterilizations and eighteen documented deaths. When the law was passed, progressive forces congratulated President Alberto Fujimori for taking on the Roman Catholic Church. Brita Schmidt, "Forced Sterilization in Peru," *Committee on Women, Population, and the Environment*, July 12, 2006. Available at: http://www.cwpe.org/resources/popcontrol/forcedster-peru. The statistic on deaths comes from "Peru: Forced Sterilization Cases Reopened; 300,000 Women Sterilized," *Latinamerica Press*, January 28, 2013, available at: http://www.eurasiare view.com/05112011-peru-forced-sterilization-cases-reopened/

[14] In Alberta, members of the Eugenics Boards traveled to and made a decision in favor or against sterilization within the mental health institutions themselves. "Return asked for by Mr Giroux respecting Members of the Sexual Sterilization Act Board," March 2, 1932, ACC, GR 1970.44, 31/1173, Provincial Archives of Alberta (PAAB). These institutions included the Provincial Training School, Red Deer, and the Provincial Mental Hospital, Ponoka.

[15] Less practical to determine is how many sterilizations occurred in private institutions and how many were voluntary. The distinction between voluntary and involuntary sterilization was, as we will see, a fuzzy one throughout the twentieth century. Some arrangements called "voluntary" with patient consent were in fact closer to coercive.

was that arbitrary. In other cases, social workers held the power of the purse: they made renewal of financial support dependent on submission to sterilization.

The relationship among agency, institutions, and ideas is a theoretical and an empirical one, and it needs to be developed. Before doing so, we position the book's argument within the scholarly literature.

THE INTELLECTUAL TERRAIN

Although it can appear at times to indulge in theorizing for its own sake, social science theory is ultimately about isolating causality, and it shares this aim with history.[16] In the latter, historians have developed, explicitly or implicitly, several theoretical frameworks for the study of eugenics and coerced sterilization. Subject to inevitable overlap, five approaches can be teased out of the existing literature.

1. Eugenics as a Religion

The first approach views eugenics a new religion,[17] which would have been a bitter irony to some of the leading eugenicists, given their hostility to religion and to religious believers. Like religion, eugenics was a comprehensive framework that gave reasons for the state of the world and also prescribed actions meant to improve it in the future. More specifically, eugenics provided an account of both the origins of human difference and the correct social and economic policies needed to deal with it. Francis Galton, as historian Daniel Kevles put it, was the "founder of the faith."[18] Like many religions, it promised salvation – in this case, from biology. "Could not," Galton wrote, "the race of men be similarly improved? Could not the undesirables be got rid of and the desirables multiplied?" Kevles takes the rhetoric further: "Could man not actually take charge of his own evolution?"[19] What made eugenics unique was the incorporation of scientific language into a religious world-view.[20] As Michael Burleigh writes in one of the best descriptions, eugenics was a "collectivist, materialist, technocratic creed which promised to conquer in a Promethean way, nature's final frontier and which, like socialism itself, had evolved from primitive utopianism into a secular religion with scientific pretentions."[21]

---

[16] There are a number of works that tell us what happened without speculating on why, but these are generally works that "broke" the story of eugenics or some part of it. See Müller-Hill, *Murderous Science*; Stefan Kühl, *The Nazi Connection: Eugenics, American Racism and German National Socialism* (New York: Oxford University Press, 1994).

[17] Kevles, *In the Name of Eugenics*, 12–13.

[18] Ibid., chapter 1.

[19] Ibid., 3. The first two questions are direct quotations from Galton. The last is posed rhetorically by Kevles. Ibid.

[20] Marius Turda, "Race, Science, and Eugenics in the Twentieth Century," in *Oxford Handbook of the History of Eugenics*, ed. Bashford and Levine, 73.

[21] Michael Burleigh, "Eugenic Utopias and the Genetic Present," *Totalitarian Movements and Political Religions* 1, no. 1 (2000): 64.

That the eugenicists held such views is indisputable, and the religion analogy is a powerful one. Galton himself directly makes this connection:

> [Eugenics] must be introduced into the national conscience, like a new religion. It has, indeed, strong claims to become an orthodox religious tenet of the future, for eugenics co-operate with the workings of nature by securing that humanity shall be represented by the fittest races.[22]

Nonetheless, defining eugenics as a religion provides little explanatory value; it is really another way of saying that eugenics was a comprehensive ideological framework with policy prescriptions. The question is why these policy prescriptions were followed. Certain religions, such as Islam and Christianity, attract billions of followers; others, such as the Bahá'í faith, attract a few million. As students of ideational approaches to politics note, the fact of policy impact cannot be read off the coherence of an ideational framework.[23]

2. Eugenics as Racialist Policy

The second approach views eugenics as a racial project designed to purge the population of nonwhite, homosexual, and/or poor people. Unsurprisingly, given the country's history of slavery, accounts of eugenics in the United States often fall into this category.[24] As one book recently put it, "in the early decades of the twentieth century, not long after the technology of surgical sterilization had been devised, state governments throughout the United States began a quest for racial purity that would change the lives of thousands of their citizens."[25] In a similar (if more jargon-ridden) vein, another scholar argues, referring to eugenicists' support for nationality-based immigration control:

> [E]ugenics ... has played a pivotal role in nationalist and racist enterprises. ... Central to the [pursuit of racial purity] is a "racism of *extermination* or elimination (an 'exclusive' racism) and a racism of *oppression* or exploitation (an 'inclusive'

---

[22] Quoted in G. K. Chesterton, *Eugenics and Other Evils: An Argument against the Scientifically Organized Society* (London: Cassel, 1922; repr., Seattle: Inkling Books, 2000), 57.

[23] Sheri Berman, "Ideas, Norms and Culture in Political Analysis," *Comparative Politics* 33, no. 2 (2001): 231–50; and Daniel Béland and Robert H. Cox, eds., *Ideas and Politics in Social Science Research* (New York: Oxford University Press, 2011).

[24] Edwin Black, *War against the Weak: Eugenics and America's Campaign to Create a Master Race* (New York: Four Walls Eight Windows, 2003); Harry Bruinius, *Better for All the World: The Secret History of Forced Sterilization and America's Quest for Racial Purity* (New York: Knopf, 2006). Black's account may be the most strident of this view: "in America this battle to wipe out whole ethnic groups was fought ... [to] create a superior Nordic race." Ibid., xv. He adds, "American eugenicists were convinced they could forcibly reshape humanity in their own image." Ibid., 21. Also see, in a more complicated way, Nancy Ordover, *American Eugenics: Race, Queer Anatomy, and the Science of Nationalism* (Minneapolis: University of Minnesota Press, 2003).

[25] Bruinius, *Better for All the World*, 9.

racism) . . ." Eugenics employed and rationalized both "inclusive" and "exclusive" racism.[26]

These analyses are not so much wrong as overblown. Many eugenicists were racist and held views of black, Asian, Latino, and indigenous people that most now regard as repugnant. In this stance they were, however, hardly unique: an unreflective assumption of Northern European racial superiority was common in North America before the Second World War. These eugenicists were products of their time. But their eugenics was not, at its core, a racist attempt to eliminate other races; the motivation was to improve the lot of white North Americans. Indeed, and rather paradoxically, racism was until the 1960s something of a shield for African Americans from eugenicists' practices. North American eugenicists viewed their society's African American citizens as so removed from the mainstream of the white society as not to warrant consideration in the development of this new public policy.

Had North American eugenics been fundamentally racist, nonwhite people would have been a primary target of the policy of coerced sterilization. They were not. On the contrary, the majority of sterilized people was white. Virginia opened the first institution for feebleminded African Americans only in 1939, and, across the state, physicians concerned with sterilization rarely if ever mention race.[27] Across the South, only a handful of institutions were opened for African Americans, and only a few white institutions admitted African Americans – to segregated sections, of course.[28] Southern eugenicists shared Northerners' overwhelming concern with the threat posed by the (white) feebleminded rather than African Americans.[29] At the same time, some leading eugenicists distanced themselves from the most overt forms of racism. In 1947, Wickliffe Draper, a wealthy racist benefactor, offered to fund Birthright, Inc., one of the United States' main pro-eugenic sterilization organizations, on the condition that the organization's research supported racial prejudice and justified Southern miscegenation laws.[30] Birthright's founder turned his offer of

---

[26] Ordover, *American Eugenics*, xv.
[27] James W. Trent, *Inventing the Feeble Mind: A History of Mental Retardation in the United States* (Berkeley: University of California Press, 1994), 177.
[28] One separate institution for African Americans was the Petersburg State Colony. Like other institutions, it began with high hopes for vocational training but soon become an overcrowded "dumping ground for delinquent black youths." Steven Noll, *Feeble-Minded in Our Midst: Institutions for the Mentally Retarded in the South, 1900–1940* (Chapel Hill: University of North Carolina Press, 1995), 100. Another institution for African Americans was North Carolina's Goldsboro State Hospital for Negroes. Ibid., 102.
[29] See ibid., 92, and chapter 5 generally.
[30] Dowbiggin, *Sterilization Movement*, 49–50. The suggestion came from Paul Popenoe. Popenoe to Marian S. Olden, January 20, 1947, 2/14, Association for Voluntary Sterilization Records (AVS), University of Minnesota, Minneapolis, MN.

$100,000 down flat and settled instead for a figure one-tenth of that amount that came free of such conditions.[31]

African Americans did become the majority of those sterilized in some instances, such as in North Carolina after – but not before – the Second World War. This outcome resulted, however, not from eugenic policy but, rather, welfare policy: African Americans on welfare became the targets of coerced sterilization. Social workers played a key role in this process. Moreover, in the American states regarded as the bastion of unreconstructed racism – the Deep South – there was less eugenic sterilization than in other areas of the country traditionally regarded as more liberal. As Mark Largent writes,

> [O]f the six founding members of the Confederate States of America – Alabama, Mississippi, South Carolina, North Carolina, Florida, and Georgia – only North Carolina and Georgia had high per capita numbers of sterilization. South Carolina and Georgia were the last two states to pass compulsory sterilization laws, and Florida never passed a eugenically based compulsory sterilization law.[32]

Only when "racial" is understood not in terms of skin color but rather in terms of genes can the eugenicists be said to have pursued an overarching racial project. And this was little more than a semantic trick that served only to restate the obvious: eugenicists believed in biological determinism. As one scholar has recently, and rightly, concluded, "arguing that eugenics was 'racist' tells us very little."[33] Indeed, doing so fundamentally misunderstands the project. American eugenicists were quintessentially progressive and advocated measures designed to improve the lives of white American citizens. The fact that Southern legislators were not terribly interested in improving the lives of African Americans saved these citizens from widespread institutionalization and therefore from sterilization in the prewar decades.[34] Racism, in other words, spared African Americans from systematic coerced sterilization before the war.

3. Eugenics as National Socialism

The third school views American eugenics as the stepchild, or at least a close cousin, of German eugenics.[35] American and German eugenicists were members

---

[31] Dowbiggin, *Sterilization Movement*, 50; Marian Olden to Mrs. Bradford, February 12, 1947, 2/14, AVS.

[32] Largent, *Breeding Contempt*, 81.

[33] Turda, "Race, Science, and Eugenics," 63.

[34] Although some superintendents were racist. See Noll, *Feeble-Minded in Our Midst*, 95.

[35] See in particular Black, *War against the Weak*. Black is an enthusiast for this perspective adumbrating about the "sad truth of how the scientific rationales that drove killer doctors at Auschwitz were first concocted on Long Island at the Carnegie Institution's eugenic enterprise at Cold Spring Harbor." Ibid., xvii. For Black, the connection is a major part of the explanation of Nazi eugenics: "the Nazi principle of Nordic superiority was not hatched in the Third Reich but on Long Island decades earlier." Ibid., xviii. For an earlier and less strident treatment of the argument, see Kühl, *Nazi Connection*.

of the same academic associations, attended the same conferences, exchanged congratulatory correspondence on legal and other accomplishments, and shared some of the same intellectual outlooks and policy views. These facts have led some scholars to view American eugenics as heavily influenced by German and National Socialist eugenics.[36] Although the connections and similarities between the two groups of people are of course worth explaining, the comparison is overdrawn.

Above all, eugenics was a complex movement, and, like any complex movement, it contained numerous strains. Overplaying the "Nazi connection" tends to associate all of American or North American eugenics with the larger movement's most extreme version.[37] Eugenicists were behind many things that we find objectionable today, but they also endorsed projects that many of us find admirable: environmental protection or, more centrally to eugenics, reproductive choice for women. Similarly, German eugenics, or "racial hygiene" as it was known vernacularly, itself contained diverse strains. It is inaccurate to see an unbroken line between, for instance, the eugenics of Alfred Ploetz (the father of the subject in Germany) and the murderous eugenics of the National Socialists. Recent scholarship has emphasized that Ploetz's worldview had more in common with his British counterparts than has been supposed.[38] It is even less tenable to view Hermann Muckermann – a German Jesuit, social reformer, and supporter of a moderate eugenics program – through the lens of National Socialism; indeed, he was sacked the moment the Nazis took power.[39] In fact, scholars interested in understanding the full complexity of German eugenics have concluded that Hitler represented a radicalization rather than a logical extension of German eugenic thought. "Had the Nazis not forced a drastic change of course in 1933," writes Sheila Weiss, "there is every reason to believe that the [German eugenic] movement would have become even more similar to its counterpart in Britain."[40]

4. Eugenics as Anti-Feminism

The fourth school cuts partly across the second (racialist policy) but views eugenics as a reaction to early feminism, to women's demands for entitlements enjoyed by men and to their rejection of fertile motherhood as their sole role. "Women," writes one observer, "were becoming masculine just as men were

---

[36] Kühl, *Nazi Connection.*

[37] On this, see Stepan, *Hour of Eugenics*, 4.

[38] See the fine essay by Turda, "Race, Science, and Eugenics."

[39] Nils Roll-Hansen, "Norwegian Eugenics: Sterilization as Social Reform," in *Eugenics and the Welfare State: Sterilization Policy in Denmark, Sweden, Norway, and Finland*, ed. Gunnar Broberg and Roll-Hansen (East Lansing: Michigan State University Press, 1996), 154.

[40] Sheila F. Weiss, "The Race Hygiene Movement in Germany, 1904–1945," in *The Wellborn Science: Eugenics in Germany, France, Brazil, and Russia*, ed. Mark B. Adams (New York: Oxford University Press, 1990), 222.

becoming increasingly weak and effeminate. Home and family were the corner-stones of society, and if women abdicated their domestic duties, what was to become of moral order?"[41] This development was bad enough, but it was rendered far worse by the fact that feebleminded women were picking up the reproductive baton:

> Beginning in the 1910s ... feeblemindedness and, in particular, the "moron" category became almost synonymous with the illicit sexual behavior of the woman adrift. Eugenic ideology provided a language and rationale for linking the sexual and reproductive behavior of women with the deterioration of the race.[42]

Gender certainly was relevant to the history of eugenics – how could it not be, as a core eugenic concern was reproduction? Certain superintendents held the deeply sexist view of feebleminded women as a greater threat than the male equivalents because they would conceive the feebleminded child. As Superintendent Bell of Virginia's Lynchburg State Colony put it, "[t]he female defective is, generally speaking, more dangerous eugenically than the male ... and it is, therefore, evident that if all mentally defective women were sterilized, there would be but little reproduction of feeble-minded persons from these sources."[43] This "danger" was simply their reproductive power.

Eugenicists also associated female sexual licentiousness generally with feeble-mindedness. For example, studies regularly confirmed that prostitutes were disproportionately feebleminded.[44] Added to these views was a wish to protect women from male sexual aggression. Because the feebleminded woman was "an easy prey to the sexual aggressions of males of superior intellect as well as those of her own mental level," she should be sterilized.[45] As a result, women made up the majority of sterilizations in many institutions. In North Carolina, women made up 80 percent of those sterilized from 1929 to 1940, whereas in Virginia the figure was 60 percent for essentially the same period.[46] The fear and con-struction of feebleminded women as a threat to the nation's core was absolutely a part of the story of eugenics.

It was not, however, the whole story, for three reasons. First, the pattern of disproportionate female sterilization did not generally obtain. In Alabama and Kansas, the majority of those sterilized was male, and men and women were sterilized in roughly equal numbers in California, Indiana, and Delaware.

---

[41] Kline, *Building a Better Race*, 11.

[42] Ibid., 29.

[43] Quoted in Noll, *Feeble-Minded in Our Midst*, 76.

[44] Thus, a 1913 Virginia study concluded that approximately 50 percent of prostitutes were feeble-minded (ibid., 24–25), whereas a 1920 Georgia study arrived at the figure of 43.5 percent (ibid., 41). See also ibid., chapter 6 and Mark T. Connelly, *The Response to Prostitution in the Progressive Era* (Chapel Hill: University of North Carolina Press, 1980).

[45] Noll, *Feeble-Minded in Our Midst*, 76.

[46] Ibid., 75.

Second, although some advocates of sterilization argued for targeting women as the carriers of babies, others argued for the sterilization of men because the operation was less intrusive, faster, and cheaper. Finally, a complete analysis of gender's role would have to account for the large number of middle- and upper-class women who endorsed eugenics and, above all, eugenic sterilization as a basis for increased female autonomy. As Alexandra Minna Stern argues,

> Throughout Europe and the Americas, female eugenicists regularly worked to bolster their own authority and professional stature by drawing a sharp line between themselves – the "fit" – and those they considered "unfit." Emblematic of this impulse was Margaret Sanger, whose tireless advocacy of contraception was tied always to a desire to lower birthrates among the laboring classes, immigrants, and racial minorities, whom she deemed to be biologically inferior. *Yet beyond high-profile actors like Sanger, there were hundreds if not thousands of professional, usually white, women who represented the early twentieth century eugenic creed by participating in local eugenics societies and mental hygiene campaigns and discouraging rural and urban poor women from reproduction.*[47]

Specific examples are numerous. In 1948, the legislative chairwoman of the Alabama Federation of Women's Clubs noted that the organization had "tried for years to get legislation passed on sterilization of mental defectives, but so far has been unsuccessful." Its members had, however, "not given up hope yet.... We are planning to put on an educational campaign, to get the press to write articles on it."[48] In South Carolina, the Federation of Women's Clubs was "especially active, the securing of this institution being one of the two things of a State-wide legislative nature" that the federation was "primarily working for."[49] In Georgia, state women's clubs, along with the state federation of labor and business associations, supported the creation of a state institution for the feebleminded.[50] Similarly, the English radical thinker and social reformer Marie Stopes championed eugenics precisely on the grounds that "our race is weakened by an appallingly high percentage of unfit weaklings and diseased individuals."[51] Sociologist Véronique Mottier expresses the dynamic by explaining that it "would be a mistake to assume that women were only the *victims* of eugenics: they were also important agents in the implementation of eugenic policies."[52]

[47] A. M. Stern, "Gender and Sexuality: A Global Tour and Compass," in *Oxford Handbook of the History of Eugenics*, ed. Bashford and Levine, 177–8. Emphasis added.

[48] Birthright, Inc., *Report to Members, January 1 to April 29, 1948*, 2/14, AVS.

[49] Quoted in Noll, *Feeble-Minded in Our Midst*, 23.

[50] Ibid.

[51] Quoted in Ross McKibbin, introduction to *Married Love* by Marie C. Stopes (Oxford: Oxford University Press, 2004), xv.

[52] Véronique Mottier, "Eugenics and the State: Policy-Making in Comparative Perspective," in *Oxford Handbook of the History of Eugenics*, ed. Bashford and Levine, 146; Mottier, "Reproductive Rights," in *The Oxford Handbook of Gender and Politics*, ed. Georgina Waylen et al. (Oxford: Oxford University Press, 2013).

5. Eugenics as Progressivism

The fifth and final school understands eugenic sterilization as the intellectual inheritance of the American Progressive movement. Those American states with strong Progressive movements – California and New York, for example – had compulsory sterilization laws. Progressivists, with their strong belief in science and the related capacity for rational, expert-based state intervention tended to support eugenic sterilization. Some of America's most famous Progressives, such as Yale University economist Irving Fisher and President Theodore Roosevelt, supported eugenic sterilization.

It is, however, a step too far to claim that "sterilization laws in most states were adopted as part of the broader Progressive movement."[53] The correlations needed to support this claim are simply absent. Ohio, Illinois, and Colorado, all of which had strong Progressive movements, never adopted compulsory sterilization laws. Three of the most ambitious eugenic sterilization programs – Georgia, North Carolina, and Virginia – were developed in the South, which was hardly the most fertile ground for American Progressivism. It is more accurate to say that Progressivists tended to lend support to campaigns for eugenic sterilization, but the success or failure of those campaigns was a function of more than the strength of the Progressive movement. More determinative in that regard were the resources of other lobbying groups, the strength of organized Roman Catholic opposition, the views of institutions' superintendents on the question of sterilization, and the links between him and the state governor.[54]

The critical reflections raised about these five schools of thought do not invalidate the works grouped under them. On the contrary: all schools uncover evidence and raise issues that are basic to understanding eugenics. Some, notably Kevles, are foundational texts to the study of the eugenics movement's (or movements') study. The discussion of these schools highlights the inherent difficulty in accounting for a set of ideas and political movement as complex as eugenics. To this end, this book does not attempt a general account of eugenics. Instead, it offers an account of why coerced sterilization had such appeal before the war and why it continued so long after.

INSTITUTIONS, IDEAS, AND ACTORS

There is longstanding debate among political scientists about the relative influence of institutions, ideas, and individuals in determining policy outcomes. One group of scholars argues that institutions exercise an independent effect on public policy: they determine preferences, encourage some actors to pursue their aims while discouraging others, and structure the political process in a

---

[53] Largent, *Breeding Contempt*, 81.
[54] Largent notes that few Southern states established mental health institutions and, thus, "compulsory sterilizations were not performed even when the laws were adopted." Ibid.

way that makes certain outcomes far more likely than others. Institutions create regularities and incentives. Westminster-style parliamentary institutions, characterized by a strong executive, a weak legislature, and a closed bureaucracy, make government less accessible to social actors, magnify the power of those who do enjoy access, and concentrate extensive power in the hands of the prime minister. The result is a tendency toward dramatic policy shifts: massive nationalization in Britain under the 1940s Labour Party, punitive tax increases under the 1970s' Labour Party, massive privatization under the 1980s' Conservative Party, and a comprehensive reduction in public spending and state responsibility under the current Conservative-Liberal Democratic coalition.

In decentralized systems with multiple levels of government and multiple veto points, the number of actors involved in the policy process expands, and the institutions are porous to interest group pressure. The United States and the Federal Republic of Germany are, in different ways, ideal types in this regard.[55] The result in policy terms is a bias toward incremental change: radical policy reforms are blocked by interest-group opposition (President Clinton's 1992 health care proposal), gridlock in Congress (President Bush's Comprehensive Immigration bill in 2007), or gain passage through the one legislative house only to be blocked by the courts and/or the other house (radical citizenship reform under Germany's Social Democratic-Green coalition in 1998). In the United States, the states often gain increased policy-making autonomy from this national infirmity.[56]

This volume focuses on two sets of institutions, the one intuitively obvious, the other less so. In the former, both Canada and the United States are federations, which allowed the regional governments to adopt legislation at the

---

[55] Alfred Stepan and Juan J. Linz, "Comparative Perspectives on Inequality and the Quality of Democracy in the United States," *Perspectives on Politics* 9, no. 4 (2011): 841–56.

[56] Gary Gerstle, "The Resilient Power of the States across the Long Nineteenth Century: An Inquiry into a Pattern of American Governance," in *The Unsustainable American State*, eds. Lawrence Jacobs and Desmond King (New York: Oxford University Press, 2009); Kimberley S. Johnson, *Governing the States: Congress and the New Federalism, 1877–1929* (Princeton, NJ: Princeton University Press, 2007). For their part, rational choice institutionalists take institutions seriously, but they reverse the causal relationship: institutions are means to reducing transaction costs and are designed and used by individual actors to maximize their preferences. Peter Hall, *Governing the Economy* (New York: Oxford University Press, 1986); Sven Steinmo, Kathleen Thelen, and Frank Longstreth, eds., *Structuring Politics: Historical Institutionalism in Comparative Analysis* (New York: Cambridge University Press, 1992), especially chapters by Hall, Immergut, King, and Rothstein; George Tsebelis, *Veto Players: How Political Institutions Work* (Princeton, NJ: Princeton University Press, 2002); Ellen Immergut, *Health Politics: Interests and Institutions in Western Europe* (New York: Cambridge University Press, 1992); Kenneth Shepsle, "Rational Choice Institutionalism," in *Oxford Handbook of Political Institutions*, ed. Sarah A. Binder et al. (Oxford: Oxford University Press, 2006), 23–38; Kathleen Thelen, "Historical Institutionalism in Comparative Politics," *Annual Review of Political Science* 2, no. 1 (1999): 369–404; James Mahoney and Kathleen Thelen, eds., *Explaining Institutional Change: Ambiguity, Agency, and Power* (New York: Cambridge University Press, 2010).

state/provincial level without federal support and then to force the issue to the federal level through the courts, specifically in the *Buck v. Bell* decision (the subject of Chapter 6). In the latter, we are particularly concerned with the role of a substate institution: homes for the feebleminded, which states and provinces established and funded but which were governed semi-autonomously by superintendents. Understanding these institutions is basic to understanding coerced sterilization.

A related theoretical approach, insofar as historical institutionalists take it seriously, concerns the role of ideas in politics. There has been a robust debate within the field of political science since the late 1980s about how ideas determine politics. Responding to the claims that institutions are decisive determinants of comparative variations in public policy outcomes, some scholars argue that the way in which ideas are formulated and used in political debate is either equally important or is of sufficient significance to warrant study.[57] The causal challenge for such a position is to show how ideas are more than instruments of material-driven political actors, including, of course, politicians on the campaign trail. In this context, eugenics is an important case study that could potentially provide further evidence of the centrality of ideas in determining politics. Eugenics constituted a comprehensive framework, drawing on the prestige of science and nailing itself to the mast of progressive social reform, and it had a concrete policy recommendation as well: compulsory sterilization. This recommendation was in turn implemented, although in a more limited way than its protagonists hoped.

Despite this strong correlation, eugenics did not simply lead to coerced sterilization. Few scholars think that policies can be read directly off the ideas.[58] Indeed, the fact that a doctrine is comprehensive is a poor predictor of its implementation. Marxist ideas were nothing if not comprehensive, but they made only limited headway in the West. New Right ideas in the 1970s and 1980s had a substantial influence in the United States and the United Kingdom but little impact in continental Europe. Similarly, Fascism and Nazism offered comprehensive programs, yet these creeds took root in some countries but not others.[59]

To understand eugenic ideas' importance, one must reflect on how they intersected with individual interests. There are two aspects to this influence.

First, sterilization was justified and/or implemented by men and women with a broad array of material interests: superintendents' interest in bigger hospitals, more staff, larger budgets, and more power; legislators' interests in passing

---

[57] On this, see Sheri Berman, "Ideas, Norms and Culture"; Mark Blyth, "Any More Bright Ideas? The Ideational Turn of Comparative Political Economy," *Comparative Politics* 29, no. 2 (1997): 229–50.

[58] Jal Mehta, "The Varied Roles of Ideas in Politics: From 'Whether' to 'How,'" in *Ideas and Politics in Social Science Research*, ed. Béland and Cox, 30.

[59] Identifying such comparative patterns is the stuff of macro-political sociology in the tradition of Barrington Moore, Theda Skocpol, and many others. See, for example, Daron Acemoglu and James A. Robinson, *Why Nations Fail: The Origins of Power, Prosperity, and Poverty* (New York: Crown Publishers, 2012).

high-profile legislation and in lowering costs to the public purse; women's groups' interests in carving out a larger role in public and policy debate; and, above all, the interests of hundreds of mental health superintendents, institution caregivers and workers, and social workers in ridding the gene pool of the feebleminded. Eugenic ideas, however, mattered precisely because they served these interests. Consistent with the scholarly theory about the political role of ideas, eugenic arguments served these interests by strengthening the case in favor of sterilizations: they extended prestige to pro-sterilization arguments,[60] defined the problem of the fecund feebleminded and framed the solution as sterilization,[61] and helped forge coalitions in favor of coerced eugenic sterilization.[62] Without the decisive interventions of mental health superintendents, state legislators, and pro-sterilization lobby groups, eugenic ideas may never have left the laboratory. Without the strength and prestige of eugenic ideas, these interventions may have led to nothing. The dichotomy, often implied in the literature, between interests and ideas is overdrawn: ideas are most powerful when they are congruent with and attached to individual interests.[63]

Second, the way eugenicists used ideas to justify their policy aims often proved to be malleable. From the 1930s, scientists grew increasingly skeptical of the hereditarian arguments advanced in its justification. Later, Nazi mass sterilization, medical experimentation, and murder presented a major challenge for the proponents of eugenic sterilization. Pro-sterilization advocates, both within and outside institutions for the insane, feebleminded, and socially deviant, did not simply capitulate. Rather, they reformulated and restated coerced sterilization in new ideational frames: the rights of the child, welfare, and world population growth. With these new anchors, coerced sterilization continued throughout the 1950s, 1960s, and 1970s. The practice only ended when the culture and language of individual rights became hegemonic following the civil rights reforms in the 1960s, which rendered the coerced sterilization of individuals a violation of inalienable rights.

Eugenics was not, particularly in the postwar period, the whole story of sterilization by any means, but it provided the most important argument in favor of coerced sterilization during the prewar period. In short, coerced

---

[60] Randall Hansen and Desmond King, "Eugenic Ideas, Political Interests, and Policy Variance: Immigration and Sterilization Policy in Britain and the US." *World Politics* 53, no. 2 (2001): 247.

[61] Martin Rein and Donald A. Schön, "Problem-Setting in Policy Research," in *Using Social Research in Public Policy*, ed. Carol H. Weiss, 235–51 (Lexington, MA: D.C. Heath and Co., 1977); Frank R. Baumgartner, *Conflict and Rhetoric*; Baumgartner and Bryan D. Jones, *Agendas and Instability in American Politics* (Chicago: University of Chicago Press, 1993); Donald Schön and Martin Rein, *Frame Reflection: Toward the Resolution of Intractable Policy Controversies* (New York: BasicBooks, 1994); Mehta, "Varied Roles of Ideas."

[62] Hansen and King, "Eugenic Ideas, Political Interests"; Vivien A. Schmidt, "Does Discourse Matter in the Politics of Welfare State Adjustment?" *Comparative Political Studies* 35, no. 2 (2002): 168–93; Mehta, "Varied Roles of Ideas."

[63] Hansen and King, "Eugenic Ideas, Political Interests." See also Blyth, "Any More Bright Ideas?"

sterilization more than meets the methodological requirements for showing how ideas influence politics, namely, by establishing that ideas (a) shape actors' actions and (b) are not reducible to some other nonideational force.[64]

VARIANCE

Why was eugenic policy implemented in some countries and in some U.S. states and some Canadian provinces but not in others? The question is important because the factors commonly cited to account for coerced sterilization either applied generally across the United States (eugenic ideas were at least known everywhere) or varied in a way that does not consistently correlate with coerced sterilization (e.g., local strength of the Progressive movement).

We argue that whether a state adopted coerced eugenic sterilization policy was a function of: (a) the role played by superintendents at institutions for the insane, feebleminded, criminal, and otherwise socially deviant and their capacity to command resources for such expenses as legal defenses; (b) the superintendents' connections with state legislators and the state's governor; and (c) the strength and effectiveness of the anti-sterilization lobby, which was often spearheaded by the Roman Catholic Church. Coerced sterilization laws were adopted in states in which superintendents were able to convince legislators of the eugenic case for coerced sterilization, where legislators either became true believers of these arguments or saw their interests furthered by pretending they believed them, and where the Roman Catholic Church was unable to launch an effective counter strategy. Scientific opponents were few and ineffective until the second half of the twentieth century.

In the history of coerced sterilization, institutions for the mentally ill and handicapped – variously called state colonies, schools, training schools, state hospitals, and so forth – were crucial sites. Sterilizations occurred within these institutions, and the variance that exists between the numbers of state sterilization trades on variance between institutions – that is, states with high numbers of sterilizations contained institutions that sterilized in large numbers. Within these institutions, the superintendent raised and dispersed funds, hired and fired staff, determined formal rules and informal norms governing the institutions and, most importantly, decided whether, how often, and by which means the mentally handicapped would be sterilized. When state eugenic boards were established beginning in the 1930s to oversee sterilization decisions, the same superintendents seamlessly became influential board members.

Some superintendents had patients sterilized without the benefit of the law, but, generally, sterilization required legal backing. In the United States and Canada, sterilization was under state or provincial jurisdiction, respectively. Superintendents in both countries who wished to instigate sterilization lobbied

---

[64] Berman, "Ideas, Norms and Culture"; Hansen and King, "Eugenic Ideas, Political Interests"; Mehta, "Varied Roles of Ideas."

governors (in Canada, premiers) and state (provincial) legislators. Because the superintendent was a state official, and an influential one at that, he enjoyed privileged lines of communications to powerful state actors and often to the governor himself. Sterilization bills were commonly drafted with direct input from mental health superintendents; sometimes, superintendents were the sole authors. As trained experts, superintendents claimed expert knowledge at a time when, partially thanks to the popularity of Progressive and socialist Fabian ideas, experts and their knowledge enjoyed deference. As heads of institutions, they were directly responsible for the targets of sterilization and were uniquely able to marshal in a single document the biological, social, and financial arguments in favor of coerced eugenic sterilization. Harry C. Sharp of the Indiana State Reformatory directly appealed to the state legislature and Indiana's Progressive governor, James Franklin Hanly, to adopt legislation on eugenic sterilization. Indiana became the first state to do so in 1907. In Mississippi, H. H. Ramsey achieved fewer sterilizations than he wanted because his institution was under-resourced. In Pennsylvania, two successive superintendents of the Pennsylvania Training School for Idiotic and Feeble-Minded Children, Isaac Newton Kerlin and Martin W. Barr, lobbied for eugenic sterilization legislation, and Barr wrote one of the state's sterilization bills. Albert S. Priddy, the first superintendent at Virginia's State Colony for Epileptics, urged his friend, Senator Aubrey Strode, to draft a coerced sterilization bill. Strode did, and the Progressive governor, Andrew Jackson Montague, signed it into law. Across the state, a cadre of former and current superintendents became proselytizers for "racial improvement" and sterilization: William Drewry, superintendent of the Central State Hospital in Petersburg and later director of the State Bureau of Mental Hygiene; Lynchburg superintendents J. H. Bell and G. B. Arnold; Petersburg State Colony Superintendent J. H. Henry; and Western State Hospital Superintendent Joseph DeJarnette.

Once a law was passed, superintendents determined its implementation: who would be sterilized, how many people would be sterilized, by what means the procedure would be carried out. They made their recommendations to state eugenic boards. Into the 1940s, promoters of eugenic sterilization saw the superintendents as the key to the robust implementation of eugenic sterilization.[65]

That superintendents played such an important role in the history of coerced eugenic sterilization relates to another institutional variable: the way in which mental health institutions themselves were structured. The institutions housing the mentally handicapped had a crystal-clear hierarchical structure in which the

---

[65] Clarence J. Gamble to the Scientific Committee of Birthright, May 9, 1949, 2/16, AVS. "If a Medical Director is secured for Birthright, I suggest that he may well undertake a study of the percentage of discharges from state hospitals which have been sterilized and of the reasons for nonsterilization of the remainder. This will have the advantage of drawing the matter to the superintendent's attention and encouraging him to give greater consideration to the desirability of sterilization in future discharges." Also see Clarence Gamble to Fred O. Butler, December 17, 1949, 2/16, AVS.

superintendent was a monarch within his (and occasionally her) realm. The rules governing the institution's operation allowed him to determine hiring and firing, spending allocations, the rules governing the handling of patients, the medical treatments applied to them, and, as part of the last, whether they would be sterilized. The superintendent, in short, was a carrier of ideas and a policy advocate outside the institution and a policy implementer within it.

LOBBY GROUPS

Naturally, superintendents and pro-sterilization legislators did not operate in a political vacuum. Throughout the seven decades of North America's experiment in coerced sterilization, two sets of groups squared off against each other. Regional and national organizations lobbied for coerced sterilization (the Human Betterment Foundation in California, the New Jersey Sterilization League (NJSL), the American Eugenics Society, smaller pro-sterilization organizations, and women's groups across the country). Christian churches – principally, the Roman Catholic Church – lobbied against it. For seven decades, the Catholic Church orchestrated a rearguard action against sterilization bills and, later, against abortion and contraception. The Christian doctrine on the sanctity of life and of reproduction as the ultimate expression of life informed Roman Catholic opposition to abortion and sterilization, and a version of the "slippery slope" argument underpinned its opposition to birth control.[66] Sanctioning contraceptives, a Catholic journal editorialized in 1931, "would be a long step on the road toward state clinics for abortion, for compulsory sterilization of those declared unfit by fanatical eugenicists, and the ultimate destruction of human liberty at the hands of an absolute pagan state."[67] Later buttressed with a papal denunciation of coerced sterilization, the Church was the most outspoken defender of what we would now call the rights of the mentally handicapped to freedom from coerced sterilization. Its effort infuriated the pro-sterilization lobby, just as it later infuriated the pro-contraception and pro-choice lobbies, but its influence was often decisive.[68] In the United States and Canada, and to some degree in Britain, the adoption or failure of a sterilization bill was a function of the success of the Catholics' anti-eugenic and anti-sterilization campaigns. Where the state or country was majority

---

[66] Dowbiggin, *Sterilization Movement*, 77.

[67] Quoted ibid., 77–8. Empirically, the Church was right about contraceptives and abortion, although wrong about contraceptives and eugenic sterilization.

[68] In a typical remark, a doctor making the case for eugenic sterilization wrote, "selective sterilization offers a safe and sane scientific form of preventative medicine whose full application is now thwarted by a misinformed and misguided religious bloc." E. A. Whitney, "Presenting Mental Deficiency to Students," *American Journal of Mental Deficiency* 50, no. 1 (July 1945): 54–8. Margaret Sanger also viewed the Roman Catholic Church as her greatest enemy, a view from which she never deviated. See Jean H. Baker, *Margaret Sanger: A Life of Passion* (New York: Hill and Wang, 2011). See also Connelly, *Fatal Misconception*, chapter 4.

Catholic (e.g, Quebec, Ireland, or France), eugenic sterilization was a nonstarter. It is no surprise that, despite a powerful eugenics movement deeply concerned with racial health, only one Latin American region, the Mexican state of Veracruz, adopted a eugenic sterilization law. Even there, however, the law was the product of the then-governor's secularism and radical anti-clericalism; eugenic sterilization did not survive his departure from office.[69] Motivated by the neo-Larmarckism (the idea that acquired characteristics could be inherited) that predominated in Roman Catholic countries, Socialist Governor Adalberto Tejeda's eugenics was deeply concerned with racial degeneration. He was highly interventionist in his assault on the supposedly degenerative effects of alcohol and prostitution and in his encouragement of good mothering and healthy babies.[70] These efforts were, in turn, bound up with Tejeda's vision of a scientific socialism involving nationalization, secular education, agrarian reform, and wealth redistribution.[71] But the program was not based on the goal of mass mutilation: it is unclear if any sterilizations, although legislatively authorized, were performed.[72]

In those cases in which the Roman Catholic Church and/or other churches did not hold the majority, their success in mobilizing anti-sterilization pressure depended on the extent to which they influenced public opinion. Throughout the entire period under study, pro-sterilization advocates complained about the public's ignorance of sterilization, its confusion of sterilization with bodily mutilation,[73] and its visceral hostility to the idea of coerced sterilization.

To summarize our argument for the pre-Second World War period, eugenic sterilization policies were made possible by the active support of mental health institutions and by the actions and authority of their superintendents. States adopted the policy and practice when superintendents and pro-sterilization lobby groups convinced legislators of the case in favor of coerced sterilization and, importantly, when the Roman Catholic Church was unable to organize effective opposition. Once the law was on the books, the superintendents determined which patients would be sterilized and thereby how many sterilizations occurred. This configuration remained part of the post-1945 policy framework, which is why any account of the persistence of eugenic sterilization into the

---

[69] Stepan, *Hour of Eugenics*, 132–4. Also see Alexandra Stern, "From Mestizophilia to Biotypology: Racialization and Science in Mexico, 1920–1960," in *Race and Nation in Modern Latin America*, ed. Nancy Applebaum, Anne S. MacPherson, and Karin Alejandra Rosemblatt, 187–210 (Chapel Hill: University of North Carolina Press, 2003).

[70] See Alexandra Stern, "'The Hour of Eugenics' in Veracruz, Mexico: Radical Politics, Public Health, and Latin America's Only Sterilization Law," *Hispanic American Historical Review* 91, no. 3 (2011), 431–43.

[71] Ibid., 437.

[72] Ibid., 441.

[73] See, for example, Robert L. Dickinson, "Sterilization without Unsexing," *Journal of the American Medical Association* 92, no. 5 (1929): 373.

second half of the twentieth century needs to understand the first half of that century as well.

POSTWAR STERILIZATION POLICY

Scholars and journalists alike now recognize that coerced sterilization continued after the Second World War, the Holocaust, and Nazi Germany's coerced sterilization of approximately 400,000 people and murder of 70,000 more. That sterilization did continue is, on its face, curious: one might expect that the mass murder of some of society's most vulnerable citizens, to say nothing of the millions of others murdered by the Nazis, would have fatally undermined both eugenics and coerced sterilization.

History was not the only strike against eugenics: science was also increasingly hostile. As Kevles notes in his authoritative study of eugenics,

> By the mid-thirties, mainline eugenics had generally been recognized as the farrago of flawed science. Jacob Landman summarized the failings of the creed: "It is not true that boiler washers, engine hostlers, miners, janitors, and garbage men, who have large families, are necessarily idiots and morons. It is not true that college graduates, people in 'Who's Who,' and some 'successful' people, such as racketeers and bootleggers, are necessarily physically, mentally or morally superior parents .... It is not true that celebrated individuals necessarily beget celebrated offspring ... [or] that idiotic individuals necessarily beget idiotic children .... It is not true that, by any known scientific test, there is a Nordic race, or that the so-called Nordic race is superior to any other race." ... and it was not true, most geneticists had come to understand, that eugenic sterilization could rapidly rid society of the eugenically undesirable.[74]

Other scholars advance similar arguments.[75] The problem, as Wendy Kline observes, is that "declinist" historians of science have assumed rather than established that attacks such as Landman's on eugenics convinced the public or politicians to abandon their beliefs.[76] They did not.[77] Indeed, in 1938, as eugenics was supposedly in precipitous decline and leading intellectuals like

---

[74] Kevles, *In the Name of Eugenics*, 164.

[75] Carl N. Degler, *In Search of Human Nature: The Decline and Revival of Darwinism in American Social Thought* (New York: Oxford University Press, 1991), 150–1; Linda Gordon, *Woman's Body, Woman's Right: A Social History of Birth Control in America* (New York: Grossman Publishers, 1976), 277.

[76] Kline, *Building a Better Race*. Although the debate on trends in genetics is best left to historians of science, it is significant that a 1936 paper by the Rockefeller Foundation lists hard-line eugenicists such as Tage Kemp (Copenhagen), Charles Davenport, and Irving Fisher as leading geneticists along with critics of eugenics such as J. B. S. Haldane and H. S. Jennings. "Actions in General Physiology," May 1936, RF 3 915,1,8, Rockefeller Foundation Archives, Rockefeller Archive Center, Sleepy Hollow, NY (RAC).

[77] This endurance is a recurring issue. As sociologist Ann Morning notes, "biological race theories have remained standing when their empirical legs have repeatedly been kicked out from under them." She adds that, in biology textbooks, "the share of texts to teach race declined from the

Columbia University anthropologist Franz Boas publicly decried the validity of race as a scientific concept, less than 50 percent of the ultimate total number of known compulsory sterilizations in the United States had occurred.[78] In 1958, the Biennial Report of South Dakota's State Commission for the Feebleminded reported that state institutions pursued an "aggressive program" for identifying the feebleminded who, if they wanted children, were sterilized because doing so "prevents feebleminded parenthood, and secondly, it prevents the transmission of the defective strain."[79] In the same year, the Director of Social Services at the Utah Training School, John F. Pero, confidently concluded "mental deficiency ... has grown to be one of the major problems in every state. When one sees the strain of mental deficiency, insanity, and the like, running thru family after family we wonder ... why there can be any objection to sexual sterilization in any section of our country."[80] Scientific skepticism was not the driving motivation of American policy makers – indeed scientific doubts may not have percolated much beyond the technical community. Members of the public, politicians, bureaucrats, and (some) doctors took it as unproblematic that sterilization was justified on economic and social grounds well into the 1960s and even later.[81]

But why was this belief tenacious? One view is that American eugenicists simply saw their project as being fundamentally distinct from that of National Socialism.[82] This might be part of the explanation, but it cannot explain why a general public revulsion against National Socialism did not also envelop eugenics and coerced sterilization. This revulsion deepened several decades later. Roman Catholics and others opposed to eugenics did not hesitate to remind people of the history of Nazi eugenics.[83] Inertia and path dependence may have

1950s to the 1990s, when their proportion rebounded. In short, race appears to be returning, not disappearing, as a topic of biological instruction." Ann Morning, "Reconstructing Race in Science and Society: Biology Textbooks, 1952–2002," in "Exploring Genetics and Social Structure," ed. Peter Bearman, special issue, *American Journal of Sociology* 114, no. S1 (2008): 108.

[78] Largent in *Breeding Contempt* records 16,066 sterilizations by 1933 and 38,087 by 1942; by 1980, the total was 63,841. Ibid., 79–80. See also Hartley F. Peart, "Vasectomy and Salpingectomy under California Law," *California and Western Medicine* 6 (May–June 1941): 1–8, Box 73, AVS.

[79] State of South Dakota, *Seventeenth Biennial Report of the State Commission for the Control of the Feeble-Minded*. SW015/73, AVS.

[80] John F. Pero, *The Problem of Mental Deficiency in Utah* (1958), SW015.1/73, AVS.

[81] The United States was not alone in this respect. In Denmark, attitudes among eugenicists in fact hardened during the 1930s: some of the most prominent eugenicists drew heavily on the radicalizing German literature, including race-based biology and the superiority of the white race. Germany's sterilization law was widely praised, as was the castration of sex offenders, and eugenicists argued that the welfare state necessitated eugenics if the poor were not to overwhelm it. See Bent Sigurd Hansen, "Something Rotten in the State of Denmark: Eugenics and the Ascent of the Welfare State," in *Eugenics and the Welfare State: Sterilization Policy in Denmark, Sweden, Norway and Finland*, ed. Gunnar Broberg and Nils Roll-Hansen (East Lansing: Michigan State University Press, 1996): 9–76.

[82] Kline, *Building a Better Race*, 6.

[83] See "The Catholic Position," *Redbook Magazine*, March 1956.

played a role: once legislation is on the books, it tends to stay there, even when the substance of the law is not enforced (hence amusing stories of long-forgotten laws such as one in California banning animals from mating near taverns). In many states, however, sterilization increased after the Second World War. In these cases, sterilization laws were not only on the books – they were enthusiastically implemented.

We argue that coerced sterilization continued, and in some cases increased, after the war for two reasons. First, there was often institutional continuity: superintendents did not resign en masse in 1945. There was also obviously no effort – as there was, however imperfectly, in Germany – to purge the medical establishment of eugenics. Significant numbers of North American superintendents remained committed to sterilization as a tool for treating the mentally handicapped well into the 1950s, 1960s, and even 1970s. The independent variable – superintendents exercising bureaucratic authority and control – that explained the implementation of sterilization policy before the war helps explain its continuance in the postwar era.

Second, pro-sterilization activists re-anchored their case for sterilization in new ideas. To understand why this was possible, it is necessary to focus on the pro-sterilization organizations that were active in the United States from the 1920s until the 1960s: the Human Betterment Foundation in Pasadena, the NJSL, Birthright, Inc. (successor to the NJSL), the Human Betterment Association/Human Betterment Association of America (HBAA, successor to Birthright), and the Association for Voluntary Sterilization (successor to the HBAA). As the scientific and moral case in favor of eugenics weakened from the early 1930s, these organizations, bolstered by their prominent academics and health profession members, developed new arguments to justify coerced sterilization: to protect the right of the unborn child to proper parents (hence the name "Birthright"), to avoid excessively high population growth, and to prevent welfare abuses. For three decades after the war, superintendents in institutions for the mentally handicapped continued to sterilize coercively; for three decades after the war, these organizations recycled old and provided new arguments to justify those sterilizations. As they did, these purely eugenic organizations worked most closely with the main American birth control organizations, which were themselves populated by eugenicists. In the postwar years, there were close institutional, personal, and intellectual links between the eugenic, euthanasia, birth control, and pro-choice movements.

PLAN OF THE BOOK

The book is divided into fifteen chapters. In Chapters 2 and 3, we introduce the topic biographically and thematically, with profiles of early influential American, British, and German eugenicists and the anxieties vitiating their agenda. Chapter 4 examines the origins and organization of the crucial institution in our account, the home for the feebleminded. Chapter 5 reviews state sterilization policy in the pre-Second World War period by discussing the first state laws on coerced sterilization and the role of mental health superintendents in urging those laws' adoption. In Chapter 6, we examine in detail the U.S. Supreme Court decision in *Buck v. Bell* (1927), which gave constitutional ballast to involuntary sterilization, thereby emboldening supporters of sterilization and galvanizing its Roman Catholic critics. Furthermore, it is important to consider cases in which eugenic sterilization was rejected and legislation was thwarted. In Chapter 7, therefore, we investigate and explain the failure of pro-sterilization advocates in enacting legislation in the state of Ohio.

The next chapter moves to the complex case of German eugenic extremism. Chapter 8 examines sterilization in National Socialist Germany, reflecting on its relationship to American eugenics and American eugenicists. This chapter also addresses the postwar collapse of eugenics in Germany, as well as the often-successful efforts of the main German eugenicists to evade responsibility for Nazi crimes against the mentally handicapped.

The book then concentrates on the post-Second World War era. Chapters 9 and 10 examine the activities of the eugenic movement in the postwar period, focusing on how its advocates regrouped and reenergized the pro-sterilization argument by linking it with the emergent "human rights" culture and with the fears of spiraling world population growth. In Chapter 11, we take a close look at the stories from the United States and Canada of individuals who were victims of coerced sterilization. In Chapter 12, we examine in greater detail the abuses suffered by such victims in the context of institutions for the mentally ill and mentally disabled. Chapter 13 explores the links between welfare and coerced sterilization by focusing on the example of North Carolina. In Chapter 14, a closer examination of the sterilizers themselves is taken up. Finally, Chapter 15 concludes the volume.

2

The Eugenicists

Short Portraits

> I am inclined to agree with Francis Galton in believing that education and environment produce only a small effect on the mind of any one, and that most of our qualities are innate.
>
> Charles Darwin[1]

The founding fathers of eugenics had much in common, and yet there was much that differentiated them. When thinking about their ideas, the task for both writer and reader is to do justice to the differences among them while highlighting what bound them together in order to isolate the core tenets of eugenics. These two goals occupy, respectively, this chapter and the next. One way to approach the former objective is to learn of the eugenicists themselves, their histories, and the evolution of their thinking. Once these biographical and intellectual differences are highlighted, we can then bring out the commonalities among eugenicists.

Whereas the bulk of this book will focus on coerced sterilization in liberal democracies – and therefore most attention is given to the United States and Canada – ideas originated on and traveled in both directions across the Atlantic. The intellectual and personal histories of British and German, as well as of American and Canadian eugenicists, are relevant to this end.[2] Throughout, it is important to keep in mind that eugenic ideas would not have been translated into practice without doctors, psychiatrists, superintendents (many of whom were doctors in homes for the feebleminded), and social workers. At the same time, these actors would have found their efforts to develop, adopt, and implement sterilization policies much more difficult, if not impossible, without the

---

[1] Francis Darwin, ed., *The Life and Letters of Charles Darwin* (New York: D. Appleton and Co., 1887), 21.

[2] Intellectually, Canadian eugenics was part of a broader North American pattern. On this, see McLaren, *Our Own Master Race*.

justificatory strength of eugenics. As argued in the introduction, (eugenic) ideas and (medical, institutional, psychiatric) interests reinforced each other and, in so doing, effected substantial policy change.

ORIGIN OF THE THESIS: FRANCIS GALTON

Although there were certainly those who worried about the social effects of reproduction before him, Francis Galton was the first to develop what we would now think of as a policy-relevant theory of human fertility. Galton was born in 1822 to a family of Birmingham entrepreneurs. The Industrial Revolution had been good to them: the family moved from farming, through gun manufacture, and into banking.[3] Galton's parents – Samuel Tertius Galton, a Quaker who converted to Anglicanism under the influence of his wife, Violetta Darwin – hoped to use their new money to secure the intellectual status possessed by his mother's biological family. She was the daughter of Erasmus Darwin, a gifted physician, inventor, and naturalist whose scholarly work was translated into German, French, and Italian and was widely read in the United States.[4] Her nephew was none other than Charles Darwin, who developed the theory of evolution. Matching the Darwins' intellectual achievements proved to be difficult. Of Violetta Darwin's eight children, five were girls, and the ambition of Francis's two brothers were limited to enjoying a life of ease among the landed gentry.[5] The entire weight of family expectations landed, therefore, on the youngest Galton's shoulders. Fortunately, he was a prodigy. His older sister taught him Latin, Greek, and French and had him recite English verse.[6] At two and a half, he could read, and, by four, he could do arithmetic.[7] Francis knew what was expected of him. When his mother asked him why he was saving his pennies, Galton replied, "why, to buy honours at the University."[8]

After King Edward's School in Birmingham, Galton moved to London and enrolled in medical school at King's College. He excelled in his first year but hated the study of medicine and suffered constant headaches.[9] He left London for Trinity College, Cambridge, where he enrolled in mathematics. He worked hard but never secured anything above a high second in university exams. Galton returned briefly to studying medicine until the death of his father in 1844, when he was bequeathed an inheritance large enough to free him from the need for study or paid work.

---

[3] Kevles, *In the Name of Eugenics*, 5.
[4] Nicholas Wright Gillham, *A Life of Sir Francis Galton: From African Exploration to the Birth of Eugenics* (New York: Oxford University Press, 2001), 13–17.
[5] Kevles, *In the Name of Eugenics*, 5.
[6] Gillham, *Life of Sir Francis Galton*, 24.
[7] Kevles, *In the Name of Eugenics*, 5.
[8] Gillham, *Life of Sir Francis Galton*, 24.
[9] Kevles, *In the Name of Eugenics*, 5.

Galton spent the next two years wandering the globe, sailing up the Nile, and visiting Beirut, establishing himself briefly near Damascus (where he learned Arabic), and visiting Jerusalem. In 1846, Galton returned to London and lived the life of an independently wealthy scientist. Like many Victorian scientists, Galton's interests were eclectic, and he left lasting contributions to several fields of enquiry.[10] In statistics, he pioneered work in correlation, regression, and biometrics. In meteorology, he discovered the anticyclone. He was also an early explorer and travel writer – and a reasonably open-minded one at that. He took the existence of inferior and superior races as given, which is hardly surprising given his time, but work on class, national, and racial differences comprises only a fraction of his writing.[11] Indeed, given his range of interests and his love and aptitude for statistics ("whenever you can, count," he liked to say), it is not obvious when and why Galton turned to eugenics and heredity. Scholars of his life cite Galton's distaste for the origins of his wealth, a desire to overcome his insecurities about his own successes by confirming that success was hereditary, and, finally, an obsession with upper- (intellectual) class fertility born of his own inability to produce a child (his cousin, Charles Darwin, produced ten).[12] Galton's complex personality makes it hard to know. As Kevles expresses it, Galton was

> a rough-cut genius, a pioneer who moved from one new field to the next, applying methods developed in one to problems in another, often without rigor yet usually with striking effectiveness. Galton's innovativeness in science was intimately bound to his relative intellectual solitude – a propensity that arose from a measure of doubt in his ability combined with a compulsion to excel.[13]

Whatever the reason, Galton's eventual work on eugenics linked his obsession with heredity to his mathematical bent. In papers published in the popular *Macmillan's Magazine*, and more fully in a book entitled *Hereditary Genius*, Galton laid out the case for nature over nurture. Relying on an 1865 edition of the *Dictionary of Men of the Time* and obituaries compiled from the *Times*, he estimated that 1 in 4,000 men in Britain qualified as "eminent." Against this baseline, he analyzed the pedigrees of well-known statesmen, peers, military commanders, judges, and men of similar ilk.[14] Making some assumptions about the number of brothers and sisters of eminent men, he provided data showing that close relatives of the eminent are much more likely to be eminent than are distant ones.[15] In a move for which he is not always given credit, Galton also attempted to control for environmental influences on eminence, if somewhat unsystematically. He did so partly through assertion, partly through comparison – education and

[10] See Gillham, *Life of Sir Francis Galton*, prologue.
[11] Kevles, *In the Name of Eugenics*, 8.
[12] See Gillham, *Life of Sir Francis Galton*, prologue; see also Kevles, *In the Name of Eugenics*, 8–9.
[13] Kevles, *In the Name of Eugenics*, 10.
[14] Gillham, *Life of Sir Francis Galton*, 161.
[15] Ibid., 160–7.

culture, he argued, were more widely accessible in America, but that country did not create more eminent men – and partly through statistical control – the distant kin could be assumed to be subjected to the same environmental influences, but they failed to achieve the same eminence. Through good luck and the intrinsic interest of its content, the book achieved a wide readership. An anonymous critic, reacting to Galton's argument that religious men were below average in vigor and longevity, penned a bitter assessment in *The Spectator*, which in turn led to a torrent of letters to the magazine. His cousin, Darwin, himself convinced that state interventions would have dysgenic effects in encouraging the survival of the unfit, found the uproar immensely entertaining.[16] He sent Galton a note congratulating him on "the tremendous stir-up your excellent article on 'Prayer' has made in England and America."[17]

Galton's argument up to this point was hereditarian – that is, it held that certain key traits were inherited – but not yet eugenic. The last chapter added the eugenic element by relating his statistical work to social policy and to the hereditary vitality of the nation. As his biographer put it,

> [Galton's] theme was straightforward. . . . Since those marrying young have larger families, produce more generations in a given period of time, and more generations are alive at the same time, the wisest policy is one that retards "the average age of marriage among the weak, and . . . hastens . . . it among the vigorous classes; whereas unhappily for us, the influence of numerous social agencies has been strongfully and banefully asserted in the precisely opposite direction."[18]

Galton continued his work on heredity by gathering data on and studying correlations among human (schoolboys) and natural (sweet pea) subjects, but there was throughout a tension between his hereditarian and his biological arguments. All the statistical work seemed to confirm two things. First, positive and negative characteristics, such as talent, vigor, and zeal, were distributed according to the bell curve: the best and worst at the extremes, with most people crowding in the middle. Worse still from a eugenic point of view, any effort to interfere with this distribution was hopeless: if only population members from the extremes were chosen for production (say, the heaviest peas), over time, their progeny would eventually regress toward the center of the bell curve (the mean).[19] This rendered the eugenic project of better breeding pointless.

Galton could not reconcile his statistical results and his eugenic aims. It would be left to his successors, the next generation of British eugenicists, to do so.

---

[16] On Darwin's eugenics, see Dowbiggin, *Sterilization Movement*, 19–20.
[17] Quoted in Gillham, *Life of Sir Francis Galton*, 167.
[18] Ibid., 169.
[19] Kevles, *In the Name of Eugenics*, 18.

KARL PEARSON

Galton's work found its continuation in his protégé, Karl Pearson, who became a lifelong friend and eventual biographer of Sir Francis. Like Galton, Pearson's father was a man of the liberal professions – a lawyer and eventual Queen's Counsel (a British and Commonwealth designation for distinguished lawyers) – who expected, indeed demanded, that his son follow in his footsteps. Pearson was able to study his passion, mathematics, at King's College, London, and as graduate student in Berlin and Heidelberg (where he balanced a hatred for the Germany of the Kaiser with a romanticized love for the Germany of Goethe). On his return to London, however, he dutifully entered Lincoln's Inn to prepare for the bar.[20] Pearson hated the law and desperately applied for mathematical posts in an effort to escape it and his father's pressure to become a criminal lawyer. On his fourth attempt, he landed a position at the prestigious University College, London (UCL).

Pearson has been described as misogynist, reactionary, and racist.[21] Such a reading is selective. He was, in fact, a man of the left and a proto-feminist. His worldview was a mix of German romantic – a Germanophile, he had changed the spelling of his first name from "Carl" to "Karl" and aspired to find a German wife – and British technocratic. From Germany, he inherited the Fichtian idea that the state embodies the values of its people, as well as (from the German left) an anti-imperialist bent.[22] From Britain, he acquired a radical edge and an interest in social reform. Pearson developed associations in London with Karl Marx's daughter, Eleanor; with George Bernard Shaw; Fabian social reformers Sydney and Beatrice Webb; and the radical sexual libertarian, Havelock Ellis.[23] The result was what one critic calls scientific socialism,[24] in which the intellectual elite (of which he would naturally be a part) would bring about a socialist state and ensure that material goods were divided as equally as possible within it.[25] Imbued with a strong authoritarian streak, Pearson's socialism did not involve enfranchising the working class.[26]

After socialism, Pearson thought the most pressing question was the "women question."[27] Hardly a feminist by postwar standards, he nonetheless was inspired by his experiences with his mother, whom he saw as financially imprisoned in marriage, to believe that genuine freedom for women required economic

---

[20] Ibid., 22.
[21] Stephen Trombley, *The Right to Reproduce: A History of Coercive Sterilization* (London: Weidenfeld and Nicolson, 1988), 40–4.
[22] Kevles, *In the Name of Eugenics*, 23.
[23] Ibid., 24.
[24] Trombley, *Right to Reproduce*, 40.
[25] Kevles, *In the Name of Eugenics*, 23.
[26] Trombley, *Right to Reproduce*, 41.
[27] Kevles, *In the Name of Eugenics*, 24.

independence, which in turn required socialism.[28] Pearson founded the Men and
Women's Club in 1885 to tackle this very issue. The Club devoted itself to
discussing prostitution, contraception, marriage, and women's economic independence as well as intellectual abilities. Although Pearson's motives may have
been suspect – Kevles suggests that he created the Club to meet rather than to
emancipate women – his was, by the standards of the day, a progressive stance
toward still-disenfranchised women citizens.[29]

The most important relationship in Pearson's life did not develop at his club
and was not with a woman. It was, rather, with a graduate student who in most
respects was the opposite of Pearson. In 1892, Pearson began working with
Walter F. R. Weldon, a Cambridge graduate, a brilliant biologist, and a convert
to Galton's mathematical techniques (he had read Galton's *Natural Inheritance*
when it appeared in 1889).[30] After a failed attempt to make London University
into a metropolitan institution with uniform standards, the two men turned to
the statistical study of evolution and heredity.[31] The relationship had a transformative effect for Pearson. He felt like "an adventurous roamer . . . a buccaneer
of Drake's days."[32] During the pair's fourteen-year collaboration, they published a hundred scientific papers. Most important for the history of eugenics,
Weldon reconciled Galtonian statistics with Galtonian eugenics. Undertaking
new observations and reworking Galton's statistical analysis, Weldon and
Pearson concluded that a population would not regress toward the mean;
selective breeding could, after a few generations, alter the distribution of characteristics.[33] Eugenic breeding was possible after all. Galton, who had encouraged Pearson and Weldon from the start, was ecstatic.

Pearson and Weldon's laboratory, along with their journal, *Biometrika*,
sought to give the statistical measure of heredity (biometrics) authenticity.
Galton, sensing a new openness to eugenics in Britain, established and endowed
a Eugenics Record Office (ERO) with the remit to pursue the study of "National
Eugenics" in 1904.[34] Seven years later, at his death, Galton's estate left an
additional 45,000 pounds sterling to endow a Galton Professorship of
Eugenics. Following the will's suggestion, Pearson was the first incumbent.

Rather paradoxically, the new British receptiveness to eugenics resulted in
part from a bitter debate that developed at the turn of the century. The forgotten
work of an Austrian monk named Gregor Mendel, who used experiments with
peas in his monastery's garden to develop a theory of "unit characteristics"
(genes), won fresh attention.[35] For Mendel, genes were not gradations along a

[28] Ibid.
[29] Ibid., 25.
[30] Soloway, *Demography and Degeneration*, 26.
[31] Kevles, *In the Name of Eugenics*, 29.
[32] Quoted in Soloway, *Demography and Degeneration*, 27.
[33] For the details, see Kevles, *In the Name of Eugenics*, 30–1.
[34] Soloway, *Demography and Degeneration*, 28.
[35] Kevles, *In the Name of Eugenics*, 41–4.

curve. They were either present or they were not: a person was male or female, colorblind or not.[36] When different genes blended, the resulting "hybridization" could be predicted. Based on observations of his plants, Mendel concluded that planting a tall plant with a tall plant would produce a tall plant, and a short plant with a short plant would produce a short plant. He had expected a tall and short plant to produce an intermediate plant, but when he noticed that such mixing regularly yielded a tall plant, he concluded that some unit characteristics were dominant (the tallness gene), whereas others were recessive (the shortness gene).[37] The prevailing assumption of hereditarianism meant no variation across time. The methodological implication was that testing for heredity required experimental breeding, not statistical enquiries.[38] The ERO, in other words, could close its doors. When British geneticist William Bateson began championing Mendel's theories, he earned Pearson and Weldon's undying enmity. The controversy that raged between biometricians and Mendelian geneticists was legendary.

In the midst of this conflict, Galton, Pearson, and Weldon received a visit from a young American scientist named Charles Davenport. The four men dined together in London, and Davenport returned to the United States imbued with the zeal of a convert.[39] He would leave a lasting imprint on the development of American eugenics.

CHARLES DAVENPORT

Davenport was born at his family's summer home in Stamford, Connecticut, in 1866. His father, a former teacher turned real estate and insurance broker, had been an early abolitionist and was an ardent puritan. Under his father's influence, he first studied engineering, taking a degree in the subject from Brooklyn Collegiate and Polytechnic Institute near the family's home in Brooklyn Heights. Plucking up enough courage to defy his father (the lot of aspiring eugenicists, it seems), he left Brooklyn for Harvard and enrolled to study biology. His early academic years were undistinguished; he spent seven of his twelve years in Cambridge, Massachusetts, as an instructor without tenure, prestige, or much of a salary.

It was, however, at Harvard that he came into contact with the work of international eugenicists by reading Karl Pearson's papers on mathematical evolution.[40] Davenport's book, *Statistical Methods with Special Reference to Biological Variation* (1899), was heavily influenced by Galton's statistical theories and those developed by Pearson. In that year, Davenport went to the

[36] Soloway, *Demography and Degeneration*, 28.
[37] Kevles, *In the Name of Eugenics*, 41–2.
[38] Soloway, *Demography and Degeneration*, 28.
[39] Kevles, *In the Name of Eugenics*, 45.
[40] Ibid.

University of Chicago, where he spent the next five years, securing tenure and rising to the position of associate professor.[41] While at Chicago, Davenport showed the first signs of what would become an impressive talent for fundraising. In May 1902, when the Carnegie Institution was only a few months old, he wrote a grant proposal for the endowment of a "Station of Experimental Evolution at Cold Spring Harbor" on Long Island, naming himself as director.[42] Davenport was already directing popular courses as director of the summer Biological Laboratory of the Brooklyn Institute of Arts and Sciences located at Cold Spring Harbor. To elicit support, he wrote to Henry Fairfield Osborn, who was director of the American Museum of Natural History and an enthusiastic supporter of eugenics.[43]

Carnegie signaled its approval, and the Station for Experimental Evolution was officially opened in 1904. In its early years, the Station employed a small staff of able students who conducted breeding experiments on animals and plants. Davenport's interest, however, was captured primarily by human heredity. As he could not experiment with human breeding, he turned instead to studying family pedigrees. The method was similar to Galton's, except that Galton showed a greater interest in "phenotype" (human traits that could be seen), whereas Davenport showed a greater interest in "genotype" (the individual's genetic makeup).[44] It was here that Davenport's work acquired its eugenic thrust. Davenport devised a "Family Record" questionnaire for high schools, colleges, and many individuals.[45] Hundreds were returned. Davenport recorded the information, and, whenever family pedigrees showed a high incidence of a particular trait, he concluded that it was hereditary.[46] From these observations, he concluded that insanity, epilepsy, alcoholism, "pauperism," and especially feeblemindedness were all hereditary conditions.

As his interest in human heredity overtook all others, Davenport laid out a more ambitious task: mapping out the genetic health of the United States. He decided to try to establish a second institute at Cold Spring Harbor. The new institute would be financially and intellectually independent of the Station of Experimental Evolution and would concern itself solely with research on human heredity. Such a task required an enormous amount of money. Here, Davenport was able to exploit an earlier contact. In the summer of 1906, he had taught genetics to Mary Harriman, daughter of railway magnate E. H. Harriman.[47]

[41] Reilly, *Surgical Solution*, 18.
[42] Charles E. Rosenberg, "Charles Benedict Davenport and the Beginning of Human Genetics," *Bulletin of the History of Medicine*, 35 (1961): 267.
[43] Ellen Condliffe Lagemann, *The Politics of Knowledge: The Carnegie Corporation, Philanthropy, and Public Policy* (Middletown, CT: Wesleyan University Press, 1989), 162.
[44] Kevles, *In the Name of Eugenics*, 45.
[45] Ibid., 46.
[46] Ibid.
[47] "Aims, Work, and Results of the Eugenics Record Office," n.d., BD 27 CD, American Philosophical Society Archive (APSA), Philadelphia.

The latter died in 1909, leaving his wife a substantial estate of $70 million. Attending closely to the sickly rich is in a serious fundraiser's job description, and Harriman's death was followed by a flurry of special pleas. Davenport patiently waited this out and used his connection to Mary to arrange a meeting with her mother. Mrs. Harriman, affected by Progressivist optimism about the scope for social reform, was seduced by the idea that the key to solving great social problems lay in isolating and eliminating defective members of the human gene pool.[48] In 1910, he convinced the widow to establish the ERO, with Davenport as director, near the Station. In 1920, she transferred the building and land (worth $183,671.15) along with an endowment of $300,000 to the Carnegie Foundation.[49] The Station's yearly budget in 1906 was already $21,000, which was twice that of Pearson's comparable institute at UCL. Its expanded size and resources in 1910 made it the United States' – and probably the world's – key foundation for research on eugenics.

Davenport's papers were left at Philadelphia's American Philosophical Society Library. They attest to Davenport's tireless research on a range of scientific and sociological questions: heredity, agriculture, race, physical anthropology, feeblemindedness, and alcoholism.[50] The ERO served as a repository of information on family histories, with records numbered in the tens of thousands; established an "analytical" index of nearly 250,000 cards containing information on families and family traits; and trained dozens of field workers in the methods of data collection on families. It also ran summer schools for budding eugenic scientists.

During this period, Davenport worked with Van Wagenen of the American Breeders Association to set up a committee to study the results of sterilization. The group began compiling data in 1911, and its first results were presented at the First International Congress of Eugenics in London in 1912.[51] The group drew on a wide range of consultants including geneticist Raymond Pearl and anatomist Lewellys Barker (both of Johns Hopkins), as well as Yale economist Irving Fisher.[52] We will meet all three again in subsequent pages.

Although Davenport is credited with producing respectable work in genetics, historians of science have not been kind in their judgments of his scientific contribution. "Greatly given to oversimplification," Kevles writes, "and little to self-critical reflection, Davenport possessed neither Galton's idiosyncratic imagination nor Pearson's formidable intellectual power. Like Pearson, he was

---

[48] Reilly, *Surgical Solution*, 19.

[49] "Aims, Work, and Results of the Eugenics Record Office," BD 27 CD, APSA.

[50] For a nonexhaustive list, see BD 27 CD, APSA: "About Eugenics," n.d.; "Agriculture and Eugenics," n.d.; "Heredity of Disease," n.d.; "A Biologists View of the Negro Problem," (rough notes), n.d.; "Battle Creek lectures," (on heredity), n.d.

[51] Paul A. Lombardo, *Three Generations, No Imbeciles: Eugenics, the Supreme Court, and Buck v. Bell* (Baltimore: Johns Hopkins University Press, 2008), 42.

[52] Ibid.

blinded by eugenic prejudice. But, unlike Pearson, Davenport did not base his eugenics on any political world view; he had none."[53]

Davenport did have, however, a decisive influence on the history of the eugenics and of America's experiment in coerced eugenic sterilization. There is an element of irony in this, since Davenport only gave half-hearted support to the idea. A fuller role was played by someone who, were it not for Davenport, history would have forgotten.

HARRY H. LAUGHLIN

On February 25, 1907, a young schoolteacher from the Midwest named Harry H. Laughlin wrote to Davenport for advice on the classification of animals.[54] Born in Oskaloosa, Iowa in 1880, Laughlin later moved with his family to Kirksville, Missouri. He had four brothers, all of whom studied osteopathy. Laughlin rejected the family's calling and became a schoolteacher instead, graduating from the North Missouri State Normal School in 1900. Although he took no degree from his subsequent state studies at Iowa State College, Laughlin became a big fish in a small pond: he was a high school principal at the age of twenty, and, by twenty-five, he was superintendent of schools.[55] At the age of twenty-seven, he received, in response to his letter to Davenport, an invitation to spend six weeks at Cold Spring Harbor. Laughlin accepted, and the experience proved transformative for him. Davenport was sufficiently convinced of the young man's talents to invite him to become superintendent at the newly established ERO in 1910.

Laughlin made, in Davenport's words, the issue of sterilization his own.[56] Laughlin became a tireless worker for and promoter of eugenic sterilization over the next thirty years, until the Carnegie Institution, increasingly embarrassed by Laughlin's obsession with race, drove him out. His preferred method involved collecting family histories, and he developed reams of family charts that can still be found at the American Philosophical Society Archive in Philadelphia and at the Social Welfare History Archives in Minneapolis.

Laughlin would later have a major influence on United States immigration policy unequalled by any other eugenicist. Before and after his appointment as eugenics expert to the House of Representatives Committee on Immigration, Laughlin provided the restrictionist Republican Congressman Albert Johnson with the scientific arguments and respectability needed to guide the race-based 1924 Johnson-Reed Immigration Act through Congress. Some of his ideas may have made him an extreme eugenicist, but his influence on policy was decisive.

[53] Kevles, *In the Name of Eugenics*, 49.
[54] Harry Laughlin to Charles Davenport, February 25, 1907, BD 27 CD, APSA.
[55] *Dictionary of American Biography*, suppl. 3 (1941–1945), s.v. "Laughlin, Harry Hamilton."
[56] "Aims, Work and Results of the Eugenics Record Office," BD 27 CD, APSA.

PART-TIME EUGENICIST IRVING FISHER

Laughlin was, in every sense of the term, a zealot for eugenics: it defined his purpose, consumed his days, and was his only legacy. For every person like Laughlin, there were dozens of others who believed in eugenics and devoted some amount of time to see it implemented but who counted it as only one interest among many. Irving Fisher was such a person. Eugenics was in no way his consuming interest; that, rather, was economics. He provided an excellent example of how an interest in eugenics could intersect with, and partly be driven by, academic and social interests that we would now regard as unobjectionable, even admirable.

Fisher was born in 1867. In contrast to many other eugenicists, he had few class insecurities. He was born into a stable but financially strapped East Coast family. His father was a Protestant pastor, and the household in which Fisher was raised combined a loving environment with a stern Protestant commitment to public service.[57] This home environment interacted with Fisher's own natural inquisitiveness and thirst for knowledge, and, very soon, scholarship became the basis for Fisher's economic advancement. During his last two years of high school, his father sent him to the Smith Academy in St. Louis.[58] Throughout 1883 and 1884, Irving Fisher and his father corresponded at length regarding which university the young man should attend.

Then Irving Fisher's world fell apart. During the spring of his last year in high school, when he was seventeen, his father fell grievously ill with tuberculosis. After writing his last college entrance exam, Fisher rushed back east. Within two weeks, his father was dead.[59] With his father gone, Fisher's family was now staring poverty in the face. The dual assault would have overwhelmed most men, but Fisher was determined to attend his father's university – Yale – and to provide for his mother and younger brother while doing so. The family moved into a three-room apartment in New Haven, and everyone threw their energy into keeping the family fed and Fisher at university. The burden fell chiefly on Irving himself: he tutored for extra money; won prizes in Latin, Greek, and mathematics; and tried to make extra income through inventions, a pursuit that he would follow throughout his life.[60] His mother earned what she could through dressmaking and renting out rooms, while Irving's eleven-year-old brother worked as her errand boy.[61] Irving went through Yale on the poverty line but still rubbed shoulders with some of America's richest sons (Henry L. Stimson was a peer and a friend/competitor of Irving, and the two boys were both in Yale's Skull and Bones society). The experience seems to have

---

[57] Robert Loring Allen, *Irving Fisher: A Biography* (Cambridge, MA: Blackwell, 1993), chapter 1.
[58] Ibid., 21–2.
[59] Ibid., 22.
[60] Ibid., 30–1.
[61] Ibid., 30.

drawn him closer to his mother, with whom Fisher exchanged warm, even gushing, letters until the end of her days.[62]

Fisher graduated from Yale in 1888. Encouraged by his professors and, as ever, supplementing his income through tutoring, he continued his studies at Yale. In 1891, he earned the university's first Ph.D. in economics, and, the following year, he was appointed to an assistant professorship in mathematics. He transferred to the new field of political economy within a year. In 1893, his financial security and personal happiness increased greatly when he married Margaret Hazard, the daughter of a wealthy Rhode Island industrialist.[63] The following year brought a fourteen-month sabbatical and a honeymoon in Europe. During his tour throughout the old continent, Fisher visited some of the world's most renowned economists in London, Oxford, Cambridge, Paris, Vienna, Lausanne, and Berlin, renting an apartment for a longer stay in the last.[64]

On his return from Europe in 1896, Fisher published his second book, *Appreciation and Interest*, and was elected to a full professorship two years later at the young age of thirty-one. Then, disaster struck again. Fisher came down with tuberculosis, the disease that had killed his father and that few survived in that day. He did recover, although he needed three years to do so, and it unremarkably altered his life. Always a physical fitness enthusiast,[65] his near-death experience fueled in him an enduring concern for personal as well as public health. He later wrote in the 1920s,

> My illness and the enforced idleness for several years greatly changed my point of view from that of an academic student of supply and demand as a mathematical problem to that of a partaker in public movements for the betterment of mankind. While I have never given up my narrow specialty of mathematical economics I have added to that central interest a keen interest in the world around me.[66]

It was a concern that would form his support for temperance movements, pacifism, health, and – not least – eugenics.

Unlike Davenport, Laughlin, and the German Ploetz, Fisher was not a natural (as distinct from social) scientist, and, unlike Laughlin, he was not solely consumed by the eugenic cause. A review of his private correspondence produces no references to coerced sterilization, and the bulk of his papers concerns

---

[62] See the letters between the two contained in the Yale archives.

[63] Allen, *Irving Fisher*, 62.

[64] In the midst of this busy schedule, a daughter arrived. Much of the correspondence in the archives gives evidence of a doting father, one who fired a harsh and unsmiling nurse after six days with his newborn. See Irving Fisher to Ella Fisher, May 2, 1894, and Mrs. Rowland Hazard to Ella Fisher, May 5, 1894, Irving Fisher Papers (IFP), Yale University Library Manuscripts and Archives, New Haven, CT.

[65] Allen, *Irving Fisher*, 75.

[66] Quoted in William J. Barber, "Irving Fisher (1867–1947): Career Highlights and Formative Influences," in *The Economics of Irving Fisher: Reviewing the Scientific Work of a Great Economist*, ed. Hans-E. Loef and Hans G. Monissen (Cheltenham: Edward Elgar, 1999), 8.

economics, chiefly monetary policy.[67] He is much more known in that field than in eugenics. After losing his fortune in the 1929 stock market crash (it "hit me between the eyes," he wrote with some understatement),[68] he wrote the definitive paper on debt deflation. Indeed, when he secured an audience with Mussolini two years earlier, it was on the subject of currency stabilization, rather than degeneration, mental illness, race, or breeding that he chose to speak.[69] On August 31, just before his meeting with the Italian dictator, he had attended a lunch at a Paris population congress, which was also attended by Raymond Pearl (a professor at Johns Hopkins University, an early critic of eugenicists' upper-class bias, and an anti–population-growth activist), Roswell Hill Johnson (co-author with Paul Popenoe of *Applied Eugenics*, a 1918 publication), and Charles Davenport, all leading U.S. eugenicists.[70]

Fisher was, in his day, both a public intellectual and respected academic figure. In politics, Fisher exchanged correspondence with Presidents Harding, Coolidge, and Franklin D. Roosevelt; in the academy, he exchanged letters with Keynes and debated his work with Einstein. Documents on eugenics might have been weeded from his correspondence, but Fisher clearly had a broader range of interests than most eugenicist zealots. He nonetheless had a lifelong interest in degeneration – of himself, his country, and his race. He was an active lobbyist for the creation of a federal department of health, and he saw a great loss to the United States in its unnecessarily high mortality rate.[71] Fisher was a founder of the American Eugenics Society and one of three American delegates (with Davenport and Laughlin) to the International Federation of Eugenic Organizations, founded in New York in 1922.[72] Fisher is an ideal example of a moderate eugenicist of the sort that journalistic histories, which gravitate naturally toward the shocking, tend to ignore. Like many people of his class and generation, he saw natural and uncontroversial links between the health of the person, the class, the race, and even the planet.[73]

---

[67] For a brief discussion of Fisher's contributions to economics, see Loef and Monissen, *Economics of Irving Fisher*. Also see William R. Allen, review of *Irving Fisher: A Biography* by Robert Loring Allen, *Journal of the History of Economic Thought* 17, no. 1 (1995): 153–5.

[68] A few days before the crash, Fisher had predicted that the stock market had reached a "permanently high plateau." Quoted in Allison Berry, "Top 10 Failed Predictions," *Time*, October 21, 2012, http://www.time.com/time/specials/packages/article/0,28804,2097462_2097456_2097459,00.html.

[69] Fisher's note on his meeting with Mussolini, September 2, 1927, IF 212/I/6/92, IFP.

[70] Note from Irving to Margaret H. Fisher, September 5, 1927, IF 212/I/6/92, IFP.

[71] Irving Fisher, *Hygienic Essays*, vol. VI, 1924–1931, reel 6, "Conservation of Human Life"; "A Department of Dollars vs. a Department of Health," *McClure's Magazine*, July 1940.

[72] Mrs. Cora B. S. Hodson – Eugenics, October 4, 1932, International Federation of Eugenics Organizations, RF 176–89,3,2,8, RAC.

[73] Irving Fisher, "Eugenics," *Good Health Magazine* (November 1913), 1–21; "The Importance of Hygiene for Eugenics," Address to the First National Conference on Race Betterment, Battle Creek Sanatorium, Battle Creek Michigan, January 8–12, 1914, IFP.

THE EMINENCE OF GERMAN EUGENICS: ALFRED PLOETZ

Germany's roughly equivalent figure to Francis Galton was Alfred Ploetz. In 1895, after a sojourn in the United States, he published a treatise on racial hygiene.[74] In 1905, he helped found the Society for Racial Hygiene, a term he had invented.[75] After a brief hiatus – the war demoralized and desensitized German eugenicists, and politicians in the early years of the Weimar Republic were more concerned with increasing population than augmenting its genetic quality – German eugenicists were able to regroup.[76] A sympathetic president of the Imperial Bureau for Public Health (*Reichsgesundheitsamt*) shared fears with Ploetz about American and Swedish advances in heredity and eugenic research. This situation led to the creation in 1927 of the Kaiser Wilhelm Institute for Anthropology, Human Heredity and Eugenics (KWI, for *Kaiser-Wilhelm-Institut für Anthropologie, Menschliche Erblehre und Eugenik*) under the directorship of Eugen Fischer.[77]

Ploetz's racial hygiene was driven by a fear of the "counter-selective" forces.[78] They ranged from war and revolution, which destroyed the race's youngest and healthiest, to alcohol and venereal disease, which damage human germ plasma. Procreation among the very old and very young was also to be avoided. Above all, well-intentioned medical interventions would have the consequence of saving the weak and allowing them to procreate when, in the absence of medicine, nature would have finished them off.

So far, this sounds like typical social Darwinism. But Ploetz was indifferent neither to the suffering of the poor nor to the cruelty entailed in allowing, still less encouraging, people to die. Ploetz was sympathetic toward insurance for unemployment, health, and accidents; old age pensions; reduced working hours; the elimination of child labor; minimum wages; and a more equitable sharing of profits between capital and labor. At the same time, however, he feared that such policies would have the unintended consequence of allowing the weak to survive.[79] There was thus a contradiction between the demands of welfare and the demands of nature.[80] Eugenics, Ploetz believed, would overcome the

---

[74] Alfred Ploetz, *Die Tüchtigkeit unserer Rasse und der Schutz der Schwachen* (Berlin: S. Fischer, 1895).

[75] Robert N. Proctor, *Racial Hygiene: Medicine under the Nazis* (Cambridge: Harvard University Press, 1988), 17.

[76] Weingart, "German Eugenics between Science and Politics," 261–2. Fischer tipped the Institute in an anti-Semitic direction: an anthropologist, he defined an issue of particular importance to the discipline as "the question of the Jewish population living among a non-Jewish one."

[77] Ibid., 263–4.

[78] The next three paragraphs are based on Proctor, *Racial Hygiene*, 14–16.

[79] Ploetz's empiricism directed much of his thinking. Citing Jesus, Spinoza, and Marx, he included Jews with Aryans at the pyramid of the races (that there was such a pyramid, with "Negros" at the bottom, was obvious to him). Science, he also believed, would sweep away European anti-Semitism. Proctor, *Racial Hygiene*, 20–1.

[80] Ibid.,16.

contradiction. By introducing selection at the right time, in the germ cells, there would be no need for a harsher, real-world process of survival of the fittest.

Although he was the founder of eugenics in the country where it reached a peak of cruelty and brutality, Ploetz is counted among the moderate German eugenicists. He was a "cautious advocate" of social reform, and progressives in turn found his brand of racial hygiene appealing.[81] Among them were Wilhelm Schallmayer,[82] who argued for "race hygiene" rather than "races hygiene" in order to make clear that the health of the whole and not simply the Nordic race was his concern, and even Alfred Grotjahn, the architect of Weimar German's progressive health reforms. Others in Germany, above all after the First World War, would give eugenics its violent and anti-Semitic thrust.

EUGEN FISCHER

In 1944, Otmar Freiherr von Verschuer penned a tribute to Eugen Fischer on the latter's seventieth birthday:

> The name Eugen Fischer has become one of a scientific program. He has single-handedly linked together an older method of anthropology with the study of heredity and transformed both into race biology [*Rassenbiologie*]. He was one of the first who grasped the discoveries of heredity and race studies and ... saw through their implementation. Through his contributions to present racial theory [*Rassengedanken*], he has contributed to the intellectual foundations of National Socialist race policy.[83]

Fischer was born on June 5, 1874, in Karlsruhe but was raised and educated in Freiburg, a city he would love until his death in 1967.[84] Fischer's father was a successful businessman, and the family was upwardly mobile: his paternal grandfather had worked himself up to the imposing title of grand-ducal senior forestry inspector (*großherzoglicher Oberforstinspektor*), making the family socially "off the land" even if professionally still very much on it. When Fischer was two, his father's health led him to resign from his firm, but he had sufficient resources to retire comfortably. The elder Fischer moved his family to the (then-idyllic) university town of Freiburg, where he purchased a large, attractive house. His son led a charmed but lonely childhood, without financial worries and surrounded by nature. Nonetheless, like Galton in England, he

---

[81] Ibid., 22.

[82] Sheila F. Weiss, *Race Hygiene and National Efficiency: The Eugenics of Wilhelm Schallmayer* (Berkeley: University of California Press, 1987).

[83] Otmar von Verschuer, "Eugen Fischer zum 70. Geburtstag am 5. Juni 1944," *Der Erbarzt* 12, nos. 5–6 (May/June 1944): 57–9. Verschuer was a teacher of the infamous Nazi doctor at Auschwitz, Josef Mengele.

[84] The next two paragraphs draw on Bernhard Gessler, *Eugen Fischer (1874–1967): Leben und Werk des Freiburger Anatomen, Anthropologen und Rassenhygienikers bis 1927* (Frankfurt: Peter Lang, 2000), 2–3.

would never be free of social insecurity. Talented but friendless,[85] Fischer completed his *Abitur* (academic high school diploma) in Freiburg and went on to study, partly under his father's influence, anatomy and (his choice) zoology at the Freiburg University. There, he attended the lectures of August Weismann, a proponent of Mendelian arguments: the theory that germ plasma is inherited from one generation to another without outside influence. Through an acquaintance with Weismann's son, Fischer was brought into the great man's circle. After obtaining his diploma, he committed himself to anatomy, finishing his doctorate in 1898 and, two years later, his *Habilitation*.[86] Academic success in Germany, then as now, depends on securing the support of a senior professor, who then acts as a patron with whose fate that of a young scholar is intimately tied. The director of the Institute for Anatomy at Freiburg University, Robert Wiedersheim, fulfilled such a role for Fischer. Wiedersheim's predecessor, Alexander Ecker, kept an extensive collection of skulls for anthropological research. They remained unanalyzed in glass jars at his retirement, and Wiedersheim gave them to Fischer with the instruction to incorporate them into his teaching. The possession of these primary materials, along with another quirk of German academe, set Fischer on his eugenic path. The department refused to allow Fischer to deliver any lectures (something incomprehensible to overburdened departments and professors today), and Fischer was forced to give his first set of lectures on *"Einführung in die Anthropologie"* (Introduction to Anthropology).[87]

When war came to Europe in 1914, it robbed Fischer of much: his brother, three cousins, and, during the Allies' single air raid on Freiburg on April 17, 1917, the Institute for Anatomy. On the night of the raid, Fischer observed the rubble that remained of the Institute, retired to his study, and wept. The experience of war and the loss of his family members – he had also lost a son in 1911 – embittered Fischer and reinforced his nationalist, monarchist sympathies. He joined the hard-line nationalist and anti-Weimar German People's Party (*Deutschnationale Volkspartei*) in 1919.[88] Fischer remained in his beloved Freiburg, oversaw the rebuilding of the Institute, and rose further in the university hierarchy.

On June 10, 1926, Fischer was granted an opportunity that altered his career and the development of German eugenics. The Kaiser Wilhelm Gesellschaft in Berlin-Dahlem, the prosperous and leafy suburb southwest of the city, offered him the opportunity to found and lead an anthropological research institute. After a comical attempt to have the Institute moved to Freiburg, this

[85] Ibid., 10–11.
[86] A *Habilitation* is second major work required by successful doctoral students who wish to become professors.
[87] Gessler, *Eugen Fischer*, 15.
[88] Robert Proctor, "From *Anthropologie* to *Rassenkunde* in the German Anthropological tradition," in George W. Stocking, Jr., ed., *Bones, Bodies, and Behavior: Essays on Biological Anthropology* (Madison: University of Wisconsin Press, 1988), 157.

quintessentially southern German prepared to move to Prussia. Fischer articulated his hopes in an emotional letter to Karl Ludwig Schemann, a race theorist and translator of Arthur de Gobineau: "I want to try to render racial theory and eugenics relevant and fruitful for our entire German people. I sacrifice my love for anatomy and the teaching of medicine, and my work for the Baden *Heimat* [his home region] to this greater goal."[89]

FRITZ LENZ

The two-volume magnum opus of German eugenics was *Grundriss der menschlichen Erblichkeitslehre und Rassenhygiene* (*Foundations of Human Heredity and Race Hygiene*). The work's junior co-author, after Fischer and Erwin Baur, was Fritz Lenz.[90] Lenz was born in Pflugrade, then part of Pomerania, today in Poland. His origins were humble and provincial: his family had farmed the land for generations, and he spent his first seven years in a town where the school was unable to prepare children for a selective academic high school (*Gymnasium*). When he was sent to relatives in Stettin for school, he suffered severely from homesickness. After completing his *Abitur* in 1905, he moved to Berlin to study medicine. He was neither intellectually nor personally comfortable in the capital and, after one semester, he made a decision that would profoundly shape the course of his professional development: he moved to Freiburg. There, he attended Weismann's lectures and became Eugen Fischer's student. As a reflection of how quickly the study of "race hygiene" had advanced in Germany, Lenz's doctorate had a boldly eugenicist title: *On the Abnormal Hereditary Dispositions of Males and the Determination of Gender in Humans* (*Über die krankhaften Erbanlagen des Mannes und die Bestimmung des Geschlechts beim Menschen*). It recommended coping with hereditary diseases through a "negative selection of the affected strains (*Stämme*)."

---

[89] Gessler, *Eugen Fischer*, 27, citing a letter from Fischer to L. Schemann, July 11, 1926.

[90] Erwin Baur played only a small role in German eugenic policy, chiefly because he died just after the Nazis came to power. He also made the mistake of picking a fight (or allowing it to be picked) with the Nazi Ministry for Agriculture and its director, R. Walther Darré. Darré also used his position to settle old scores. Baur had previously been a bitter critic of Darré, and Darré returned the favor by blocking funding renewal for Baur's institute and placing control of its research agenda in the hands of the Farmer's Union (*Bauernverband*), a thoroughly Nazi-infiltrated organization. Baur traveled to Berlin in December 1933, ostensibly to attend an awards ceremony for Carl Correns, but perhaps also to secure sympathy for his plight. On December 2, he suffered a heart attack and died. See Heiner Fangerau, *Etablierung eines Rassenhygienischen Standardwerkes 1921–1941: Der Baur-Fischer-Lenz im Spiegel der zeitgenössischen Rezensionsliteratur* (Frankfurt: Peter Lang, 2001), 36–7; Hans-Peter Kröner, *Von der Rassenhygiene zur Humangenetik: Das Kaiser-Wilhelm Institut für Anthropologie, menschliche Erblehre und Eugenik nach dem Kriege* (Stuttgart: Gustav Fischer Verlag, 1998), 16, 26; and Paul Weindling, "Weimar Eugenics: The Kaiser Wilhelm Institute for Anthropology, Human Heredity and Eugenics in Social Context," *Annals of Science* 42, no. 3 (1985): 303–18.

After completing his studies, Lenz moved from Freiburg to Munich, where he worked as an assistant to Max von Gruber, a co-founder in 1907, with Ploetz and Ernst Rüdin, of the Munich Society for Race Hygiene. Lenz worked with Eugen Fischer to found a Freiburg branch of the society in 1910. In Munich, Lenz became editor (1913) and then co-editor (1922) of the *Archiv für Rassen- und Gesellschaftsbiologie* (*Journal of Racial and Social Biology*, or *ARGB*). Alfred Ploetz offered him both posts.[91]

In 1923, Lenz was offered the first Race Hygiene Chair in Munich, and his fame grew throughout the decade. It was, however, exactly a decade later that Lenz reached the peak of his powers. When the Weimar Republic collapsed, Lenz welcomed Hitler and the Nazis with hope and joy. "The Nationalist Socialist state," he intoned, citing *Mein Kampf*, "has, following Adolf Hitler, placed race at the center of everyday life"[92] Gushing with the enthusiasm of a late convert, he suggested that "National Socialism is a political lesson on the foundations of a particular world view, a world view with its roots in the belief in race."[93] His enthusiasm was soon rewarded. Hermann Muckermann, a Jesuit, was pushed out of his position as head of the Institute for Race Hygiene at the Berlin Kaiser Wilhelm Institute,[94] and Lenz stepped into his place. He joined Walther Darré, Alfred Ploetz, Ernst Rüdin, Heinrich Himmler, and Fritz Thyssen on the Expert Committee for Population and Race Policy (*Sachverständigenausschuss für Bevölkerungs- und Rassenpolitk*). The committee drafted Germany's 1933 sterilization law, the Law for the Prevention of Hereditary Offspring. In 1937, possibly under pressure from Arthur Gütt, Lenz joined the National Socialist the Party.[95]

MARGARET SANGER

Back on the other side of the Atlantic, Margaret Sanger (née Higgins) was born in 1883 in Corning, New York. Her childhood – her mother gave birth to eighteen children, eleven of whom survived – and her later experience with her own children imbued a life-long commitment to the cause of birth control and the need to empower women in this key reproductive sphere. Margaret Higgins married a Jewish radical named William Sanger in 1902, and, after a happy but brief period of motherhood in the suburbs, she moved to the Upper East Side and began working illegally as a nurse on the Lower East Side.[96]

There, she came into contact with the poor mothers of large families who, between bitter asides about wealthy female New Yorkers who managed to

[91] Fangerau, *Etablierung eines Rassenhygienischen Standardwerkes*, 42.
[92] Fritz Lenz, *Die Rasse als Wertprinzip. Zur Erneuerung der Ethik* (Munich, 1933), 5.
[93] Ibid.
[94] Muckermann had allegedly referred to Hitler as an "idiot." Weindling, "Weimar Eugenics," 316.
[95] Kröner, *Von der Rassenhygiene zur Humangenetik*, 65.
[96] See Baker, *Margaret Sanger*, chapter 2.

secure birth control, asked Sanger how they could avoid another pregnancy.[97] A defining moment for her came in 1912, when she nursed a twenty-eight-year-old mother of three, Sadie Sachs.[98] Sachs had paid a backroom abortionist five dollars to terminate a pregnancy; he botched the job, and she became infected. After recovering, she asked Sanger and the presiding doctor one question: what could she do to avoid another pregnancy?[99] Ignoring the question, the doctor replied, "Any more such capers, young woman, and there will be no need to send for me." When Sachs persisted – "what can I do to prevent it?" – the doctor laughed, accused her of wanting to have her cake and eat it too. "It can't be done," he replied. "Tell Jake [her husband] to sleep on the roof."[100] Sadie made one last appeal to Sanger: "He can't understand. He's only a man. But you do, don't you? Please tell me the secret, and I'll never breathe it to a soul. *Please!*"[101]

The next time Sanger saw Sachs was three months later. Sachs was suffering from septicemia from a self-inflicted abortion; her husband and three children wept as they and Sanger watched death take Sadie.

As is often the case in such stories, it is unclear if Sadie Sachs ever existed. Sanger was described as being "dramatic," possessed of "fabulist predilections," and Sachs may have been a composite, a conflation in Sanger's mind of the recollection of several medical histories.[102] But there can be no doubt about the effect on Sanger of needless deaths among women seeking to avoid pregnancy. "I walked and walked through the hushed streets," she later wrote,

> [And w]hen I finally arrived home . . . I looked out my window and down upon the dimly lighted city. Its pains and griefs crowded in upon me, a moving picture rolled before my eyes with photographic clearness: women writhing in travail to bring forth little babies; the babies themselves naked and hungry, wrapped in newspapers to keep them from the cold; six-year-old children with pinched, pale, wrinkled faces, old in concentrated wretchedness, pushed into gray and fetid cellars crouching on stone floors, their small scrawny hands scuttling through rags, making lamp shades, artificial flowers; white coffins, black coffins, coffins, coffins interminably passing in never-ending succession.[103]

Sanger was, despite or because of her upbringing by an Irish anticlerical contrarian who liked a drink, singularly convinced of her own destiny. She saw herself as the city's salvation.

> The sun came up and threw its reflection over the house tops. It was the dawn of a new day in my life also . . . I was resolved to seek out the root of the evil, to do something to change the destiny of mothers whose miseries were as vast as the

---

[97] Ibid., 49.
[98] Details from Margaret Sanger, *An Autobiography* (New York: Dover Publications, 1971), 89–92.
[99] Baker, *Margaret Sanger*, 49.
[100] Quotations from Sanger, *Autobiography*, 90–1.
[101] Ibid., 91. Emphasis in original.
[102] Baker, *Margaret Sanger*, 51.
[103] Sanger, *Autobiography*, 92.

sky. . . . I was now finished with superficial cures, with doctors and nurses and social
workers who were brought face to face with the overwhelming truth of women's
needs and yet turned to pass on the other side. . . . I resolved that women should
have the knowledge of contraception . . . I would tell the world what was going on
in the lives of these poor women . . . . I *would be heard*. No matter what the cost, *I
would be heard*.[104]

Sanger was the key figure in the American birth control movement; indeed, she
invented the term. But Sanger's activism extended beyond enhancing women's
reproductive choices to include an embrace of standard eugenic goals: improving
national breeding through positive and negative policy measures. The journal
she founded in 1917, *Birth Control Review*, opened its pages to numerous
eugenic advocate articles, some penned by Sanger herself.[105] Her book, *The
Pivot of Civilization*, published in 1922, rehearsed standard pro-sterilization
arguments from eugenicists in respect to the feebleminded and others whom she
thought ought not to be allowed to procreate. One theme that she emphasized in
particular was the dangerous effect of benevolence and charity in helping those
whom natural selection processes would have eliminated. This applied particularly to very poor women in urban ghettos who then went on to procreate in a
way that increased fiscal charges on the public purse.

   Like so many other social reformers, Sanger was deeply sympathetic to
eugenicists' anxieties about who should and should not be allowed to procreate.
Like others who kick away social ladders once they have climbed them, she
decided that her own history – as the daughter of a poor, Irish Catholic immigrant with a large family – was a cautionary tale rather than evidence of people's
opportunity to rise socially given the right opportunities: she had inherited her
mother's tuberculosis, and her family's weak racial inheritance explained its
poverty.[106] In addition, she saw in the impoverished slums of the Lower East
Side daily proof of how disease, criminality, and stupidity passed from one
wretched generation to another.[107] "Tentatively in the [1910s] and wholeheartedly in the 1920s, Margaret Sanger became a fellow traveler and then promoter
of the eugenics movement," writes her sympathetic biographer, Jean
H. Baker.[108] Baker suggests that Sanger's support for eugenics was tactical – a
means of increasing support for birth control – but the author provides no
evidence in favor of this assertion. Indeed, the very passages cited by Baker
suggest that birth control could help eugenics rather than the other way around.
"Birth Control," Sanger wrote in response to an article published by Paul
Popenoe in her *Birth Control Review*, "opens the way for the eugenicist and

---

[104] Ibid.; portion from "I resolved" quoted in Baker, *Margaret Sanger*, 50.
[105] Examples include "Some Moral Aspects of Eugenics" (June 1920) and "Birth Control and
      Positive Eugenics" (July 1925).
[106] Baker, *Margaret Sanger*, 146.
[107] Ibid.
[108] Ibid.

preserves his work. By freeing reproduction from its present chains it will make a better race."[109]

Sanger was an ardent opponent of what she took to be excessive fertility individually – poor women were having too many children – and collectively – there were too many poor, nonwhite people on the planet. Sanger was an organizer of the first World Population conference, held in Geneva in 1927. She died in 1966, passionate, formidable, and thoroughly convinced of her cause until the end.

We will meet all of these figures again in subsequent chapters to bring into view their lobbying efforts for sterilization, as well as their success in getting policy implemented in the postwar era. In the case of the Germans, we will examine their attempts – at times unapologetic, at times sycophantic, and at times deceitful – to reestablish themselves and their causes in Germany after 1945.

WE WERE ALL EUGENICISTS THEN

These individuals – Galton and Pearson in Britain; Davenport, Laughlin, and (to a far lesser degree) Fisher and Sanger in the United States; and Ploetz, Fischer, Verschuer, and Lenz in Germany – would leave a permanent mark on eugenic ideas and (Britain excepted) eugenic policies. They by no means exhaust the list of influential eugenicists. Many others will be introduced in the course of this book, but it is reasonable to view those others as successors to these intellectual pioneers and policy entrepreneurs.

More generally, all high-profile eugenicists fed both into and off of a broader popularization of eugenics in the early decades of this century. It comprised conservatives and social democrats, doctors, psychiatrists, and social workers, policy makers, and activists. In the United Kingdom, eugenicists grouped in the Eugenics Education Society (later, the Eugenics Society), which published the *Eugenics Review*. Its first president was Montague H. Crackenthorpe, author of *Population and Progress*.[110] Major Leonard Darwin, the son of Charles Darwin, led the organization from 1911 to 1928. Subsequent presidents included C. P. Blacker and Julian Huxley, the first Director General of the United Nations Educational, Scientific, and Cultural Organization (UNESCO). Illustrious figures such as George Bernard Shaw, John Maynard Keynes, Winston Churchill, and Sydney and Beatrice Webb lent varying degrees of sympathy to eugenics.[111] But this in itself means very little. Eugenics could

---

[109] Quoted in ibid.

[110] Montague H. Crackenthorpe, *Population and Progress* (London: Chapman and Hall, 1907).

[111] Searle, *Eugenics and Politics in Britain*; Michael Freeden, "History and Progressive Thought: A Study in Ideological Affinity," *Historical Journal* 22, no. 3 (1979): 645–71; Harvey G. Simmons, "Explaining Social Policy: The English Mental Deficiency Act of 1913," *Journal of Social History* 11, no. 3 (1978): 395.

mean so many different things, and its appeal was so broad that different people could support eugenics for varied reasons and with varying levels of enthusiasm. Furthermore, there was a world of difference between the passing interest in eugenics displayed by John Maynard Keynes or Winston Churchill, the greater but still-secondary involvement of Irving Fisher, and the unqualified ardency of Charles Davenport or Harry Laughlin.

The only general point that can be made was that, for almost all of the early adherents, their commitment to eugenics was genuine. They believed that support for eugenics was support for progress. They did not perceive of eugenics or eugenic policies as punitive or retributive, employed to the disadvantage of those targeted by its negative implications, but, rather, as a means to building a better race.

Three macro-historical and social changes swept across what historian Daniel T. Rodgers calls the transatlantic community in the last decades of the nineteenth century.[112] The first was rapid industrialization and scientific discovery. The second was the uncovering, through early social scientific research, of widespread social ills: poverty, disease, and ill health. The third, which flowed directly from the first two, was a determined effort to secure, through increasingly emboldened national and state governments, scientific solutions to social problems. To understand how eugenics, and above all eugenic sterilization, could once seem to hold so much promise, we must turn to the problems and concerns that the eugenicists themselves faced when devising their ideas.

[112] Daniel T. Rodgers, *Atlantic Crossings: Social Politics in a Progressive Age* (Cambridge, MA: Belknap Press, 1998).

3

Eugenic Anxieties

The word "eugenics" derives from the Greek and denotes "good in birth" or "noble in heredity." Francis Galton understood by it the science of improving the human gene pool by granting "the more suitable races or strains of blood a better chance of prevailing speedily over the less suitable."[1] His American counterpart, Charles Davenport, put it slightly differently: "Eugenics may be defined as the science and art of social advancement by better breeding, or of improving the population by increasing the number of those with valuable racial (heredity) traits."[2] What distinguished eugenics from many existing academic branches of science was that it was a program with a set of prescriptions for law-makers – it was science with an agenda. Eugenicists in fact liked to compare this policy implication to the application of a newly discovered vaccination. But, in contrast to a medically prescribed preventive vaccination, eugenics demanded political action.

HEREDITY AND MENTAL ILLNESS: THE JUKES

Eugenics depended on two principles: heredity and differential fertility. Absent these two precepts, a eugenic program would be pointless. From the eugenicists' point of view, eugenics was not only policy-relevant but was urgently so. This resulted from the assumption that the majority of mental illnesses were heredi-tary, increasingly pervasive, and degenerative of a society's overall human stock.

This assumption was based partly on personal prejudices and partly on a desire to show that existing inequalities had a biological and thus immutable basis, but also on what might be called the eugenic research method (although those prejudices and desires may also have influenced that method). Following

[1] Quoted in Kevles, *In the Name of Eugenics*, xiii.
[2] Charles Davenport, "Eugenics," in *Body Build: Its Development and Inheritance* (Bulletin No. 24), February 1924, BD 27 CD, APSA.

Galton, researchers in the late nineteenth century traced family histories. As parents' wealth and social class were (and are) strong predictors of the wealth and social position that their children will be likely to occupy, the family histories method had an intrinsic hereditarian premise. The most infamous study, a flawed one that still haunts the mental health profession, was that of the Jukes.

The Jukes were a New York family that came to the attention of a transplanted Englishman, Richard Dugdale. Dugdale had no university training, although he had a smattering of courses in sociology and business from Cooper Union and had read Thomas Malthus, who famously predicted that population growth would inevitably outstrip food production. In his commitment to civic involvement, Dugdale was more a model Tocquevillian American: he was secretary of the Section on Sociology of the New York Association for the Advancement of Science and the Arts; treasurer of the New York Liberal Club; vice-president of the Society for the Prevention of Street Accidents; a member of the New York Social Science Society and of the New York Sociological Club; and, from 1868, a member of the Executive Committee of the New York Prison Association. In 1874, Elisha Harris, a physician and corresponding secretary of the Prison Association, presented some preliminary results of work on the recurrence of family names in county prisons.[3] This finding caught the attention of Dugdale, and he undertook a more extensive survey.[4] Armed with a Malthusian concern for overpopulation and imbued with a Lombrosian belief that criminals owed their fate to their genes,[5] Dugdale headed north. While interviewing inmates in one prison, he was struck by the fact that six of them were related. Dugdale redirected his efforts to retracing the family's lineage, and he eventually traced the lives of 709 people related by blood (540) or marriage/cohabitation (169) to five sisters who had married the sons of "Max," a descendent of Dutch settlers born around 1730. Coining the name "Jukes" for them, he traced the family "with more or less exactness for five generations." The Jukes were to be found in the brothels, prisons, and almshouses of New York. Dugdale suggested a complicated mix of heredity and environment as the cause for the Jukes' asocial behavior – a subtlety that subsequent readers would appreciate – but it was only the hereditarian element that received immediate attention.

When the Jukes study was published in 1877, it was an instant success, although it was likely more talked about than thoroughly read. It went through three printings in several months, inspired a new social science interest in heredity, and would forever be associated with a strong claim of heredity as

---

[3] Elof Axel, *The Unfit: A History of a Bad Idea* (Cold Spring Harbor, NY: Cold Spring Harbor Laboratory Press, 2001), 163.

[4] Ibid.

[5] Dugdale refers to the Italian criminologist Cesare Lombroso (1835–1909), whose work is credited with approaching the study of criminals as a science and not just within a legal framework.

the cause of social ills. In the late 1930s, mainstream, pro-eugenic organizations were citing it as an argument in favor of coerced sterilization.[6]

THE KALLIKAKS

In the spring of 1908, the head of the research department at the Training School for Feebleminded Girls and Boys in Vineland, New Jersey, Henry H. Goddard, left for a research trip to Europe. He met some of the most distinguished researchers of the day, but his most important meeting was with a Belgian doctor and special educator, Ovide Decroly.[7] Decroly showed Goddard tests on children's intelligence that had been conducted by a prolific experimental psychologist, Alfred Binet, and a physician, Théodore Simon. Employing progressively difficult questions, the tests were meant to measure attention, memory, and verbal skill. Goddard brought the tests back with him to the United States and translated them into English.[8] Binet and Simon viewed their tests with a certain degree of detachment and saw them as a means to identifying students in need of extra attention or alternative teaching methods. For Goddard, their purpose was quite different. He used them to identify and delineate the feebleminded, whom he divided into three categories: the "idiot" (with a mental age of one or two), the "imbecile" (mental age of three to seven), and the high-grade defective labeled the "moron" (mental age between eight and twelve).[9] Of these, the moron was the most threatening, because he or she was most likely to go unnoticed and to reproduce. "The idiot is not our greatest problem. He is indeed loathsome; he is somewhat difficult to take care of; nevertheless, he lives his life and is done. He does not continue the race with a line of children like himself.... It is the moron type that makes for us our great problem."[10] By the 1920s, intelligence tests such as the Binet-Simon test co-opted by Goddard were the standard means for identifying the "feebleminded."[11]

Back at the Vineland Training School, Goddard came on an unusual set of data about a feebleminded family spanning six generations – much like the Jukes – who lived in the Pine Barrens area of central and southern New Jersey. The story began with Martin Kallikak (the surname was a pseudonym),[12] an upstanding New Jersey Quaker who had fought in the Revolutionary War, contrary to his family's religious beliefs. While serving in the Continental Army, Kallikak had a brief tryst with a country girl from the Pine Barrens (in

---

[6] *Selective Sterilization in Primer Form* (Princeton: Birthright, Inc., 1937), 26, 9/9, AVS.
[7] Leila Zenderland, *Measuring Minds: Henry Herbert Goddard and the Origins of American Intelligence Testing* (Cambridge: Cambridge University Press, 1998), 92.
[8] Trent, *Inventing the Feeble Mind*, 158–66.
[9] Kevles, *In the Name of Eugenics*, 78; Trent, *Inventing the Feeble Mind*, 159–62.
[10] Henry H. Goddard, *The Kallikak Family, a Study of the Heredity of Feeblemindedness* (New York: Macmillan, 1912), 101–2.
[11] Noll, *Feeble-Minded in Our Midst*, 27.
[12] Derived from the Greek *kalos* ("good") and *kakos* ("bad").

some accounts she is referred to as a "feeble-minded tavern wench").[13] This union produced a son, but Goddard had no contact with him or the girl. After the war, he settled on a large farm in New Jersey and married a respectable young woman with whom he started a legitimate line of offspring.

According to Goddard's research, the first union produced a long line of paupers, criminals, and degenerates, whereas the second, legitimate line resulted in doctors, professors, politicians, and eminent citizens, all members of a prominent New Jersey family.[14] In contrast with the Jukes, Goddard's study, *The Kallikak Family*, was a short book written in a simple, dramatic tone with numerous photographs. It was directed at a lay audience and became immensely popular: Macmillan issued reprints in 1913, 1914, 1916, and 1919.[15]

The studies of the Jukes and Kallikaks were widely read and absorbed. Their success both contributed to and benefited from the new interest across northern Europe and North America in hereditarian explanations. In North America, the studies of the Jukes and Kallikaks were followed by state commission reports on the mentally ill. Massachusetts, Michigan, New Jersey, New York, Pennsylvania, Virginia, Arkansas, Florida, Indiana, California, and Maryland all commissioned comparable investigations.[16] Each study reached pessimistic conclusions, and several were at times alarmist about the scale of the problem.

Described as the "family studies," the reproving tone taken by the researchers' reports made for more popular reading and wider dissemination than usually enjoyed by academic studies. This reflected both the new era of Progressivism, when scholars were still interested in writing for general audiences, and the novelty of the results, which appealed to readers. The studies were also consumed by specialists, welfare and social workers, and Progressive-era (and later New Deal) administrators committed to improving both society generally and the conditions for the feebleminded by providing better institutional care. But it is the role of family studies in making the case for eugenics that the scholar Nicole Rafter rightly emphasizes.[17] Such was their success that the Eugenics Record Office run by Davenport published a booklet, *How to Make a Eugenical Family Study*, which helped to explain to general readers how such accounts were constructed and could be undertaken, a do-it-yourself guide for eugenic research.

Fifteen family studies were completed and published between 1877 and the mid-1920s, beginning with *The Jukes* and ending with a study of the offspring of interracial marriages in Virginia.[18] The authors were principally trained eugenic

---

[13] Stephen Jay Gould, *The Mismeasure of Man* (New York: Norton, 1981), 198.

[14] Bruinius, *Better for All the World*, 204

[15] Ibid.; Nicole Rafter counts five additional reprintings in 1922, 1923, 1927, 1931, and 1973. Rafter, *White Trash: The Eugenic Family Studies, 1877–1919* (Boston: Northeastern University Press, 1988), 3n5.

[16] "Memorandum on proposed survey to determine the number of mentally defective persons in Nassau County," n.d., RF 11/200/32/367, RAC.

[17] Rafter, *White Trash*, "Introduction," 1–2.

[18] Ibid., 4.

field workers, products of the Eugenic Record Office summer schools, who were joined in part by social workers and a few sociologists: eight of the twelve authors were in the first grouping. This new professional class, engaged in what Rafter characterizes as the "business of social control – welfare workers, eugenic field workers, institutional superintendents, mental testers," was animated by a shared anxiety about the "survival of the unfit."[19] They were convinced by the studies that individual traits and defects, not environmental factors such as poverty, explained the type of degeneration manifest in the family studies. Negative traits identified by the studies' authors were piled upon each other to depict a picture of familial degeneration.

Undertaken at a time when Mendelian genetics was the ascendency, these studies seemed to confirm the hereditarian conclusion made salient by Goddard, but they also highlighted what was for the eugenicists another worrying trend: high fertility among the mentally incompetent. The large Jukes and Kallikak clans were evidence of this fear, but research more broadly "confirmed" their findings. The development and application of new eugenic-based census techniques led to the supposed discovery of large numbers of poor, mentally ill, and cognitively impaired (although the last two categories were lumped together at this time). This research coincided with a shift in public policy toward the mentally ill as the number of institutions for such people continued to grow. The state began for the first time to keep systematic track of the mentally handicapped and mentally ill, and policy makers were shocked by the numbers of people and the conditions under which they lived. In the southern United States, the introduction of extensive medical and psychological testing of military conscripts and volunteers revealed high levels of "feeble-mindedness" as well as preventable physical ailments among recruits.[20] In Canada, the Canadian National Committee for Mental Hygiene was created and later upgraded to a Royal Commission.[21] Through correspondence with provincial hospitals across the country, it painted a picture of a large and growing population of mental defectives.[22] In 1904, the United Kingdom appointed a Royal Commission on the Care and Control of the Feeble-minded. When it reported back in 1908, it concluded that the feebleminded were so large in number that segregation in special institutions was necessary to prevent their further growth. Finally, from the 1920s, a series of scientific studies – notably those by Cyril Burt and Raymond Cattell in Britain – was meant to have confirmed an inverse relationship between intelligence and family size. The result

---

[19] Ibid., 5, 15.
[20] Noll, *Feeble-Minded in Our Midst*, 18; Gould, *Mismeasure of Man*, 192–9.
[21] Secretary, Canadian National Committee for Mental Hygiene to Department of Provincial Secretary, May 4, 1926, GR 865 1/1, British Columbia Archives (BCA), Victoria, BC. Its American namesake had been founded in 1909. See Gerald N. Grob, *Mental Illness and American Society, 1875–1940* (Princeton, NJ: Princeton University Press, 1983), chapter 6.
[22] Psychopathic Hospital, Winnipeg to E. J. Rothwell, December 2, 1925, GR 865 1/1, BCA. See also C. A. Baragar et al., "Sexual Sterilization: Four Years Experience in Alberta," *American Journal of Psychiatry* 91, no. 5 (1935): 909.

of all this was that, within the space of a few decades, the mentally ill and mentally handicapped were transformed in the public's eye from a marginal, numerically insignificant group to a mass, threatening horde.

The apparent growth of the number of feebleminded was set against the backdrop of an overall decline in the national birthrate. The decline was first detected in France from the 1870s, then in Britain, and finally, after the First World War, in Germany.[23] In an argument that anticipated contemporary debates in Europe, eugenicists contended that these demographic developments would bankrupt state finances, above all because of spiraling health costs.[24] If the overall birth rate was declining while that of the feebleminded was increasing, the only explanation could be that the upper-class birth rate was dropping precipitously.[25] In Britain and Germany, eugenicists argued that members of the upper classes were more likely to remain single, that their lengthy academic studies meant delayed marriage and procreation, and that they made more frequent use than the lower classes of birth control.[26] At the same time, for Lenz and Fischer (although not for Pearson), the upper-class propensity to educate its daughters in addition to its sons granted women financial

---

[23] On Germany, see in the secondary literature Gisela Bock, "Racism and Sexism in Nazi Germany: Motherhood, Compulsory Sterilization and the State," *Journal of Women in Culture and Society* 8, no. 3 (1983): 405–6; John E. Knodel, *The Decline of Fertility in Germany, 1871–1939* (Princeton, NJ: Princeton University Press, 1974); and Annette F. Timm, *The Politics of Fertility in Twentieth-Century Berlin* (Cambridge: Cambridge University Press, 2010). For primary sources, see "Massregeln gegen den Geburtenrückgang," R/1501 109344, Bundesarchiv (BArch), Berlin; Fritz Lenz, "Über die idioplasmatischen Ursachen der physiologischen und pathologischen Sexualcharaktere des Menschen," *Archiv für Rassen- und Gesellschaftsbiologie* (ARGB), Bd. 9, Heft 5 (1912): 545–603; and Alfred Grotjahn, *Geburten-Rückgang und Geburten-Regelung im Lichte der individuellen und der sozialen Hygiene*, 2nd ed. (Berlin: Colblentz, 1921), 153. On France, see William H. Schneider, *Quality and Quantity: The Quest for Biological Regeneration in Twentieth-Century France* (Cambridge: Cambridge University Press, 1990), chapter 2. On the United Kingdom, see Soloway, *Demography and Degeneration*; "Minutes of the 6th meeting of the Brock Committee [into sterilization]," November 14, 1932, 3, MH51/217, UK National Archives (UKNA), Kew, London; "Minutes of the first meeting of the Brock Committee," June 23, 1932, 6–7, MH51/212, UKNA.

[24] Ludwig Jens, "Was kosten die schlechten Rassenelemente den Staat und die Gesellschaft?" *Archiv für Soziale Hygiene* 8 (1913): 213–37, 265–322, cited by Weindling, *Health, Race and German Politics*, 239n322.

[25] Erwin Baur, Eugen Fischer, and Fritz Lenz, *Grundriss der menschlichen Erblichkeitslehre und Rassenhygiene, Menschliche Auslese und Rassenhygiene* (Munich: J. F. Lehmanns Verlag, 1923), 2:8–10; Eugen Fischer, "Aufgaben der Anthropologie, menschlichen Erblehre und Eugenik," *Die Naturwissenschaften* 14, no. 32 (1926): 753; Theodor Fürst and Fritz Lenz, "Ein Beitrag zur Frage der Fortpflanzung verschieden begabter Familien," *ARGB* Bd. 17, Heft 4 (1925): 353–9; Alfred Grotjahn, *Die Hygiene der menschlichen Fortpflanzung. Versuch einer praktischen Eugenik* (Berlin: Urban & Schwarzenberg, 1926), 113.

[26] Eugen Fischer, "Sozialanthropologie," in E. Korschelt et al., eds., *Handwörterbuch der Naturwissenschaften*, Bd. 9 (Jena: Fischer, 1913), 172–88; Fritz Lenz, "Individualistische Hemmnisse der Bevölkerungspolitik," *ARGB* Bd. 12, Heft 1 (1916–18): 124; Fritz Lenz, "Vorschläge zur Bevölkerungspolitik mit besonderer Berücksichtigung der Wirtschaftslage nach dem Kriege," *ARGB* Bd. 12, Heft V (1916–18): 440–68.

independence, thus removing one of the chief motives for marriage. When educated women did marry, they had fewer children.[27]

These foundations of the eugenic argument – the inheritance of mental illnesses, differential fertility rates between the genetically fit and unfit, and the scope for good breeding – crystallized in the years between 1890 and 1914. A 1900 study showed that Harvard graduates had on average fewer than two children, a figure that deeply worried that famous university's president.[28] Margaret Sanger was able to retain both humor and perspective in noting that this development was just as well since Harvard men made lousy fathers, but the majority of eugenicists saw in the Harvardians' apparent infecundity a metonymy for American national decline.[29] "There are evil forces at work among us in America," declared future U.S. president and father of six Theodore Roosevelt in 1899. "The diminishing birth rate among the old native American stock, especially in the north east ... I should consider the worst."[30]

The First World War cemented these fears. The cost of war, measured in the deaths of millions of young men, shocked the European consciousness profoundly, particularly in elite circles. Eugenicists were no exception. Although some had looked forward to war as a great cleansing exercise out of which the country would emerge stronger, the slaughter of the Great War shattered this illusion.[31] In Germany, it lent eugenicists a harder edge that would mark their views throughout Weimar and into the Nazi period. Industrialist Hans Werner Siemens wrote in 1917:

> We are in the midst of a disastrous situation in which we allow the progressive eradication of our best hereditary strains; it is draining our blood and life power. Unless we end it, we are heading indelibly towards our certain destruction. ... When the fertility of those hereditarily superior is less than that of the hereditarily inferior, the fitness of a people seeps irrevocably away.[32]

---

[27] Anette Herlitzius, *Frauenbefreiung und Rassenideologie. Rassenhygiene und Eugenik im politischen Programm der "Radikalen Frauenbewegung" (1900–1933)* (Wiesbaden: Deutscher Universitäts-Verlag, 1995), 58. See also Fritz Lenz, review of *Frauenarbeit und Rassenhygiene*, by J. Kaup, *ARGB* Bd. 11, Heft 5 (1914–15): 680–2.

[28] Baker, *Margaret Sanger*, 40.

[29] Ibid.

[30] Ibid.; Roosevelt to Cecil Spring Rice, August 11, 1899, in *The Letters of Theodore Roosevelt*, ed. Elting E. Morison, John Morton Blum, and John J. Buckley (Cambridge: Harvard University Press, 1951), 2:1053.

[31] Some still found something positive in the experience. Lewellys F. Barker, a prominent Baltimore physician, told the American Psychiatric Association in June 1924, "The Great War, responsible for so many tragedies, was not wholly devoid of benefits, and among those may be counted the awakening of the medical profession, of military authorities and of the public, to the extraordinary prevalence of mental defects and disorders of the emotions and the will among the young men who were drafted." "Psychiatry and Public Health," June 1924, GR 865/2/2, BCA. On Germany, see Kröner, *Von der Rassenhygiene zur Humangenetik*, 50.

[32] Hermann Werner Siemens, *Die biologischen Grundlagen der Rassenhygiene und der Bevölkerungspolitik. Für Gebildete aller Berufe* (Munich: J. F. Lehmanns Verlag, 1917), 62n and 92.

He continued the theme twenty years later: "In a biological sense, there is only one form of selection: fertility selection. ... Through it, the struggle for existence is a biological, not a social, idea, and the victory in the struggle for existence consists only of one thing: that the victor has more children than the 'vanquished.'"[33] Above all, the war – for both victors and vanquished – had itself contributed to this dysgenic trend by leading to the loss of the nation's fittest young men.[34]

OTHER EUGENIC CONCERNS

Beliefs in heredity and differential fertility were foundational in that one could not be properly called a eugenicist without subscribing to those ideas. At the same time, a number of other assumptions, arguments, and programs attracted eugenic interest. Because the movement was so broad, it is inaccurate to suggest that all eugenicists endorsed each and every eugenic argument, but they were part of the debate and, as such, warrant at least brief attention.

Race and Racism

On both sides of the Atlantic, the eugenicists were both generators and products of their time. Like most white Westerners then, they believed in a hierarchy of races and, by extension, of nations. Before 1918 (and in many cases after), Europeans took their racial superiority over Africans and other colonized peoples as given. For a few, notably Charles Davenport and Harry Laughlin, race was a central concern. Davenport, in a series of essays on race, casually refers to whites as a "dominant" and Indians/"Negroes" as "inferior";[35] that blacks are less intelligent[36] but more musical and possessed of better rhythm than

---

[33] Siemens, *Grundzüge der Vererbungslehre, Rassenhygiene, und Bevölkerungspolitik*, 8th ed. (Munich: J. F. Lehmann, 1937), 92.

[34] See Gustav Boeters, "Die Unfruchtbarmachung geistig Minderwertiger," *Wissenschaftliche Beilage zur Leipziger Lehrerzeitung* 28–9 (1924), 217; Erwin Baur et al., *Von der Verhütung unwerten Lebens – ein Zyklus in 5 Vorträgen* (Bremen: G. A. v. Halem, 1933), especially 15–62; Eugen Fischer, "An die deutschen Universitäten [Aufruf des Gesamtvorstandes der Deutschen Gesellschaft für Anthropologie, Ethnologie und Urgeschichte]," *Korrespondenz-Blatt der Deutsche Gesellschaft für Anthropologie, Ethnologie und Urgeschichte* 50 (1919): 23–4; Stefan Kühl, "The Relationship between Eugenics and the So-called 'Euthanasia Action' in Nazi Germany," in *Science in the Third Reich*, ed. Margit Szöllösi-Janze (Oxford: Berg Publishers, 2001), 185–211; Lothrop Stoddard, *The Rising Tide of Color Against White-Supremacy* (New York: Scribner, 1920), 190–1. When elected to an Advisory Council of the Eugenics Committee of the United States in December 1924, Irving Fisher put it thus: "The time is ripe for a strong public movement to stem the tide of racial degeneracy following in the wake of the War. America in particular needs to protect herself against indiscriminate immigration, criminal degenerates and the race suicide deplored by President." On Canada, see McClaren, *Our Own Master Race*, 43–5.

[35] Charles Davenport, "*Race Crossing in Jamaica*," April 15, 1928, BD 27 CD, APSA.

[36] Charles Davenport, *Body Build: Its Development and Inheritance*, Bulletin No. 24, February 1924, BD 27 CD, APSA.

whites;[37] that Jews have crooked noses;[38] and that mixing any of the above would lead to predictable and predictably substandard genetic results: "[a] population of hybrids will be a population carrying an excessively large number of intellectually incompetent persons." Racial differences in intelligence were immutable:

> In their mental traits ... different peoples are unlike. It has formerly been maintained that the obvious mental differences in races are due to differences of education and training merely, but the experience with native tribes in Australia and in Africa has shown that the children of these peoples do not respond in the same way as the white children to the same sort of education. ... [T]he army intelligence test ... showed that there is a marked difference in average mental capacity between the major races of mankind, and even between the peoples of different parts of Europe. ... In fact, it seems probable that in the same country we have, living side by side persons of advanced mentality, persons who have inherited the mentality of their ancestors of the early Stone Age, and persons of intermediate evolutionary stages.[39]

Davenport then goes on to link these "obvious" mental differences with temperament. Had it not been so harmful, the result might be amusing:

> [T]he different races of man are unlike in temperament also. Some of them, like the Scotch, are prevailingly profound,[40] thorough, somewhat somber; others like the Mediterranean peoples are prevailingly mercurial and light hearted; the trait of reserve has been developed to a high degree among the North American Indians; that of fidelity to a superior race among the Bantu Negroes; that of industry and dependability in the Chinese; and so on.[41]

Despite these comments, it is too easy to tar the eugenicists with the brush of crude racism. It was the case that they all believed in the existence of "races," but this term was so commonly used at the turn of the century that it became for many a synonym for what is now termed "nationality." At the same time, Davenport's racism did not entail an unbridled hatred for African Americans. In another piece, he reflected on their better qualities and the possibly positive results of racial mixing: "On the other hand, a population composed of hybrids between whites and negroes will contain persons better endowed in appreciation of music and in simple arithmetical or mental computations, than a pure white population."[42] In the last few lines, Davenport leads himself to the conclusion that race policy might, in some ideal world, allow for a positive racial engineering. "If only society had the force to eliminate the lower half of a hybrid population," Davenport surmises, "then the remaining upper half of the hybrid

[37] Charles Davenport, "Race Crossing in Jamaica."
[38] Ibid.
[39] Davenport, *Body Build.*
[40] An earlier draft suggested that Scots were "deep."
[41] Davenport, *Body Build.*
[42] Davenport, "Race Crossing in Jamaica."

population might be a clear advantage to the population as a whole, at least so far as physical and sensory accomplishments go." Throughout the twentieth century, the target of eugenicist fear and loathing was the feebleminded, but most of those targeted were white.

More broadly, there was a basic difference between, on one hand, those who believed that eugenics was about improving the lot of all the races and, on the other hand, those who thought that it was about improving and protecting the Nordic, white race. American and German eugenicists were to be found in both camps.[43] Charles Davenport, Eugen Fischer, Harry Laughlin, and (although this is contested) Fritz Lenz were in the latter category;[44] Wilhelm Schallmayer, Alfred Ploetz, Irving Fisher, and Sydney and Beatrice Webb were in the former category.

All eugenicists – German, American, British, and Scandinavian – did agree, however, that the state had an interest in and even an obligation to ensure the fitness of the race. Although such talk of state, race, and nation-forging has a vaguely threatening ring, these ideas were broadly shared across the West right through the interwar period. Consider the following ruling by a judge in 1936:

> [T]he state is vitally interested in the health and welfare of its citizens. It is certainly interested in preventing contamination of them by venereal disease. It may be conceded that intelligent medical science has succeeded in producing a cure for syphilis which is efficacious in the great majority of cases. However, as the trial court very properly observed, it was not so much concerned with curing the disease with which appellant was afflicted as it was with preventing appellant from transferring the disease to his possible posterity. If reproduction is desirable to the end that the race shall be continued, it is clearly desirable that the race shall be a healthy race, and not one whose members are afflicted by a loathsome and debilitating disease.[45]

The judgment could well have been from a eugenic court in Hamburg. It is, in fact, from a California judge in *People v. Blankenship*, a case concerning the statutory rape of a thirteen-year-old by a twenty-three-year-old. Politicians, scholars, and citizens across the West took as self-evident the need for race protection (against the evils of immigration, for example) and for race improvement.

Immigration

Eugenics and race concerns in North America coalesced most clearly in discussions of immigration. In both Canada and the United States, a fear of eugenic decline was bound up with intense opposition to Asian immigration on North

---

[43] On this, see Kröner, *Von der Rassenhygiene zur Humangenetik*, 61.

[44] See Renate Rissom, *Fritz Lenz und die Rassenhygiene* (Husum: Matthiesen, 1983), introduction and 32–4. Lenz oscillated between the claim that race hygiene would serve all races and the claim that the Nordic race was superior. In the 1930s, he succumbed to the latter position.

[45] *People v. Harley Blankenship*, 16 Cal. App. 2d 606, 610 (1936).

America's west coast.[46] More broadly, eugenics intersected with public fears that new waves of immigrants – from southern and eastern Europe, including many Jews in the latter – were less able to assimilate than past waves of German or British immigrants. This fear was not a new one; the same fears and suspicions had been expressed first about German and later about Irish immigrants. And this fear did not die with the eugenics movement. It has a new home and is frequently expressed in the United States today in reference to Latino migrants.[47] What was unique about the early decades of the last century was that eugenics gave scientific credibility to anti-immigration sentiment. As Charles Davenport put it:

> I know of no subject of vaster eugenic, as well as political moment than that of the genetical [sic] consequences of the union of dissimilar races of mankind. ... [W]e are justified in going slow in bringing together [in our land] very diverse races of mankind. If future research supports present suspicions as to the dangers, the mixtures cannot be unscrambled. If the suspicions of danger prove to be unfounded, then it will not be too late to throw the doors open to free intermigration of the most diverse peoples. The present safe course is to pursue the ideal of race homogeneity.[48]

Davenport's research institute and above all his colleague Harry Laughlin would be of immense assistance to the American anti-immigration movement, feeding facts and arguments to politicians for use in buttressing the restrictionist case to which they independently subscribed.

After sterilization, the adoption of quota-based immigration policies in the United States was the most important instance of eugenic ideas influencing policy.[49] There is, however, an important difference between the two. Eugenic ideas fed into and supported extant opposition to immigration; nativist sentiment would have existed and possibly even been decisive without the eugenics movement. By contrast, eugenics provided the primary arguments for forcibly sterilizing the feebleminded.

---

[46] See file RG 276, vol. 474/729921 at BCA.

[47] Leo R. Chavez, "Narratives of Nation and Anti-Nation: The Media and the Construction of Latinos as a Threat to the United States," in Michael Bass ed. *Narrating Peoplehood Amidst Diversity: Historical and Theoretical Perspectives* (Aarhus: University of Aarhus Press, 2011).

[48] Charles Davenport, "A Decade of Progress in Eugenics," *Address to the Third International Congress of Eugenics*, New York, August 21–23, 1932, available at: http://www.archive.org/stream/decadeofprogress00inte/decadeofprogress00inte_djvu.txt. A still-bolder statement can be found in Charles Davenport, "First Rough Draft of Eugenics Militant," n.d., Ms Coll 77 ERO Series X, APSA.

[49] Mae Ngai, "The Architecture of Race in American Immigration Law: A Reexamination of the Immigration Act of 1924," *Journal of American History* 86, no. 1 (1999): 67–92.

What Ever Happened to Positive Eugenics?

A basic if problematic distinction in the study of eugenics is between positive –
meaning policies designed to encourage breeding among the racially fit – and
negative – meaning policies designed to prevent the breeding of the unfit –
eugenics. Throughout the first half of the twentieth century, eugenicists in
Germany, Britain, Canada, and the United States professed support for both.
This is strikingly evident, for example, in a talk Davenport gave in 1912 to
university students in Minneapolis. In it, he informed the American young of
their duty to further "eugenical mating."[50] Early in relationships, before infat-
uation develops into love, they were "to consider whether the possible union
would be desirable from the standpoint of eugenics" and to walk away were it
not. They were to resist the call of travel, career, and money and to immediately
pursue "biological immortality" through procreation. Finally, they were to
fulfill their duty of rearing "fairsized families," defined by Galton and Lenz as
no fewer than four children, "else our best blood will be swamped by the more
fertile blood of the less effective strains." Davenport ended his talk with a ringing
call to the students to seize their genetic destiny:

> Let me urge again that you are the trustees of the best strains. See to it that they do
> not perish from our land, to the end that those fine qualities whose possession has
> given our nation such a high rank shall not be permitted to die; but that America's
> position shall be maintained through the coming ages.[51]

For eugenic sympathizers, support for good breeding had fairly obvious implications for family policy. Direct and indirect financial support were to be given
to large families of the right stock; pay scales were to be altered so that the
intellectually and genetically superior, notably professors, were to be paid more;
and, in some cases, punitive measures were to be directed at those who failed to
fulfill their eugenic duty. Thus, with an eye to the losses of young men Germany
suffered in the First World War, Fritz Lenz proposed during the Nazi period that
a law be adopted requiring all German families to have at least four children.
Those who failed to meet this obligation would have their income reduced for
each child by a sum equivalent to the cost of child rearing.[52]

The logical implication Davenport's arguments might have been positive
(direct and indirect financial support for not having children) and punitive

---

[50] "Family and the Nation," November 7, 1912, BD 27 CD, APSA.

[51] On the eugenicists' commitment to good breeding, see also Charles Davenport, "How a
Eugenicist Looks at the Matter of Marriage," n.d., BD 27 CD, APSA. The lecture contains an
implicit appeal for multiple sex partners: "control by some means the intercourse so that it shall
not be promiscuous but regulated." No doubt a creature of his time, Davenport does not quite
come out and call for it directly.

[52] Otmar von Verschuer to Fritz Lenz, May 10, 1940, Va 16/4, Archiv zur Geschichte der Max
Planck Institut (MPA), Berlin-Dahlem. See also Lenz, "Wege weiterer Vormarsches der
Bevölkerungspolitik," June 8, 1940, Va 16/4, MPA.

measures (marriage restrictions) for the genetically unfit. As it happened, the former never got onto the table, and the latter were viewed as unreliable. As Davenport put it,

> In few, if any, of the States of the Union have the legislators or the people grasped the idea of restricting the marriage of the mentally or physically defective in order to diminish the procreation of more defectives. Laws against the marriage of the feeble minded are futile in any case. For so long as a feeble minded person is at large he will find another feeble minded person who will live with him and have children by him. It would be as sensible to hope to control by legislation the mating of rabbits. The only way to prevent the reproduction of the feeble minded is to sterilize or segregate them.[53]

The British Brock committee, established to report on the desirability of coerced sterilization, reached much the same conclusion a few years later: "there is evidence that in the poorest districts neighbour marries neighbour, and like marries like."[54] Added to these strategic arguments was a fiscal fact: positive eugenics cost more. At a moment when sterilization was promising to save money by reducing costs through the elimination of criminals and dependents on welfare aid from the gene pool, proposals to encourage the genetically fit to reproduce promised only more costs.

In summary, three assumptions – of heredity, of the scope for breeding, and of differential fertility – founded a program based on preventing the procreation of the unfit and encouraging the breeding of the genetically fit, and a migration policy that stopped the genetically inferior from slipping in through the back door. An undated Eugenics Record Office pamphlet (probably published around 1910) summarized the "principal business of Eugenics" as follows:

1. To find out what matings are fittest for society as organized. This presupposes, among other things, an understanding of the social, economic and biological factors that govern mate selection and fecundity, and also a knowledge of the method of inheritance of human traits. To these tasks scientific investigators have set themselves earnestly.
2. To disseminate a knowledge of the facts of inheritance as they are ascertained.
3. To secure such social ideals as will facilitate the mating of the fittest, – i.e., a large opportunity for acquaintanceship among young persons, and a fuller knowledge, on the part of all, of the hereditary traits carried by each.

---

[53] "Marriage Laws and Customs," 1927, BD 27 CD, APSA. The same lecture held that such measures would obviate the need for restrictions against mixed marriages: "thus will the mentally incompetent strains be eliminated and the good physical traits of some of the black races be added, as a valuable heritage to enhance the physical manhood of the South."

[54] *Report of the Departmental Committee on Sterilisation* ("Brock Report") (London: HMSO, Cmd 4485, 1934), 21.

4. To educate organized society, especially as represented by several state governments, to a point where it will act with an eye to racial progress, encouraging the reproduction of the "best blood," and discouraging or preventing the reproduction of its worst strains. Eugenical improvement means the diversification, purification and conservation of highly effective and talented human families.

5. To encourage every intelligent and patriotic family to establish an *Eugenical Family Archive* for preserving genealogical and biographical materials, and for working out as accurately and as completely as possible a record of the family distribution of natural physical, mental and temperamental traits. The establishment of this custom would aid greatly the practical application of eugenics.

The pamphlet ends with a sort of eugenic call-to-arms:

> [I]f a race is to make progress along the lines of natural abilities, those in control must see to it that there shall be fit matings and many children among those most richly endowed by nature, and that hereditary defectives and degenerates shall not be permitted to reproduce at all. ... The State can be expected to take means to bring about these ends only when due pressure is brought to bear by aroused citizens.[55]

It would be tempting to conclude that this was a call to which, over the next forty years, researchers, social workers, mental health superintendents, and a few policy makers would energetically and enthusiastically respond. And there is an element of truth to this suggestion. But, in the same way that ideas do not drive policies in any simple, unidirectional manner, those who lobbied for and implemented sterilization policy were not following direct marching orders from Davenport, Goddard, or Dugdale. Rather, they too were both contributing to and taking inspiration from a broader culture that was newly concerned with issues of heredity, degeneration, and how states might intervene to encourage the former and block the latter. Whereas Davenport and Dugdale drew their conclusions from early social science research, others based theirs in more practical experience: work in homes for the feebleminded and mentally ill, and medical experiments on patients and inmates.

---

[55] Eugenics Record Office, "Eugenics Seeks to Improve the Natural, Physical, Mental and Temperamental Qualities of the Human Family" (Washington, DC: Carnegie Institution, Department of Genetics), MS Coll 77 ERO Series X, APSA. For a similar statement, see "Report of Committee of Research in Eugenics," n.d., BD 27 CD, APSA.

4

Homes for the Feebleminded

The majority of sterilizations in the United States and Canada occurred within prisons, hospitals, and – above all – homes for the feebleminded. By the heyday of the coerced sterilization movement, from the 1920s until the late 1950s, there were dozens of mental health institutions in the United States, Canada, and Scandinavia performing the procedures. Some of the chief North American institutions included the Provincial Training School in Red Deer, Alberta; the Sonoma State Home in Eldridge, California; the State Hospital at Raleigh, North Carolina; Fairview Hospital and Training Center in Salem, Oregon; the Utah State Training School in American Fork, Utah; the Virginia State Colony for Epileptics and the Feebleminded, Lynchburg, Virginia; and the Northern Wisconsin Colony and Training School, Chippewa Falls, Wisconsin. In the 1950s, all of these institutions were performing eugenic sterilizations under the procedural arrangements that concentrated the decision to sterilize in recommendations from superintendents followed by approval by boards of eugenics.[1] Because sterilization cannot be understood without an account to these public institutions, it is necessary to devote some attention to their origins and development and to the rules and norms governing them.

The distinction we make today between mental illness and mental disability was less sharply drawn in the nineteenth century, and the discipline of psychiatry itself was a by-product of the establishment of mental hospitals.[2] The treatment of both groups – the mentally ill and the feebleminded ("people with mental disabilities" now) – began with many good intentions and a desire to learn from and avoid the errors and injustices associated with past efforts. The institutions in which these people were housed were to a degree segregated, although many institutions continued to mix groups of the mentally ill and the mentally

---

[1] See the files in 8/71 (1951), AVS.

[2] Gerald N. Grob, *The Inner World of American Psychiatry, 1890–1940: Selected Correspondence* (New Brunswick, NJ: Rutgers University Press, 1985), 2.

handicapped, along with epileptics, the crippled, those suffering brain tumors, severe alcoholics, young criminals, and sexual deviants.[3] And, in any case, the care of all these patient groups followed a similar trajectory. The professionals responsible for their care were, from the first quarter of the nineteenth century onward, optimistic about their capacity to care for their patients, and they saw institutions as the setting for doing so. Their optimism was encouraged by initially generous state funding. The superintendents governing the institutions were convinced that residence at the institution would be temporary or cyclical and that the purpose of the institution would be educational. Because people both entered and left the institutions regularly, numbers in them would be small and manageable. Within a few decades, all of this changed. Funding became tighter, and the medical and psychiatric professions to which institutional super-intendents belonged became pessimistic about their patients' curability. Both the need for money and therapeutic pessimism encouraged making more decisions to commit patients on a permanent basis, and the institutions themselves became ends rather than educational stepping stones. As far fewer patients left the institutions than envisaged, the numbers moved onto an upward trajectory that would last into the second half of the twentieth century.

CURING THE INSANE AND FEEBLEMINDED

Institutions for the insane and the feebleminded grew up alongside each other. Before the advent of the asylum, the insane were left in the care of their families – often hidden in a cellar or locked in a pigpen – or to wander the streets.[4] Where there were medical and psychiatric treatments, they could be worse than the relevant mental condition. Doctors throughout the first half of the eighteenth century cut and bled patients, blistered their skin, and forced them to sweat and vomit.[5] Institutions, such as they existed, were glorified prisons: England's workhouses, Germany's *Zuchthäuser*, and France's hospitals.[6] In all of these countries, paupers, the "mad," criminals, people with mental disabilities, and the feebleminded were indiscriminately mixed together, if they made it into the institutions at all. The nineteenth-century American mental health institution was created to end such abuses. Moreover, it was intended to provide a place in which the insane would not only be housed but also in which they might be cured.

The intellectual foundations for this shift were laid out by Parisian psychia-trist Philippe Pinel. Pinel combined empiricism – recognition of the obvious fact

[3] Gerald N. Grob, *The Mad Among Us: A History of the Care of America's Mentally Ill* (Cambridge, Harvard University Press, 1994), chapter 5.
[4] Roy Porter, *Madness: A Brief History* (Oxford: Oxford University Press, 2002).
[5] Dowbiggin, *Keeping America Sane*, 5.
[6] Porter, *Madness*. See also Kathleen Jones, *Asylums and After: A Revised History of the Mental Health Services: From the Early 18th Century to the 1990s* (London: Athlone Press, 1993).

that bleedings, blistering, and induced vomiting were ineffective – with optimism – the conviction that insanity could be cured.[7] On this basis, he devised in
1790 a new method: moral treatment. Moral treatment involved employing
kindness, reason, and discipline to cure insanity.[8] Physicians believed that
moral treatment necessitated control over their patients' lives, and the treatment
of insanity became "inseparable from the asylum itself."[9] Inspired by Pinel and
similar ideas developing in Italy, Massachusetts planned the first asylum for the
insane in Charlestown, which opened in 1818.[10] It became the McLean Asylum
for the Insane in 1926. Philadelphia opened its first insane asylum in the suburb
of Frankford in 1817, and Connecticut followed with the Hartford Retreat in
1822.[11] Within three decades, sixteen state, one federal, and four municipal
institutions had been opened in the United States.[12]

Progressive ideas circulated within the North Atlantic community by travel
and scholarship,[13] and the new interest in moral treatment of the insane shaped
the thinking of British and American physicians caring for the feebleminded. In
1844, John Conolly, the chief physician at the county lunatic asylum at Hanwell,
in outer London, visited two well-known Parisian institutions: the Salpêtrière
and the Bicêtre.[14] The head of these institutions, Edouard Séguin, had developed
the theory that idiots suffered from an arrested development of the will at
childbirth and could be trained through physical exercise to be functioning
individuals.[15] Conolly's enthusiasm was limitless in a report he published in
1845. Whereas English towns left idiots "in total innocence and apathy,"
Séguin's tutelage meant that even the most severe cases of idiocy showed educational and moral improvements.[16] Samuel Howe, head of the Massachusetts
Asylum for the Blind, and Samuel Woodward, superintendent of the
Massachusetts Lunatic Asylum, read and circulated Conolly's piece. In 1846,
Howe was appointed chair of a state commission on idiocy. Marshaling evidence of the horrors that resulted from leaving idiots to their own "brutishness,"
he convinced the Massachusetts legislature to appropriate funds for an idiot
school for the blind in South Boston in 1847.[17] New York followed
Massachusetts by hiring a young physician named Hervey B. Wilbur, whose
research on the Bicêtre had inspired him to open private schools for idiots in

[7] Grob, *Mad Among Us*, 27.
[8] Dowbiggin, *Keeping America Sane*, 6; Grob, *Mad Among Us*, 27–49.
[9] Dowbiggin, *Keeping America Sane*, 6.
[10] Grob, *Mad Among Us*, 32.
[11] Ibid., 33.
[12] Dowbiggin, *Keeping America Sane*, 6.
[13] Daniel T. Rodgers, *Atlantic Crossings*.
[14] Trent, *Inventing the Feeble Mind*, 13.
[15] Ibid., chapter 2.
[16] Ibid., 13.
[17] Ibid.

western Massachusetts and near Albany, New York.[18] Thereafter, institutions opened in Germantown, Pennsylvania in 1853 (directed by a former teacher at Howe's school); Ohio in 1857; Connecticut in 1858; Kentucky in 1860; and Illinois in 1865.[19] New York opened a second school on Randall's Island in the borough of Manhattan.[20] As in regular schools, pupils could expect to return home for the summer and holidays, and they could be released from the school to take up employment.

While the states redesigned the institutions housing the feebleminded, Ohio took the lead in recasting the bureaucratic structures that governed them. The Ohio State Board of Charities was created in 1867, and the new body initiated what would become a trend: a further emphasis on custodial care rather than education.[21] Other states followed, with governors appointing the officials occupying positions on the boards, and legislatures expanding funding for them.[22]

Howe, Wilbur, and other early reformers were self-confident men. Their self-assurance rested in the belief, inspired by their tutor Séguin, that idiots could learn,[23] and they infused their institutions with optimism. Isaac Kerlin, who would later become a supporter of coerced eugenic sterilization, was among the most enthusiastic. In an 1858 book, he published with pride a letter written by one of his students, Grub, about a trip Kerlin organized to Trenton. Grub wrote:

> I have been to Trenton. . . . We showed the people what we could do; all the boys and me sung, and did the dumb-bell exercises; sung geography, and did some sums. A whole lot of people was in, and ladies, and they stamped their feet. The Governor of New Jersey talked to us, and I made him a present of a smoking-cap, that Lizzie M – made.[24]

DESIGNING THE INSTITUTIONS

The treatment of the feebleminded could not be separated from the way in which they were housed. The intellectual and architectural father of the modern insane asylum was a Quaker psychiatrist, Thomas S. Kirkbride (1809–83).[25] Kirkbride,

---

[18] Ibid., 14–5.

[19] Ibid., 15.

[20] Ibid.

[21] Ibid., 66.

[22] Ibid.

[23] On this, see Grob, *Mad Among Us*, 1–6 and Grob, *Inner World*, chapter 1.

[24] Isaac Newton Kerlin, *The Mind Unveiled; or, A Brief History of Twenty-Two Imbecile Children* (Philadelphia: U. Hunt and Son, 1858), 54–5 cited in Trent, *Inventing the Feeble Mind*, 15. On the confidence of early reformers, see also Grob, *Mental Illness and American Society*, prologue and Grob, *Inner World*, chapter 1.

[25] Kirkbride's 1854 book, *On the Construction, Organization, and General Arrangements of Hospitals for the Insane*, was, according to one scholar, the "bible of asylum construction in the United States." Trent, *Inventing the Feeble Mind*, 91.

who worked as head psychiatrist at the Pennsylvania Hospital for the Insane in Philadelphia from its opening in 1841, believed that only through institutionalization would patients have any hope of recovery.[26] The very architecture of the institution was to be part of the cure. Kirkbride outlined, in a series of propositions adopted at 1851 and 1853 meetings of the Association of Medical Superintendents of American Institutions for the Insane, the physical attributes of the ideal institution.[27] It would be made up of attractive buildings located outside the city on spacious, well-kept grounds. The grand center building would house the superintendent, his family, and the administrative offices. On each side of this pivotal structure, wings housing the patients would extend out.[28] Admissions would have an upper limit of 250. Each wing would contain wards subdivided in a hierarchical manner: as patients' behavior improved, they would move closer to the center building and perhaps enjoy additional privileges, such as the right to walk on the institution's grounds.[29] Kirkbride's plans were realized in mental health institutions across North America.

The Kirkbride model dominated until 1880, when the combined pressure of patient numbers and reduced resources led to a search for structures that would house more patients at lower cost. The cottage system provided them.[30] Under this system, the single schools for idiots and the Kirkbride asylums were replaced, in a move initiated in Ohio and imitated elsewhere, by a series of two-story "cottages" and "colonies" housing different classes of the feebleminded.[31] Known as the Kankakee model, advocates of the colony plan argued that it could be built for half the cost of the Kirkbride institutions and, with patient labor, operated at even less than that.[32] Following Illinois's early move in 1878, the colony plan became the most widely adopted within a decade, and, by 1910, the model was ubiquitous.[33]

As ever, the superintendents were decisive. These administrators saw patient labor as the solution to the problem of securing and retaining staff to work with custodial cases: higher-functioning patients were assigned the role of looking after them.[34] Indeed, whereas much changed in the population, aim, and

---

[26] See Nancy Tomes, *A Generous Confidence: Thomas Story Kirkbride and the Art of Asylum-Keeping, 1840–1883* (Cambridge: Cambridge University Press, 1984). The institution was later renamed the Institute of Pennsylvania Hospital.

[27] Grob, *Mad Among Us*, 71.

[28] Ibid., 72.

[29] Ibid.

[30] The system prevailed across the northern United States and in parts of the South. Southern states aimed to imitate the northern cottage system, but shortage of funds sometimes led to cheaper designs. Noll, *Feeble-Minded in Our Midst*, 108.

[31] Trent, *Inventing the Feeble Mind*, 65. A competing model, only rarely adopted, was the "Willard" system based on separate and cheaply built facilities for chronic cases. The goal would be purely custodial, without even a pretense to education or medical treatment.

[32] Ibid., 92.

[33] Ibid., 98.

[34] Ibid., 93.

organization of the asylums and homes for the feebleminded before 1890, one feature of them remained constant: the pivotal role of a superintendent, who enjoyed near complete control of the entire operation.[35] The pattern lasted well into the twentieth century, and it applied as much to institutions devoted to the care of the feebleminded as it did to those designed for the insane.

Beyond managing the institution, overseeing the staff, and setting the daily routine, the superintendent performed another key function: he decided which cases would be presented to Eugenic Boards for mandatory sterilization. Thus, in a typical procedure, the superintendent selected those individuals who would be sterilized and presented the case in favor of the operation for each individual to the board of his institution.[36] The board informed the subject or his or her guardian and gathered evidence. If the board concluded that the person

> is a habitually sexual criminal, or is insane, idiotic, imbecile, feebleminded, or epileptic, or is afflicted with degenerate sexual tendencies and by the laws of heredity is the probable potential parent of socially inadequate offspring likewise afflicted and that the inmate or person may be sexually sterilized or asexualized without detriment to his or her health, and that the welfare of the inmate or person and of society will be promoted by such sterilization or asexualization, [then] the board may order such superintendent or warden to have performed by some competent surgeon ... the operation of vasectomy, if the inmate or person is male, or of salpingectomy, if the inmate or person is female, or asexualization of either.[37]

With gravitas and a touch of the divine, a superintendent would move silently through wards, inspecting the smooth functioning of his kingdom. His position would remain almost untouched well into the twentieth century, through two world wars, until the deinstitutionalization movement of the 1960s and 1970s.[38] As historian Stephen Noll writes, "[a]side from continual concern over lack of funding, many superintendents ran their institutions as virtual private fiefdoms, given almost free rein by state agencies concerned only with removing mentally handicapped individuals from public sight."[39] That their role was decisive can be seen in the sterilization statistics. In California, superintendents Dr. John A. Reilly and Dr. G. M. Webster at the Patton State Hospital (a home for the insane, inebriates, and drug addicts) were "vocal advocates" of eugenic sterilization; between 1909 and 1950, 4,500 sterilizations occurred at the

[35] Grob, *Mad Among Us*, 72–7.
[36] This example is drawn on the procedure in Utah but is representative of the protocol followed in institutions in many states. *Utah Government Report on State Institutions*, chapter 10: Sterilization, 1958, SW015.1/73, AVS.
[37] Ibid., 304–5.
[38] It was not a coincidence that the Human Betterment Association, in its studies of coerced sterilization throughout the 1950s and 1960s, sent its questionnaires directly to superintendents of state hospitals and mental health institutions. See the correspondence with mental health superintendents and other state officials from the early 1950s to the 1960s, SW015 1/73, AVS.
[39] Noll, *Feeble-Minded in Our Midst*, 52.

institution.[40] Similarly, at Sonoma (a home for the feebleminded), Fred Butler personally and proudly performed over 4,000 surgeries.[41] By contrast, Dr. Leonard Stocking, superintendent at Agnews State Hospital (for the insane) from 1913 until 1931, took a skeptical view of sterilization, replying casually to a 1916 questionnaire that he performed "practically no sterilizations."[42]

The centrality of the superintendent persisted after the Second World War. In March 1945, the recently rechristened Birthright, Inc. (previously the Sterilization League of New Jersey) launched a campaign to, in its words, foster "by educational means, a nationwide program of sterilization."[43] Marian S. Olden, Birthright's leader, wrote directly to superintendents of institutions for insane and feebleminded people right across the United States.[44]

FROM OPTIMISM TO PESSIMISM

From the 1820s until the 1860s, superintendents' hopes for dedicated institutions for the insane and feebleminded were largely fulfilled. They led an isolated existence, but they enjoyed prestige, power over a large and expanding institution, the perks of a steady salary and a large house, and a sense of moral and professional purpose. This all changed very quickly, with the Civil War marking a turning point. Most institutions had accepted private patients, and the sudden loss of patients from the South (particularly for Kerlin, who was geographically close to the Confederacy) resulted in a financial shortfall.[45] This loss created perverse effects. It encouraged a search for expanded public funds, thereby creating an incentive to admit more publicly financed pupils for longer periods of confinement. Kerlin and other superintendents began to emphasize in their annual reports the need for dealing with two types of "idiots": (1) the dischargeable idiot, who would be educated in the institution but could find employment outside, and (2) the low-grade idiot, who could not be educated and who would be safer in the institution than in a poorhouse or prison.[46] The institutions thus came to house both pupils and inmates. Year-round care was added to the existing nine-month school year, and inmates were trained with skills that were useful inside the institution rather than outside it.[47] Major economic downturns from 1873 to 1879 and again between 1883 and 1885 intensified

---

[40] Alexandra M. Stern, "From Legislation to Lived Experience: Eugenic Sterilization in California and Indiana, 1907–79," in *A Century of Eugenics*, ed. Paul A. Lombardo, 105.

[41] Stern, "From Legislation to Lived Experience," 109.

[42] Quoted in Stern, "From Legislation to Lived Experience," 110.

[43] "Sterilization in the United States," *Journal of the American Medical Association* 127 (April 28, 1945): 1131. 11/92, AVS.

[44] See the correspondence contained in 11/92, AVS.

[45] Trent, *Inventing the Feeble Mind*, 62.

[46] Ibid., 63.

[47] Ibid. The first custodial patients were private, but then superintendents, again driven by financial necessity, admitted public patients.

financial pressures, and these pressures encouraged a reframing of superintendents' arguments. Instead of funding requests being made on the basis of proven success in training "idiots," superintendents justified their requests for continued institutionalization by pointing to the inability of their charges to adapt to work or educational environments on the outside.[48] Séguin's optimism died, and mass institutionalization, which would become the dominant means to dealing with the feebleminded in the United States, was born.

The practice of mixing these very different populations, which was a basic feature of the institutions in the twentieth century, was already established. As once-private schools for affluent "idiotic" patients had become primarily houses for the poor in the late 1870s, they became increasingly dependent on often-ungenerous state funds.[49] In the insane asylums, populations also increased as the elderly were transferred from the declining almshouses, and as sufferers of syphilis, cerebral disorders, Huntington's disease, brain tumors, and other chronic conditions were transferred from general hospitals.[50] The increase in numbers became unstoppable, and overcrowding was the predictable and inevitable result. By 1915, most institutions would have more than 1,000 patients, a far cry from Kirkbride's aspirational limit of 250.

The admission of "incurables" at first served the interests of the superintendents. They could concentrate resources and energy on the most capable and curable individuals, thus creating the image of success, while a steady admission of incurables would increase funding.[51] The trick worked for a few years, but, inevitably, a growing mass of purely custodial cases overwhelmed the few successes, and the jig was up. The insane asylum's failure to reduce significantly the reservoir of the insane bred disillusionment among staff and suspicion among state governments.[52] The expansion of numbers in the homes for the feebleminded made the superintendents' original role as teacher untenable. The contradiction became increasingly clear after the Civil War, and, by the 1880s, state governments had had enough. They created charity boards that treated asylums as only one institution among many and sought to centralize control over them.[53] They pressured superintendents of schools for the feebleminded to abandon their educational role and to become full-time administrators.[54] The superintendents acquiesced and quickly came to see the admission of ever-broader classes of patients (epileptics, cripples, and "low-grade defectives") as the way to ensure continued state support and job security. In one of many changes of heart, superintendents who had extolled the merits of small institutions and temporary admission now argued for expansion and permanent

[48] Ibid., 64.
[49] Ibid.
[50] Grob, *Mad Among Us*, chapter 5.
[51] Trent, *Inventing the Feeble Mind*, 64.
[52] Grob, *Mad Among Us*, 97–102.
[53] Dowbiggin, *Keeping America Sane*, 9.
[54] Trent, *Inventing the Feeble Mind*, 83.

institutionalization. The institution's educative function did not end entirely, but access to education became inconsistent and arbitrary. In addition, the goal of teaching shifted to equipping patients with skills useful for institutional life: caring for other patients, maintaining the buildings, and working the land.[55] Many superintendents took advantage of low post-1873 land prices to acquire farms, and the patients were prepared for heavy work on them. By the early decades of the twentieth century, most institutions reduced their costs on the backs of their patients.[56]

As the institutions expanded and the superintendent's role became more managerial, the distance between superintendent and the patients grew[57]; work devolved to assistant physicians and especially to the attendants. In some cases, the managerial role – and above all, presenting a clean, orderly, efficient, and abuse-free image to the public – superseded all others and virtually became the raison d'être of the institution.[58] The attendants, never especially well trained or well paid, were overwhelmed by the pressures of large numbers, a difficult clientele, and inadequate conditions.[59] Private institutions for the insane, which were able to bill families of wealthy patients, fared better.[60] But the majority of mentally ill and mentally handicapped patients was housed in overcrowded, custodial institutions where they were treated by staff who were, at best, demoralized, and, at worst, untrained and ill equipped for the job.

Five decades after their founding, institutions for the insane and mentally handicapped had strayed from their original mission. Where numbers had been small, they were now overcrowded and expanding; where optimism and hope had inspired employees, pessimism now reigned; where resources had been ample, they were now overstretched and inadequate; and where superintendents had once had individual contact with patients, they now rarely saw those whom they administered. It was in this context that they sought a surgical solution to the problems of the feebleminded.[61]

---

[55] Ibid., 84.

[56] Fairview fell into this category. Dennis Heath, interview, July 19, 2003, Portland, OR.

[57] Residential patients were increasingly called, from the late 1850s, "inmates."

[58] Gordon Bullivant (Executive Director, Foothills Academy, Calgary), interview, October 16, 2003, Calgary, AB. Discussion on attendant salaries.

[59] Grob, *Mental Illness and American Society*, 19–20.

[60] Dowbiggin, *Keeping America Sane*, 42. These institutions too, however, had to balance the exclusivity and comfort provided to the wealthy with the requirement of accepting enough public patients to claim to be serving a public good.

[61] Reilly, *Surgical Solution*.

5

The Eugenicists' First Throw

Sterilization Policy before the Second World War

It is common in the study of eugenics to cite structural factors such as racism, misogyny, and – perhaps above all – fear as the doctrine's causes: fear of population decline, fear of polluted genetic stock, and fear of the undesired subaltern peoples.[1] Although political institutions contain, shape, and limit behavior, they themselves cause nothing. People, or "actors," in social science language, do. Without a cohort of individuals prepared to lobby for state sterilization laws, to draft them, to present them before state legislatures, and to sign them into law, there would have been no coerced sterilization in America. Many of the same actors were also needed to implement the laws.

In the majority of examples, the key actor in the history of state sterilization laws was the superintendent of homes for the feebleminded. He (and, in one instance, she) naturally needed support, and this support came from sympathetic legislators prepared to place a bill before the legislature and to guide it through the law-making process. As is possible for all bills passed by state legislatures, state governors could veto eugenic sterilization bills. The decision to do so was partly about personal conviction but was mostly about politics: then as now, governors had to weigh the political costs of signing or vetoing a law. The targets of coerced sterilization were in the vast majority of cases poor, powerless, and mostly did not vote, so the incentive structure tilted in favor of approving

---

[1] There is an impressive scholarly literature on the history of eugenics in the United States and Canada. It includes Gregory M. Dorr, *Segregation's Science: Eugenics and Society in Virginia* (Charlottesville: University of Virginia Press, 2008); Dowbiggin, *Keeping America Sane*; Largent, *Breeding Contempt*, especially chapters 3 and 4; Edward J. Larson, *Sex, Race, and Science: Eugenics in the Deep South* (Baltimore: Johns Hopkins University Press, 1995); Noll, *Feeble-Minded in Our Midst*; Christine Rosen, *Preaching Eugenics: Religious Leaders and the American Eugenics Movement* (New York: Oxford University Press, 2004); Stern, *Eugenic Nation*; and Trent, *Inventing the Feeble Mind*. For an excellent recent overview, see Wendy Kline "Eugenics in the United States," in *Oxford Handbook of the History of Eugenics*, ed. Alison Bashford and Philippa Levine, 511–22.

coerced sterilization laws. It is thus not a surprise that the majority of U.S. states adopted them. Coerced sterilization became politically costly only when articulate, determined, and moneyed interests opposed the sterilization law. Throughout twentieth-century America, this opposition came from the Roman Catholic Church.

In the early 1890s, F. Hoyt Pilcher, superintendent of the Asylum for Idiotic and Imbecile Youth in Kansas, faced what he regarded as a serious disciplinary problem: his young inmates would not stop masturbating. To tame this behavior, he castrated forty-four boys and fourteen girls, an effort that cost him his job.[2] Some of his victims were still housed at the Asylum in the late 1930s (then renamed the State Training School).[3] In Pennsylvania, Superintendent Isaac Kerlin at the Pennsylvania Training School for Feeble-Minded Children at Elwyn also initiated castrations at his institution. His successor, Martin W. Barr, continued to castrate illegally and helped guide the United States' first, albeit failed, effort at a sterilization law in 1905.[4] Men, however, were not the only victims: the first institutionalized woman was castrated in the United States in 1872, and the practice flourished throughout the 1880s and beyond.[5]

Castration would continue to be practiced throughout much of the twentieth century. In some penal institutions, homosexuality or "extreme self-abuse" (excessive masturbation) were combated with castration.[6] Women judged to be "nymphomaniacs" or who masturbated had their clitorises cut off.[7] A few states also used castration as a basic technique for sterilizing mental defectives. In Oregon, between 1918 and 1941, 68 percent of the 207 men sterilized under Oregon's eugenic law were castrated.[8] The practice continued after 1945, and in the early years of the new millennium, social workers in Portland worked with individuals diagnosed as mental defectives who had been castrated in the 1950s.[9]

[2] Dowbiggin, *Sterilization Movement*, 23–4.

[3] Superintendent, Kansas State Training School, to Paul Popenoe, October 20, 1936, 11.8, E. S. Gosney Papers and Human Betterment Foundation Records ("Gosney Papers"), Caltech Archives, Pasadena, CA.

[4] Julius Paul, "State Eugenic Sterilization History: A Brief Overview," in Jonas Robitscher, ed. *Eugenic Sterilization* (Springfield, IL: Charles C. Thomas, 1973), 30.

[5] G. J. Baker-Benfield, *The Horrors of the Half-Known Life: Male Attitudes Toward Women and Sexuality in Nineteenth Century America* (New York: Harper & Row, 1976), chapter 11.

[6] Dickinson, "Sterilization without Unsexing." Castrations may also have had a punitive component. G. H. Sears, "Abstract of Castrated Males in Utah," May 10, 1934, 11.8; letter to Frank C. Reid, Human Betterment Foundation, 11.8, both in Gosney Papers.

[7] Dickinson, "Sterilization without Unsexing."

[8] "Eugenic Cases," n.d., *Minutes of the Oregon State Board of Health, 1914–22, Eugenics Records 1918–1945* and Eugenic Records, 1918–1945/5/4/04, Oregon State Archives (OSA), Salem, OR.

[9] Interview with an official from the Oregon Council on Developmental Disabilities, July 21, 2003, Salem, OR.

THE SEARCH FOR AN ADEQUATE TECHNIQUE

Castration had its limits. The operation was brutal, it took too long, and publics in all countries were wary of the technique as a policy instrument. As a tool to prevent breeding of certain populations, it was flawed. The sort of mass castration required to tackle the problem of mental deficiency was unrealistic and had another technique not been adopted, the eugenicists likely would have had little success.

In 1899, their hour arrived. A surgeon at St. Mary's Hospital and Augustana Hospital in Chicago, Albert John Ochsner, performed vasectomies on two patients in the hope of curing prostate problems.[10] When the patients told him that the operation had led to no change in their sex lives, Ochsner began to see the eugenic possibilities in the operation.[11]

Two years after the operation, he published a report in the *Journal of the American Medical Association*. He effused over vasectomy's potential to address multiple social pathologies. Ochsner began with the proposition that "it has been demonstrated beyond a doubt that a very large proportion of all criminals, degenerates, and perverts come from parents similarly affected." He added another dimension to undesirable inheritance: "it has also been shown, especially by Lombroso, that there are certain inherited automatic defects which characterize criminals, so that there are undoubtedly born criminals."[12] From this hereditarian premise, which had certainly not been convincingly demonstrated, he then jumps to the conclusion that "if it were possible to eliminate all habitual criminals from the possibility of having children, there would soon be a very marked decrease in this class ... [a vasectomy] could [also] reasonably be suggested for chronic inebriates, imbeciles, perverts and paupers."[13]

In the Indiana State Reformatory, a penal institution for boys, a senior administrator, Harry C. Sharp, read Ochsner's piece closely. It made Sharp think of one of his charges, a boy in his early teens named Clawson who was haunted by his inability to stop masturbating. At the time, of course, blindness, insanity, an early death, and probably a journey to hell were all linked with

---

[10] Reilly, *Surgical Solution*, 30. And see Philip Reilly, "The Surgical Solution: The Writings of Activist Physicians in the Early Days of Eugenical Sterilization," *Perspectives in Biology and Medicine* 26, no. 4 (1983): 637–56; Reilly, "Involuntary Sterilization in the United States: A Surgical Solution," *Quarterly Review of Biology* 62, no. 2 (1987): 153–70. Ochsner reported his sterilization operation success in the April 1899 issue of the *Journal of the American Medical Association*; his initial focus was on the sterilization of the so-called criminal classes. A. J. Ochsner, "Surgical Treatment of Habitual Criminals," *Journal of the American Medical Association* 32, no. 16 (April 22, 1899): 867–8.

[11] William M. Kantor, "The Beginnings of Sterilization in America," 1935, folder 12.2, Gosney Papers.

[12] Ochsner, "Surgical Treatment of Habitual Criminals."

[13] Ibid.

masturbation. The boy had asked Sharp to castrate him.[14] The doctor was initially reluctant to perform such a "mutilation." He told Clawson that he could perform a vasectomy and that the operation would just as effectively cure him of the propensity to masturbate.[15]

Clawson despaired when the operation predictably had little effect, and Sharp performed another vasectomy.[16] He told the boy to come back in six months; if the urge to masturbate did not stop by then, Sharp would castrate him. When Clawson came back, he had either acquired other interests, self-control, or the sense to lie. He told Sharp that he had ceased masturbating, thought more clearly, and did better at school. Other boys began lining up for the operation.

After the operation, Sharp displayed unalloyed enthusiasm for the procedure. Fresh from and emboldened by his experience of sterilizing children, the medic made a clarion call in favor of sterilization. He presented his findings at the Mississippi Valley Medical Association in the autumn of 1901 and published them the following winter in the *New York Medical Journal*. Six years later, he addressed an annual meeting of the National Prison Association extolling the virtues of sterilization based on his treatment of 223 prisoners.[17]

Sharp became fixated on numbers and particularly on the seemingly endless expansion of the mentally unfit and the fiscal and criminal threat they posed in America. He ruminated: "in 1850 there were 6,737 criminals in the United States or one to each 3,442 of the population; while in 1890 the penal population is shown to be 83,329 or one to each 957 of the population." These percentages concerned criminals solely. But "if all dependents were considered, such as inhabit public and private insane hospitals, almshouses and institutes for the feeble-minded we should find the proportion to be in the neighborhood of one to three hundred of the population."[18] Multiplication of "the defective and diseased" through procreation would be detrimental to "our race and our nation," Sharp argued, implying eugenic sterilization measures probably on an industrial scale.[19]

These results called for "a most heroic method of treatment." Sharp had led the way: between 1894 and 1902, he had sterilized young men between the ages of eighteen and twenty-five. These operations were performed with the putative consent of the individual,[20] but absent signed state legislation, their legality was indeterminate.

[14] William M. Kantor, "Beginnings of Sterilization in America," *Journal of Heredity* 28, no. 11 (1937): 374.

[15] Ibid.

[16] Ibid.

[17] Reilly, *Surgical Solution*, 33.

[18] Harry C. Sharp, "The Severing of the Vasa Deferentia and Its Relation to the Neuropsychopathic Constitution," *New York Medical Journal* 75(1902): 411–4, quoted in Reilly, *Surgical Solution*, 31–2.

[19] Sharp quoted in Reilly, *Surgical Solution*, 33.

[20] Kantor, "Beginnings of Sterilization in America," 375.

TABLE 5.1 *Sterilization Laws Enacted up to 1921 by State*[21]

State	Year(s) of enactment	Overturns (if applicable)
Indiana	1907	Rendered unconstitutional in 1921
California	1909	–
Connecticut	1909	–
Washington	1909, 1921	–
Iowa	1911, 1913, 1915	First law repealed; second law rendered unconstitutional
New Jersey	1911	Rendered unconstitutional in 1913
New York	1912	Rendered unconstitutional in 1918, surgeries cease; repealed in 1920
Kansas	1913, 1917	–
Michigan	1913	1913–1918 (rendered unconstitutional)
North Dakota	1913	–
Oregon	1913, 1917	First law repealed by referendum in 1913; second law rendered unconstitutional in 1921
Wisconsin	1913	–
Nebraska	1915	–
Nevada	1911	Rendered unconstitutional in 1918
South Dakota	1917	–

TOWARD STATE STERILIZATION LAWS

After Ochsner and Sharp's surgical experimentations, lobbyists for drastic solutions to limit allegedly expanding feebleminded populations quickly joined with eugenicist propagandists. They set out together to achieve better breeding and, to that end, set to work convincing state legislatures to enact appropriate laws.

They succeeded.

These early coercive sterilization laws (see Tables 5.1 and 5.2) sat against other policies designed to socially engineer good births. Paralleling early legislation to permit involuntary sterilization were state laws proscribing marriage between a man and a woman if either was feebleminded, imbecilic, or epileptic. Connecticut, the vanguard state in this respect, legislated such a marriage measure in 1896. The statute was copied by most northern and western states

---

[21] Derived from Reilly, *Surgical Solution*, 49, table 3 and Lutz Kaelber, "Eugenics: Compulsory Sterilization in 50 American States," University of Vermont, http://www.uvm.edu/~lkaelber/eugenics/ (accessed January 17, 2013).

TABLE 5.2 *Enacted Sterilization Laws by State*

State	Year of enactment	Year of major subsequent measures (if any)	Number sterilized
Alabama	1919	–	224
Arizona	1929	–	30
California	1909	1913, 1917, 1951	20,108
Connecticut	1909	1919, 1965	557
Delaware	1923	1929	945
Georgia	1937	–	3,284
Idaho	1925	1929	38
Indiana	1907	1927, 1931	2,500
Iowa	1911	1913, 1915	1,910
Kansas	1913	1917	3,032
Maine	1925	1929, 1931	326
Michigan	1913	1923, 1925, 1929	3,786
Minnesota	1925	–	2,350
Mississippi	1928	–	683
Montana	1923	–	256
Nebraska	1915	1929, 1957	902
New Hampshire	1917	1929	679
New York	1912	–	42
North Carolina	1919	1929	7,600
North Dakota	1913	1927, 1961	1,049
Oklahoma	1931	1933, 1935	556
Oregon	1913	1917, 1923, 1925	2,341
South Carolina	1935	–	277
South Dakota	1917	1925	789
Utah	1925	1929	772
Vermont	1931	–	253
Virginia	1924	–	7,325
Washington	1909	1921	685
West Virginia	1929	–	98
Wisconsin	1913	1955	1,823

in the next decades.[22] According to the most recent evidence, of those states that enacted eugenic sterilization laws, 87.5 percent already had laws on their books restricting marriage on grounds of specified mental or physical conditions.[23] To illustrate how widespread the adoption of eugenic sterilization laws were across

[22] Larson, *Sex, Race, and Science*, 22–3.
[23] Largent, *Breeding Contempt*, 65.

the United States and to demonstrate the decisive role of superintendent-led initiatives, it is helpful to examine in detail the process in a number of selected states.[24]

Indiana in the Lead

As anticipated in the hypotheses in the first chapter, a particularly committed state superintendent was decisive in seeing through the passage of America's first eugenic sterilization law. Harry C. Sharp took the results of his medical operations to the Indiana state government. Six months to a year after their vasectomies, Sharp had the forty boys operated on sit down in their cells and write a record of their experience.[25] Then, when Indiana's first sterilization bill was brought before the legislature in April 1907, he took these "testimonials" to the legislators to prove that sterilization could be free of coercion and cruelty. The bill was passed by the state law makers, but it faced opposition from the state's governor.

Indiana's governor, Republican James Franklin Hanly (an ardent Prohibitionist) wanted to veto the act. He threatened to cut funding to state institutions that implemented the practice, but Sharp again intervened. He implored the governor to refrain from using his veto and promised that the institutions would not carry out coerced sterilizations while Hanly was still in office. The governor eventually relented. The law was enacted in 1907 and focused on recommending sterilization for habitual criminals, "idiots," "imbeciles," and "rapists." Indiana consequently had the distinction of being the first state with a sterilization law on its statute books.

The Indiana Supreme Court struck down the state's sterilization law in 1921 in *Williams v. Smith* on the grounds that it violated the constitutional guarantee of due process of law.[26] In a response later imitated in other state houses, the Indiana legislature crafted a new law in 1927 designed to satisfy the Court's objections. The law carefully framed and delimited the target population for sterilization by excluding confirmed criminals and rapists and by adding epileptics to the feebleminded and insane.[27] Reports from the governor-appointed Committee on Mental Defectives issued in the 1920s helped sway political support in favor of eugenic sterilization. The Committee, established in 1915 by Democratic Governor Samuel Ralston, issued family studies tracing family pedigrees in the classic eugenic fashion.[28] These studies helped diffuse the

---

[24] Derived from Kaelber, "Eugenics" website.

[25] Largent, *Breeding Contempt*, 65.

[26] L. Potter, "Medical and Legal Aspects of Sterilization in Indiana," *Proceedings of the Fifty-Eighth Annual Session of the American Association on Mental Deficiency*, New York, May 26–29, 1934, 5, SW015.1, AVS.

[27] Alexandra M. Stern, "'We Cannot Make a Silk Purse Out of a Sow's Ear': Eugenics in the Hoosier Heartland," *Indiana Magazine of History* 103, no. 1 (2007): 28–30.

[28] *Mental Defectives in Indiana: Report of the Committee on Mental Defectives Appointed by Governor Samuel M. Ralston*. Indianapolis, IN. November 10, 1916.

hereditary view of mental "defectiveness" as an inheritable and hence control-lable trait – that is, if inheritance were stopped dead in its tracks through sterilization. Efforts at promoting positive eugenics included "Better Baby" and "Fitter Families" contests at county and state fairs. The Committee's eugenic findings and recommendations were conveyed to the state legislature by one of its members, State Senator C. Oliver Holmes.[29] Holmes drafted the new steri-lization bill passed in 1927.

Like most states, the superintendent of the institution housing patients rec-ommended sterilization; unlike most states, there was no central state eugenics board required to ratify that decision. Rather, on the advice of the superintend-ent, the institutions made a sterilization recommendation to the courts, which in turn made the ultimate decision.[30] The 1927 law also introduced a thirty-day notice period before a sterilization recommendation could be carried out, during which time the inmate could prepare an appeal.[31] "We have doubted for various reasons the advisability of having a central Eugenics Board in Indiana," noted a psychiatrist from the Fort Wayne State School. "We feel that staff members in institutions will make better decisions where there has been at least a thirty day observation of the case."[32]

Further amendments to the law in 1931, 1935, and 1937 reaffirmed the authority of individual governing boards of state institutions to decide on sterilization for an inmate. This measure naturally enhanced the role and power of the superintendents. Lobbyists such as Sharp quickly built on the Indiana success by urging laws in other states. Several obliged. New Jersey enacted a bill in 1911, encouraged by the chief physician at the State Village for Epileptics, Dr. David F. Weeks. Weeks subsequently developed the steriliza-tion program at the Village.[33]

California

Of all the states that adopted eugenic sterilization laws, California was the most aggressive in carrying out its statute. Half of all sterilizations in the United States were performed in the state. The law's adoption followed the pattern established by Virginia and, especially, by Indiana.

---

[29] Jason S. Lantzer and Alexandra M. Stern, "Building a Fit Society: Indiana's Eugenics Crusaders," *Traces of Indiana and Midwestern History* 19, no. 1 (2007): 4–11.

[30] "Sexual Sterilization – Hearing – Evidence – Finding and Decree of Court – Appeal – Commitment – Act of 1931," SW015.1/73, AVS.

[31] Largent, *Breeding Contempt*, 71.

[32] L. Potter, "Medical and Legal Aspects of Sterilization in Indiana," 6, SW015.1, AVS. In 1953, the state legislature created the position of Commissioner of Mental Health, which replaced and assumed the responsibilities of the trustees and governing bodies of the institutions. Frederick M. Harrison, Deputy Attorney General, to Irene Headley Armes, Human Betterment Association of America, July 17, 1953, SW015.1/173, AVS.

[33] Reilly, *Surgical Solution*, 36.

Californian physician Dr. Frederick W. Hatch, the secretary of the State Commission in Lunacy from the turn of the century, worked in liaison with State Senator W. F. Price, a close friend, to orchestrate the drafting and introduction of a bill into the California legislature in March 1909. Hatch saw in sterilization a remedy to the state's growing population of "mental defectives," fueled in part by growing immigration numbers (particularly "Oriental" immigrants, although these were never part of the pool of those sterilized in the state). Hatch's commitment to the eugenic program stemmed from his embrace of the hereditary argument about the causes of insanity. For Price, the embrace of eugenics was rooted in alarm about California's changing demographics and the aspiration to restrict reproduction by the undesirable.[34] In common with many states, the bill passed the California legislature almost unanimously (21–1 in the Assembly, 41–0 in the Senate) and was signed into law by Governor James N. Gillett on April 26, 1909. Under the legislation, sterilization was designed for inmates in prisons (who were sex offenders) and institutions for the mentally retarded. But there was no explicit eugenic language about hereditarian motives for sterilization in the law.

Hatch later became general superintendent of California State Hospitals, in which role he applied the act's remit to identify and sterilize so-called mental defectives.[35] He used this position to appoint other eugenicists as superintendents and surgeons at state institutions and to pass over potential appointees who lacked sufficient enthusiasm for the eugenic agenda. Making these appointees consolidated Hatch's fervor for eugenic sterilization into the Californian institutional system for decades after his death in 1924.[36] By the time of his death, ten Californian institutions had sterilized 3,000 inmates,[37] an achievement accredited to Hatch's single-minded advocacy of sterilization for twenty-five years.[38]

California's early sterilization program highlights the central role played by state superintendents with privileged access to key legislators. As it developed, the eugenic accent of the language authorizing sterilization became even more pronounced than it was in the 1909 law. Subsequent reforms expanded the focus of sterilization from tackling those presently manifesting signs of mental illness or imbecility to the quintessential eugenic principle of preventing the emergence of such flaws in future generations by eliminating the capacity of patients to procreate (although this latter aim was always a consideration for eugenic

---

[34] Ibid.; Paul Popenoe, "The Progress of Eugenic Sterilization," *Journal of Heredity* 25, no. 1 (1934): 19–26.

[35] Joel T. Braslow, "In the Name of Therapeutics: The Practice of Sterilization in a California State Hospital," *Journal of the History of Medicine & Allied Sciences* 51, no. 1 (1996), 33 (drawing on Californian reports of its State Board of Charities and Corrections); Reilly, *Surgical Solution*, 36; Reilly, "Involuntary Sterilization," 155.

[36] Reilly, *Surgical Solution*, 36.

[37] Ibid.; Largent, *Breeding Contempt*, 79.

[38] Popenoe, "Progress of Eugenic Sterilization," 20.

supporters in the state). Under Hatch's prodding, and with the support of reforming legislators, the control of heredity assumed dominance in the sterilization laws.

Further legislation, passed on June 13, 1913, repealed the 1909 law and clarified the pool of citizens eligible for sterilization treatment. The new measure brought a focus on inherited illness as understood by eugenicists:

> Before any person who has been lawfully committed to any state hospital for the insane ... and who is afflicted with hereditary insanity or incurable chronic mania or dementia shall be released or discharged therefrom, the state commission in lunacy may in its discretion, after a careful investigation of all the circumstances of the case, cause such a person to be asexualized ... whether with or without the consent of the patient.[39]

The law was entitled the Asexualization Act. The new law reflected growing confidence in employing eugenic reasons for sterilization decisions. Unlike the 1909 statute, the legislation in 1913 did not sail smoothly through the state legislature: in the Assembly, twenty-four opposed and forty supported it, and in the Senate, it received twenty-one votes in favor and four against.[40]

This opposition, however, proved transient and certainly would not be a barrier to further sterilization measures in the state. Thus, California's third major sterilization law further enlarged the pool eligible for sterilization by singling out those likely to pass on a "mental disease" if permitted to procreate. The bill was approved in 1917, having received only seven votes against it in the Assembly.[41]

In requesting permission to conduct a sterilization, state hospital superintendents uniformly cited the danger of the subject transmitting the mental or other illness to any potential offspring: "we think this patient should be operated on for sterilization as he [or she] would likely transmit to descendants."[42] From 1920, the request for approval to administer sterilization went to the State Department of Institutions, which replaced the Commission in Lunacy that year.[43] A standard letter later drafted by the Department for sterilization requests, which was in use until late 1959, listed various categories of illness legally permissible for sterilization.[44]

---

[39] Quoted in Braslow, "In the Name of Therapeutics," 33–4.

[40] Jeffrey Alan Hodges, "Dealing with Degeneracy: Michigan Eugenics in Context" (Ph.D. diss., Michigan State University, 2001), 137.

[41] Kline, *Building a Better Race*, 50.

[42] Quoted in Joel T. Braslow, *Mental Ills and Bodily Cures: Psychiatric Treatment in the First Half of the Twentieth Century* (Berkeley: University of California Press, 1997), 57.

[43] The Department of Institutions was given the more eugenic title of "Department of Mental Hygiene" in 1945, which remained in place until 1973, when it became part of the California Department of Health.

[44] Quoted in Braslow, "In the Name of Therapeutics," 4.

California's legislators' early and wide embrace of eugenic sterilization should have made it a propitious setting for initiatives in the mid-1930s to create a state eugenics board. Such bills were entertained by the California Assembly in 1935 and 1937 but failed to gain sufficient support. The proposed board would have extended the sterilization law to prisons, correctional schools, reformatories, and detention camps, giving superintendents, wardens, and directors of all these institutions the capacity to petition that any patient manifesting "a tendency to serious physical, mental, or nervous disease or deficiency" be sterilized before release.[45] A broad array of professionals that included medical doctors, professors, and institution administrators joined with journalists and philanthropists to promote a deeper eugenic commitment in Californian public policy.[46]

Failure to win sufficient support to pass these bills did nothing to reduce the power of superintendents at institutions covered by earlier and existing sterilization legislation. Superintendents at state hospitals played the lead role in selecting inmates for treatment and acquiring written consent from the California Department of Institutions. The standardized permission process did not weaken the key role of each individual institution's superintendent: he specified the illness justifying sterilization – from epilepsy to "mental disease" – which was always combined with a citation of its hereditary nature. As psychiatrist-historian Joel Braslow shows, superintendents took an expansionist view of supposedly hereditary conditions in order to maximize the number of persons sterilized. They diluted or exaggerated diagnoses to fit this operational goal, and they wrote letters to patients' families seeking consent that deliberately obfuscated the underlying eugenic rationale.[47]

Since superintendents played the decisive role in recommending an inmate for sterilization, the number of sterilizations varied across California's state hospitals as a function of their enthusiasm. Of all the sterilizations carried out in the state up until mid-1926, 40 percent came from the Stockton State Hospital. It was a testimony to the centrality and dedication of its superintendent from 1929 to 1945, Dr. Margaret Smyth, and her predecessor, Dr. Fred Clark, both of whom encouraged their physicians to recommend sterilizations in considerable numbers.[48] In contrast, Superintendent Leonard Stocking of Agnews State Hospital was one of those judged relatively cautious in recommending patients' sterilization.[49] The Department of Institutions biennial report in 1926 included a disappointed assessment of some superintendents' diligence: "it is to be regretted that the superintendents of some of our state hospitals have failed to realize that

---

[45] Quoted in Stern, *Eugenic Nation*, 83.

[46] Ibid., 83–4.

[47] Braslow "In the Name of Therapeutics," 36–9.

[48] Dr. Asa Clark was superintendent from 1892 to 1906; he was then replaced by his son, Fred Clark. The younger Clark was superintendent from 1906 until 1929, when Smyth took over.

[49] Braslow, "In the Name of Therapeutics," 35.

there is another obligation laid upon them equally heavy with that of the humane ministry to the needs of these unfortunates, namely: the protection of society against a further reproduction of the unfortunates."[50]

There was no such regret concerning Stockton State Hospital. Dr. Margaret Smyth, a psychiatrist and graduate of Stanford University School of Medicine, started her professional career at the Hospital in 1900 and did not disappoint as superintendent in ensuring patients were sterilized. Smyth later praised how Nazi Germany emulated but massively expanded on American eugenic sterilization practices. She won a national and international reputation for her work in psychiatry, including the treatment of the feebleminded through eugenic means. Smyth was the first woman in the United States to be appointed as superintendent of such an institution.[51]

If Virginia's sterilization laws earned historical and political importance in producing the basis for *Buck v. Bell*, California's practices gained a distinct notoriety from their scale. Almost a third of the officially recorded sterilizations in the United States in public institutions from 1907 occurred in California. The running total of sterilizations performed by the state increased consistently: by 1921, 2,248; by 1930, 5,274; by 1941, 9,856; by 1946: 10,998; and by 1950, 11,491.[52] These numbers were divided almost equally between men and women and would eventually reach a total of 20,108.

Naturally, activist superintendents did not operate in a vacuum. While superintendent lobbyists and sympathetic legislators ensured the adoption and implementation of a sterilization law, the law's underlying eugenic doctrine justifying the law enjoyed the support of California's most prominent citizens. Many of them played a national role. David Starr Jordan, founding president of Stanford University and a respected scientist, had helped Charles Davenport persuade Mrs. E. H. Harriman to use some of her dead husband's largesse to finance the Eugenics Record Office's (ERO) establishment in Cold Spring Harbor in October 1910. Through his membership on the Eugenics Committee of the American Breeders' Association,[53] Jordan was linked with the East Coast eugenics intelligentsia. This presence gave California a leading role in the national development of eugenics as policy agenda.

As in many states, lobbying for eugenic sterilization was undertaken by medical professionals, especially those associated with state institutions for the mentally ill. Jordan was in regular contact with a fellow progressive, physician John R. Laynes, who lobbied enthusiastically for the eugenic cause throughout the 1910s and 1920s, having conducted a survey of sterilization needs in

---

[50] John R. Haynes, *Eighth Biennial Report of the State Board of Charities and Corrections from July 1, 1916 to June 30, 1918* (Sacramento: California State Printing Office, 1918), 61, quoted in Braslow, "In the Name of Therapeutics," 35.

[51] Neal L. Starr, "Stockton State Hospital: A Century and a Quarter of Service," *San Joaquin Historian* 12, no. 4 (1976): 123–8.

[52] Compiled in Joel T. Braslow, "In the Name of Therapeutics," 32.

[53] Stern, *Eugenic Nation* 84–5.

California in 1918.[54] His influential report stressed the problem of "morons," the category invented by Goddard.

Many eugenic ideas converged with other doctrines and populist causes in the state, such as hostility to Chinese immigrants in the 1870s and 1880s, and a later aversion to Mexican immigrants. According to the historian Alexandra Minna Stern, there were extensive networks of social groups and civic organizations able to circulate eugenic and other scientific racist notions among the state's elites and public intellectuals and eventually to translate those ideas into legislation. As in other parts of the country, a variety of different concerns and doctrines – racism, degeneration of the "national stock," anti-immigrant sentiment, and so forth – converged on the fulcrum of protecting particular visions of America as a nation.

California's sterilization laws also influenced other American states' efforts. The state was recognized as an early starter in eugenic practices, and its success in the realm received apparent endorsement in studies issued by the Human Betterment Foundation in Pasadena, which was directed by Paul Popenoe and Ezra Gosney. Californian success spurred superintendents in other states to lobby for their own sterilization laws. The passage of Vermont's sterilization law in 1931, for example, was directly influenced by California's achievement.[55]

Michigan

Legislators in Michigan made their first attempt to adopt a sterilization law in the aftermath of Ochsner's success with vasectomies.[56] In 1897, and twelve years before neighboring Indiana, a bill was brought before the state legislature.[57] "An Act for the Prevention of Idiocy" was punitive rather than eugenic – it authorized the sterilization of "perverts" and criminals – but it contained many of the formal procedures for making and reviewing sterilization decisions, as well as the surgical technique (in this instance, castration), that were later integral parts of later successful legislation. Lawyers in the state, however, objected to the constitutionality of the measure's punitive nature and to the scope of the new law; the bill never made it out of the legislature. But eugenic commitment and enthusiasm among the state's medical, political, and institutional elite was entrenched, and in both the 1900s and 1910s, they lobbied systematically and successfully for sterilization laws.

Michigan's next bill made it through the legislature but not much further. A group of eugenicists in the state professionally linked to the superintendents of state institutions campaigned in favor of the bill. The group included Dr. Victor

---

[54] Cited in ibid., 103, 253n126.

[55] Nancy L. Gallagher, *Breeding Better Vermonters: The Eugenics Project in the Green Mountain State* (Hanover, NH: University Press of New England, 1999), 4.

[56] For one valuable study on sterilization in Michigan, see Hodges, "Dealing with Degeneracy."

[57] Largent, *Breeding Contempt*, 66–9.

Vaughan, who was not only dean of the University of Michigan Medical School but also president of the State Board of Health. Vaughan gave lectures on eugenics at the University of Michigan in the 1910s – comparable courses existed at Columbia, Harvard, Cornell, Brown, Northwestern, and the University of Wisconsin – and published a book for the lay public, *Sex Attraction.*[58] His writings and lectures revealed a conviction that hereditary traits were a source of mental and other illnesses and cited the work and speeches of leading national figures in eugenics, such as Davenport and Goddard, as authoritative sources. Adopting the language of good and bad "stock," Vaughan emphasized the "three-generation" theory of degenerative inheritance. Exemplary of bad stock were the "unit characters" associated with "alcoholism, feeblemindedness, epilepsy, insanity, pauperism and criminality." He explained:

> [A]ll of these classes should be excluded from the list of those to whom is granted the privilege of exercising the highest, holiest, most important function of the race – parenthood. . . . [I]n order to boast of good stock it is necessary to have the history of at least three successive generations. Among these there should be none of the defective unit characters mentioned above.[59]

These undesirable illnesses were all included in subsequent Michigan legislation as grounds for eugenic sterilization.

Vaughan featured in 1914 at the statewide conference in Battle Creek funded by the Race Betterment Foundation, which included John Harvey Kellogg (the cereal magnate and philanthropist) as well as Davenport and Laughlin from the ERO. Michigan's governor was honorary president of the Race Betterment Foundation. This first Race Betterment conference heard numerous papers read on all aspects of eugenic fitness and alarmist degeneracy concerns. It concluded with the popular "Better Baby" and "Fitter Family" contests, which gave the occasion a fair-like atmosphere and carnival joviality.

The previous year, acting in his role as president of the Michigan State Board of Public Health, Vaughan set up a commission on the problem of the feebleminded and related illnesses. A state law, based on a bill introduced by State Representative Arthur Odell, was passed in the same year, 1913, which legalized sterilization of "mentally defective persons maintained wholly or in part by public expenses in public institutions."[60] Once again, a superintendent played a central role. The director of the Michigan Home and Training School at Lapeer, Dr. R. L. Dixon, ensured that those officials helped to push the legislation in the state legislature. Dixon served with Vaughan and Kellogg on the Central Committee of the Race Better Foundation and on the State Board of

---

[58] Hodges, "Dealing with Degeneracy," 82–5.

[59] Victor C. Vaughan, "Eugenics from the Point of View of the Physician," in *Eugenics: Twelve University Lectures* (New York: Dodd, Mead and Co., 1914), 60, cited in Hodges, "Dealing with Degeneracy," 85.

[60] Public Act 24, Public Acts of Michigan 1913, 52–4, cited in Hodges, "Dealing with Degeneracy," 143.

Public Health. These figures were identified both in Michigan and nationally as strong eugenic advocates and publicly endorsed state eugenic sterilization legislation.[61] The 1913 bill was passed by votes of 72–16 in the House and 21–9 in the Senate.

The 1913 law was challenged. But in a pattern common in many states, the challenger was part of the eugenic circle seeking to help establish the constitutionality of the new measure. The Medical Director of the Michigan Institute for the Feebleminded (and the future Medical Director at Vaughan's Medical School), a Dr. Haynes, brought the test case of Nora Reynolds. The lawsuit questioned the "right or constitutionality of a law to inflict such an operation upon a subject who has violated no law, and no assumption can be legally indulged in that any law will be violated."[62] In *Haynes v. Williams*, first a Michigan circuit court and then the Michigan Supreme Court concluded that the 1913 law violated due protection because the sterilization measure applied only to individuals resident in institutions and not generally.[63] Thus, five years after the law was adopted and after only one person was sterilized, the Michigan Supreme Court threw out the law on the grounds of violating equal protection.[64] Five years later, in 1923, a new sterilization law was enacted. This one stuck, and it was amended two years later. At least 3,786 sterilizations were carried out in Michigan over the following decades.

Vermont

Vermont was a latecomer to coerced sterilization in effect, although not in intention. As elsewhere, superintendents working with supportive legislators were at the fore. Opened in 1912,[65] the State School for the Feebleminded at Brandon quickly filled up and had a waiting list for new patients. It and the Vermont State Hospital for the Insane in Waterbury (founded in 1891) were the main institutions dealing with the so-called mentally ill in the state; superintendents at both were active supporters of the state's progress toward enacting eugenic sterilization laws.[66] A year later, the superintendent at Brandon strongly supported a sterilization bill, which was passed by the Vermont legislature with the unusually strong backing of Governor John A. Mead, who was also a physician. Entitled "An Act to Authorize and Provide for the Sterilization of Imbeciles, Feeble-minded and Insane Persons, Rapists, Confirmed Criminals and Other Defectives," it was vetoed by Mead's gubernatorial successor, Allen

---

[61] Hodges, "Dealing with Degeneracy," 143.

[62] This is taken from "Our Supplement," *American Lawyer* [journal of the American Bar Association] 5, no. 7 (1897): 299, quoted by Largent, *Breeding Contempt*, 67.

[63] Hodges, "Dealing with Degeneracy," 145–6.

[64] Ibid., 144.

[65] Some date its opening to 1915, but 1912 is the proper date according to Gallagher, *Breeding Better Vermonters*, 52.

[66] In 1929, it was renamed the Brandon Training School. The institution was closed in 1993.

Fletcher, on grounds that it would prove unconstitutional, an opinion provided by state Attorney General R. A. Brown.[67]

Despite this setback, the usual professional combination of institutional superintendents, medics, professors, sympathetic law makers, and public intellectuals lobbied the state to enact a new eugenic sterilization law. The assessment that army draftees from Vermont included the country's second-highest number of "defectives" in 1917 gave the issue of "stock" and breeding salience in the state.[68] Such draft board results about defectives were widely cited across the United States to signal concerns about a declining "national stock." As in other states, Vermont did get a bill through its law-making process in 1915 to place restrictions on marriage between Vermonters judged to be "mentally deficient." It took until 1931, however, to achieve a eugenic sterilization law. Although the eugenics movement by then faced more skepticism nationally about the robustness of its heredity arguments, the Vermont statute was consistent with ones enacted in other states.

From 1916, lobbying in Vermont was organized first through the state chapter of the National Conference on Charities and Corrections and then through a dedicated commission of enquiry on rural life in the state. The former organization had a presence and role in advocating sterilization in many other states. These lobbying efforts took a considerable time before achieving success. Four years after the governor's veto, another law made it through the state's House of Representatives but not the Senate. On March 31, 1931, the governor finally approved a sterilization law entitled "An Act for Human Betterment by Voluntary Sterilization." This one stayed on the books, and its focus on hereditary degeneration was at the forefront: "henceforth it shall be the policy of the state to prevent procreation of idiots, imbeciles, feeble-minded or insane persons, when the public welfare, and the welfare of idiots, imbeciles, feeble-minded or insane persons likely to procreate can be improved by voluntary sterilization."[69]

The genesis of the 1931 measure lay in a comprehensive eugenic survey of the state conducted from 1925 under the direction of Henry F. Perkins, a professor of zoology at the University of Vermont and a keen supporter of eugenics. Perkins's study followed as a response to the abysmal Army Draft Board results published for Vermont as the United States entered the First World War. Perkins consequently initiated and administered a series of major eugenic investigations from 1925.[70] They built on an earlier study conducted by a political scientist, Professor Kemp R. B. Flint, called *Poor Relief in Vermont*. Flint's report, published in 1917, focused on the difficulties of dependent children and advocated eugenic remedies to the problem of procreation by unsuitable

[67] Gallagher, *Breeding Better Vermonters*, 52–3.
[68] Ibid., 39–40.
[69] Quoted in ibid., 185.
[70] Ibid., 41, 66–8.

parents.[71] Based on extensive field research in the state's charitable institutions, including poorhouses and poor farms (a practice of putting children in agriculture-based institutions designed to exploit the state's rural structure), the study's author calibrated different forms of poverty, pauperism, and feeble-minded in Vermont. Flint sought clearer differentiation between these categories of state dependents, as well as better treatment for children. For long-term improvement, he looked to addressing the problem of inherited defection: "sterilization or segregation of the feebleminded may gradually eliminate their kind."[72] Superintendents of the state's Industrial School and the Brandon State School for the Feebleminded welcomed the Flint report and used its findings to advocate sterilization for their inmates.

Similar views came in annual reports from the state's Board of Health. The superintendent of the Vermont State Hospital for the Insane between 1918 and 1936, Dr. E. A. Stanley, served on the advisory committee for Perkins's eugenic surveys, illustrating the close ties between professionals committed to eugenic reforms.[73]

The work directed by Perkins – orchestrated under his Committee on the Human Factor – included family studies similar to that of the Jukes to identify undesirable hereditary lineages. A massive study, *Rural Vermont: A Program for the Future*, detailed the need for eugenic measures throughout the state's communities and focused eugenic attention on the state's Native Americans, the Abenakis, and the state's French-Canadian population.[74] Perkins appointed a eugenicist scholar, Dr. Henry C. Taylor, who was familiar in national eugenics circles,[75] to act as executive director of the Vermont Commission on Country Life, the body created in response to his early eugenic surveys that collected the data for Perkins's report. The statewide eugenic studies were funded with grants from the Laura Spelman Rockefeller Memorial and the national Social Science Research Council. Perkins adroitly persuaded the state's governor, John Weeks, to serve as chair of the Commission's executive committee. Perkins also skillfully convinced state civic leaders and public intellectuals to serve on various sub-committees established by the Commission. Although some of these participants and others injected noneugenic-type explanations into the Commission's deliberations to explain Vermont's pattern of poverty and child dependency, Perkins opted to embrace and implement the eugenic arguments he had absorbed through meetings with ERO members including Davenport and Arthur H. Estabrook, an ERO field worker. He also hired a psychologist to administer IQ tests and a more general eugenic survey. This latter undertaking included the

---

[71] Flint was a faculty member at Norwich University. Ibid., 53–4.

[72] Cited in ibid., 54.

[73] Marsha R. Kincheloe and Herbert G. Hunt, Jr., *Empty Beds: A History of Vermont State Hospital* (Barre, VT: Northlight Studio Press, 1989), 99.

[74] Gallagher, *Breeding Better Vermonters*, 81–2.

[75] Ibid., 92.

Key Family Study, which compared the histories of families in different Vermont towns to determine patterns of degeneracy traceable to persistent inheritable traits.[76]

One distinct feature of the Vermont studies was its focus on Native Americans insofar as most state sterilization laws targeted the white feebleminded. The effects of these studies and resultant sterilization policy were especially devastating for the Native American community, many members of which were institutionalized and sterilized after the results of Perkins's survey appeared. The Abenakis continued to be sterilized long into the twentieth century without proper consent procedures, as historian Ian Dowbiggin records.[77]

As often was the case elsewhere, the eugenics interest in Vermont was part of the state political and intellectual elite's notional embrace of Progressive ideas. Perkins enjoyed membership and respect in the national eugenics networks and served as president of the American Eugenics Society between 1931 and 1934, an influential platform from which to lead and orchestrate opinion. The Perkins survey came in this spirit and was modeled on comparable ERO studies elsewhere in the United States. The Key Family Study survey focused on families already known to have allegedly defective children and sketched out pedigree profiles of families whose progeny inherited the feebleminded and other undesirable traits. A devotee of Davenport, Perkins was bitterly disappointed that Davenport would not use ERO funds for expanding the Vermont study to include French Canadians as a distinct part of the state population.[78]

Appearing in 1929,[79] the Vermont Commission on Country Life, with its strong eugenic undercurrent, fed into the state's enactment of a voluntary sterilization law two years later. The 1931 statute permitted sterilization of the "unfit" in institutions and strengthened restrictions on granting marriage licenses between those assessed to be eugenically inferior.

Alabama

In 1901, the director of the Alabama Insane Hospitals, Dr. William Hassell Somerville, gave an influential speech to the annual meeting of the Medical Association of the State of Alabama (MASA). He had a simple but powerful message: criminal traits were inherited. The same meeting heard from another eminent participant about the way in which mental illnesses, including forms of feeblemindedness, are inherited between generations.[80] These physicians and MASA itself launched more than a decade-long lobbying effort in favor of a state eugenic sterilization law. It was the key professional body advancing this cause

---

[76] Ibid., 93–5.
[77] Dowbiggin, *Sterilization Movement*, 181.
[78] Gallagher, *Breeding Better Vermonters*, 95–7.
[79] The year of the Commission's establishment.
[80] Larson, *Sex, Race, and Science*, 50.

in the state and also supported the creation of a state institution for feebleminded youth.

In 1915, MASA initiated the Alabama Society for Mental Hygiene (later affiliated to the eugenics-advocating National Committee for Mental Hygiene), and, through it, set out detailed plans for a sterilization bill.[81] One of the leading MASA members, Dr. William Dempsey Partlow, head of state facilities for the mentally disabled and later a keen supporter of eugenic sterilization, agreed to build relations between the eugenics-centered National Committee for Mental Hygiene and a new, MASA-inspired Alabama Society for Mental Hygiene.[82] Funded by a grant from the Rockefeller Foundation, the National Committee conducted surveys of the feebleminded in the second half of the 1910s in five Southern states in which officials perceived a need for greater reform.[83] In 1919, Mississippi passed legislation that created a new home for the feebleminded in Tuscaloosa, of which Partlow was superintendent; the law included a clause that all patients scheduled to be discharged from the Home were to be sterilized.[84]

But Partlow and his professional colleagues wanted a stronger version in Alabama, a law that would be applicable to all state institutions. Partlow drafted a bill empowering all institution superintendents with authority to sterilize "any or all patients upon their release."[85] This ambition failed. Bills to broaden the terms of eugenic sterilization across the state were passed by the state legislature in 1934, but the first was found unconstitutional by the state supreme court and the second vetoed by the governor.[86] Partlow's inspired bill included the conventional arrangement to establish a three-member state eugenics board filled with medical professionals to determine whether recommendations from institution superintendents to administer sterilization should be granted. Subsequent bills foundered in 1935, 1939, 1943, and in revised form in 1945.[87]

Partlow and fellow eugenics advocates in MASA achieved passage of the 1919 sterilization law but failed subsequently to marshal sufficient statewide political and legislative support for a deeper eugenic program, which reflected a general reluctance to push eugenics in Southern states.[88] In contrast to events in Vermont, Catholics in Alabama organized to protest the eugenic principle of

---

[81] Ibid., 60.
[82] Gregory Michael Dorr, "Eugenics in Alabama," *Encyclopedia of Alabama*, October 10, 2007, http://www.encyclopediaofalabama.org/face/ArticlePrintable.jsp?id=h-1367 (accessed January 16, 2013).
[83] Steven Noll, "The Public Face of Southern Institutions for the 'Feeble-Minded,'" *The Public Historian* 27, no. 2 (2005): 30.
[84] Edward J. Larson and Leonard J. Nelson, "Involuntary Sexual Sterilization of Incompetent in Alabama: Past, Present and Future," *Alabama Law Review* 43, no. 2 (1992), 413–5, 417.
[85] Ibid., 418.
[86] Ibid., 422–3.
[87] Ibid., 399–444.
[88] Larson, *Sex, Race, and Science*, 59–63.

sterilization.[89] Weak state resources and strong conservatism – an instinctive suspicion of new radical ideas that might upset the prevailing social order (although it is not the case that Progressives wished to challenge racial segregation) – made Alabama less receptive to eugenic sterilization before the Second World War.

Like so much else, Alabama's approach changed after 1945.

Oregon

Passage of Oregon's eugenic sterilization law in 1917 was presaged by over a decade of intense lobbying for such a measure. A bill was first introduced in 1907. When it failed to acquire a majority in the state legislature, it was reintroduced in similar form every year until enacted a decade later. In some of these years, the bill was passed by the legislature, as in 1909, but the state's governor vetoed the law due to the bill's generality and its potential violation of an individual's constitutional rights.

As in other states, the advocates of using sterilization to stop criminals and "imbeciles" from procreating simply waited for a sympathetic governor to be elected. Governor Oswald West obliged in 1913, but opponents of sterilization successfully forced a statewide referendum, held on November 4, 1913, by which the voters rejected the newly minted statute by a vote of 53,319 to 41,767.[90] Undeterred, and despite the lobbying efficacy of the state's Anti-Sterilization League, Progressive reformers returned to the fray and got another bill through the legislature, which was finally signed by the next governor in 1917. Prominent among these reformers was women's rights advocate Bethenia Owens-Adair, one of Oregon's first female physicians. Owens-Adair conducted a campaign through letters to the editor of the *Portland Oregonian* that made the case for mandatory sterilization of the feebleminded.[91] From 1907 until 1917, she repeatedly lobbied for a sterilization bill to be introduced into the state legislature.[92] An American Marie Stopes in part, Owens-Adair championed eugenics as a solution to reproduction among the so-called criminal and mentally feeble, since procreation by such individuals formed the "greatest curse of the race."[93]

The 1917 law established a Board of Eugenics, composed of superintendents of the state's hospitals, the state penitentiary, and the state institution for the feebleminded, as well as officials from the Oregon Board of Health, whose members supported enactment of sterilization laws. The Board of Eugenics's

---

[89] Bishop of Mobile [Alabama] to Rev. John J. Burke, February 8, 1935, AVS.
[90] Mark Largent, "'The Greatest Curse of the Race': Eugenic Sterilization in Oregon, 1909–1983," *Oregon Historical Quarterly* 103 (2002): 188–97.
[91] Largent, *Breeding Contempt*, 70.
[92] Largent, "'Greatest Curse,'" 195.
[93] From a letter by Owens-Adair published in 1904. See Largent, *Breeding Contempt*, 70–1.

remit was to review recommendations for sterilization brought from the superintendents of the state's various confining institutions. Quarterly reports identified and reviewed individuals who showed that they, "because of inheritance of
inferior or anti-social traits[,] would probably become a social menace or a ward
of the state."[94] By 1983, 2,341 sterilizations were documented in Oregon.[95]

Mississippi

Mississippi became the twenty-sixth state to enact a eugenic measure when it
passed its sterilization law in 1928. The legislation mirrored that of Virginia.
Between 1930 and 1963, sterilization was performed on 523 women and 160
men – a modest total compared with other states.[96] The law defined eligibility for
sterilization as applying to "persons who are afflicted with hereditary forms of
insanity that are recurrent, idiocy, imbecility, feeble-mindedness or epilepsy." A
strong influence on the reform dynamic throughout the nation, already noted in
other states, was the National Committee for Mental Hygiene. The Committee
helped conduct a survey of the number and character of the feebleminded and
also worked to defuse alarm about the so-called menace posed by this expanding
category. Eugenic arguments were similarly propagated through the Southern
Sociological Congress, which had been established in 1912 for the discussion
and dissemination of eugenic analyses and to develop solutions to social problems.[97] These organizations supported reform-minded superintendents and
legislators who sought eugenic sterilization laws to be used on institutional
patients. It was part of the Progressive-era idiom, as one historian underlines,
that those designated as feebleminded should be deemed returnable to society at
some point instead of being permanently institutionalized (a confinement necessary for the insane); to overcome the characteristics of "idiots" or "morons,"
they need to be sterilized against the potential of becoming parents.[98] The
superintendent of the South Carolina State Training School expressed this
eugenic aspiration: "if our Institution does nothing more than incarcerate,
teach, and train those admitted hereto, of course, it will have failed in many of
the high purposes for which it is intended."[99]

Institutions for the feebleminded and their superintendents were procedurally
crucial to the process of sterilization. Superintendents made a recommendation

[94] Legislation quoted in R. Newton Crane, "Recent Eugenic and Social Legislation in America,"
     Eugenics Review 10, no. 1 (April 1918): 25.
[95] Lombardo, *Three Generations*, 294; Largent gives a figure of 2,269 by 1980, in *Breeding
     Contempt*, 79.
[96] Cf. Julius Paul, *"Three Generations of Imbeciles Are Enough": State Eugenic Sterilization Laws
     in American Thought and Practice* (Washington, DC: Walter Reed Army Institute of Research,
     1965), 399 and Kaelber, "Mississippi," http://www.uvm.edu/~lkaelber/eugenics/MS/MS.html.
[97] Larson, *Sex, Race, and Science*, 60–71; Noll, *Feeble-Minded in Our Midst*, 15–7.
[98] Noll, "Public Face," 28.
[99] Quoted ibid.

that one of their inmates be sterilized; a hearing was held at the institution's board within thirty days of the inmate receiving the notice. The inmate was present, with a legal guardian or counsel, to review the recommendation. The law was a compulsory one, but consent was sought from families. Inmates could appeal sometimes all the way to the state supreme court.[100] The consent requirement helps explain why the numbers of those sterilized were modest: superintendents and boards of institutions felt constrained. One superintendent, H. H. Ramsey, who was head of the Mississippi School and Colony for the Feeble Minded, founded in 1923, noted that the law's provision for consent led them "to proceed cautiously under [the law's] provisions and sterilize only such cases as consent from parents or guardians can be secured."[101]

This legalistic emphasis differentiated the Mississippian process for reaching the recommendation for sterilization from that of other states. There was more recourse to judicial appeal, because the allegedly feebleminded person or his or her family or guardian could ask for a trial to review the decision for committal to an institution, a necessary condition before sterilization.[102] Part of the reason for this more cautious approach was the greater role given to the family in Southern states and hostility or, at least, resistance to Progressivist social reforms involving an expanded role for state agencies. Commitment to a state institution in the first place did not rely solely on medical evaluations but was predicated more often on court proceedings, including sometimes the results of jury deliberation.

But confinement was often long term, and, without sterilization, inmates could not expect to leave. This policy developed over time: initially, superintendents such as Ramsey at the School and Colony for the Feeble Minded, favored segregation. From the early 1930s, though, they extolled the merits of sterilization, believing that "selective sterilization should become an ally to the parole system of the institution."[103] In other words, discharge should be conditional on an inmate being sterilized. Mimicking the system developed in Ohio under Goddard's guidance, Ramsey recommended forming "traveling clinics" composed of eugenic experts and psychiatrists who would test school children to identify the feebleminded for institutional care. This arrangement, Ramsey argued, would "enable the state to assume charge of its defectives during the formative period, before they have become a menace and social liability."[104]

Ramsey's salient role is evidence of the influence of superintendents in the eugenic sterilization process. His Colony was established in 1920 under a "Law of Mental Deficiency." Initially for white males, women were admitted starting in 1928, but the proportion of women quickly outpaced that of male inmates.

---

[100] Paul, *Three Generations*, 399.
[101] Larson, *Sex, Race, and Science*, 121.
[102] Noll, *Feeble-Minded in Our Midst*, 34.
[103] Quoted in Trent, *Inventing the Feeble Mind*, 200.
[104] Quoted in Larson, *Sex, Race, and Science*, 95.

This trend reflected the common, although not universal, view that feebleminded women, as bearers of children and as easy targets for predatory males, posed the greater threat than did feebleminded men and therefore required more attention. Ramsey did not have the resources to make sterilization a large-scale operation in his Colony because funding, especially to counter legal challenges to recommendations, was limited. But his successor, Superintendent Paul Haney, proved more effective in ensuring a significant increase in sterilizations in the 1940s.[105]

The push for eugenics in Mississippi arose in part from factors that were common in other states. Alerted by the national surveys of feebleminded populations in the United States – for example, those undertaken by the National Committee for Mental Hygiene in the 1910s[106] – superintendents sought to highlight the problem in Mississippi. Eugenic solutions had overtones of Progressivism. But, as in other Southern states, there was suspicion of Progressivist ideas apparently being imported from the North, and a religious conservatism, centered on family life, stemmed enthusiasm among some policy makers. Roman Catholic leaders in Mississippi explicitly opposed eugenic sterilization laws, but Roman Catholics were sparse in the state – less than 2 percent of the population – and theirs was a weak voice.[107]

Progressive reform of any sort – including eugenic – functioned within the limits of the Jim Crow order,[108] meaning that state "services" such as sterilization would not be extended to the state's sizable black population. A related reason for the modest number of sterilizations carried out was the limited resources available to implement the law, which generally paled in comparison to Northern states.[109] In keeping with the profoundly segregationist order of the South,[110] the putatively progressive reform regarding the issue of feeblemindedness was conceived of in reference to white needs only. There were no facilities for African Americans. The Colony for the Feeble Minded admitted African Americans only in 1968, forty-eight years after its establishment.[111]

Within state institutions and hospitals, however, sterilizations were seen as a legitimate response to the feebleminded menace. The label "therapeutic" was often assigned to sterilization operations, rather than "eugenic," thereby undercounting the number of operations recorded as eugenic.[112]

---

[105] Ibid., 153.
[106] Ibid. 60–71; Noll 16–17.
[107] See Table 7.2.
[108] See Kimberley Johnson, *Reforming Jim Crow: Southern Politics and State in the Age before Brown* (New York: Oxford University Press, 2010), 41.
[109] Noll, "Public Face," 33.
[110] Desmond King and Rogers M. Smith, "Racial Orders in American Political Development," *American Political Science Review* 99, no. 1 (2005): 76–89.
[111] Larson, *Sex, Race, and Science*, 91–2, 122–3; Noll, "Public Face," 30–1.
[112] Susan K. Cahn, *Sexual Reckonings: Southern Girls in a Troubling Age* (Cambridge: Harvard University Press, 2007), 173–4.

TABLE 5.3 *Canadian Provinces and Eugenic Sterilization Laws*

Province	Sterilization law		Year(s) enacted
	Yes	No	
New Brunswick	–	X	–
Nova Scotia	–	X	–
Alberta	X	–	1928, 1937
Ontario	–	X	–
Quebec	–	X	–
British Columbia	X	–	1933
Manitoba	–	X	–
Prince Edward Island	–	X	–
Saskatchewan	–	X	–
Newfoundland & Labrador*	–	X	–

* Joined Confederation in 1949

Eugenics in Canada

The Canadian pursuit of eugenic sterilization laws nationally and in the provinces both mirrors American practice and diverges in some important ways. The paired comparison is the subject of Ian Dowbiggin's influential work on eugenics.[113] As in other countries, eugenic policies gained support from across the ideological and political spectra, and its promotion won endorsement from progressive and reforming forces in national politics.[114] The parallel with other advanced democracies, including the United States and some European countries such as Britain, is unsurprising.[115] Across many countries, reformers of different ideological hues embraced the putatively progressive notion of alternatively "helping" or "controlling" the mentally unwell.

As in the United States, Canadian advocates of eugenic sterilization operated subfederally in the early days of their movement. Unlike their U.S. counterparts, however, they remained essentially at that level and failed to achieve a universal eugenic standard on the federal level of the sort granted by the U.S. Supreme Court in 1927 (discussed in the next chapter). Officials in two provinces – British Columbia and Alberta – sterilized (Table 5.3).

Alberta's law, initially based on consent when it was passed in 1928,[116] included a provision for involuntary sterilization when it was amended in 1937. Five other provinces commissioned "mental hygiene reports" on eugenic sterilization that were conducted by the energetic and eugenics-proselytizing

[113] Dowbiggin, *Keeping America Sane.*
[114] McLaren, *Our Own Master Race.*
[115] Michael Freeden, "Eugenics and Progressive Thought: A Study in Ideological Affinity," *Historical Journal* 22, no. 3 (1979): 645–71.
[116] Dowbiggin, *Keeping America Sane*, 180.

Canadian National Committee for Mental Hygiene.[117] The organization was created in 1918 by Drs. Clarence M. Hincks and Clifford W. Beers, both psychiatrists, and was deeply influenced by renowned psychiatrist and Hincks's mentor, Dr. Charles K. Clarke.[118] It was not affiliated with any national professional bodies but nonetheless enjoyed authority due to the renown of its founding medical experts.[119] Echoing state-level studies conducted in the United States, these reports identified growing populations of "feeble-minded" or "defectives" in the 1920s. The reports' recommendations included the remedy of eugenic sterilization or deportation for new immigrants in response.[120] The Committee had a decisive influence in British Columbia and Alberta, effectively aiding and abetting the two provinces' enactment of eugenic sterilization laws. The United Farmers of Alberta (UFA) political party seized on the recommendation for eugenic sterilization[121] and lobbied the province's lawmakers to pass an act. New immigrants were deemed inferior to genuine Canadian stock, which was, of course, an argument also circulating among eugenics advocates in the United States.[122] The Eugenics Society of Canada advanced the cause of eugenic sterilization as a component of national policy to address the dilution of the "core" Canadian population resulting, its members maintained, from a recklessly open and nonselective immigration policy between the 1890s and 1910s.

Eugenic sterilization in Alberta, and especially in British Columbia, was adopted in the context of extreme public hostility against immigrants and above all East Asian immigrants. In British Columbia, women's groups supported negative eugenics, that is, the prevention of future procreation by the mentally incompetent. After adoption, the British Columbian law placed the superintendent at the core of the sterilization regime: he recommended sterilization to the province's Eugenics Board, which in turn had to approve the sterilization unanimously.[123]

Alberta passed its Sexual Sterilization Act in 1928, under which patients reviewed for discharge from mental institutions would be routinely assessed for eugenic sterilization. The bill for the law was introduced into the Alberta legislature by Minister for Health George Hoadley and was passed on March 7,

---

[117] McLaren, *Our Own Master Race*, 60, 99. The CNCMH was the forerunner of today's Centre for Addiction and Mental Health (CAMH).
[118] Dowbiggin, *Keeping America Sane*, 167–8; McLaren, *Our Own Master Race*, 109–10.
[119] Dowbiggin, *Keeping America Sane*, 168–9.
[120] Carolyn Strange and Jennifer A. Stephen, "Eugenics in Canada," 530–2.
[121] Dowbiggin, *Keeping America Sane*, 180.
[122] Julio Decker, "The Immigration Restriction League and the Political Regulation of Immigration, 1894–1924" (Ph.D. diss., University of Leeds, 2012). This study of the Immigration Restriction League (IRL) demonstrates how the League's influence fed into the eugenic turn in U.S. immigration policy more deeply than previously considered. It also shows that the IRL played a crucial role in defining a conception of whiteness in Anglo-Saxon America.
[123] "Canada: Sterilization," n.d. [likely late 1940s], 4/32, AVS.

1928. The UFA's support of eugenic sterilization was reinforced by the lobbying efforts of the complementary United Farm Women of Alberta, whose president, Margaret Gunn, urged in 1924 the enactment of a government policy of "racial betterment" for "weeding out ... undesirable strains."[124] Introducing the sterilization bill, Hoadley cited the perennial argument that persons with mental problems and unable to look after children were increasing in number and posed therefore an ever-greater burden on taxpayers. Efforts to oppose the legislation, mobilized under the People's League to Act, garnered publicity but insufficient support among law makers to thwart the law's passage.

The statute permitted sterilization of an inmate in a mental health institution if it could be demonstrated that "the patient might safely be discharged if the danger of procreation with its attendant risk of multiplication of evil by transmission of the disability to progeny were eliminated."[125] Like their American counterparts to the south, superintendents at individual mental health institutions in Alberta had considerable discretion and power in deciding which inmates might be recommended for sterilization and presented those names to the Board of Eugenics.[126]

The institutional procedures for assessments fell within the broad comparative practice, with two physicians or other medical experts and two lay persons serving as a review committee. Together, they adjudicated on the suitability of individual patients for release, given the danger of their procreating and passing on their incompetence. The decision to sterilize was made compulsory in 1937, with the four-member board having absolute power over each case.[127] The chair of Alberta's Eugenics Board between 1928 and 1965, Dr. J. M. MacEachran (a professional university philosopher), personally approved recommendations to sterilize more than 2,000 individuals.[128] The Board was disbanded in 1972, with a physician having served as chair for its last seven years.[129] Of the 4,785 cases reviewed by the Eugenics Board, it recommended sterilization for 99 percent of them and deferred decisions about the remaining 1 percent – not once did it vote "no."[130] In the case of Alberta, the pattern of decisions over the lifetime of the Eugenics Board shifted increasingly toward the sterilization of women judged to

---

[124] Quoted in Jana Grekul "Sterilization in Alberta, 1928 to 1972: Gender Matters," *Canadian Review of Sociology* 45, no. 3 (2008): 250. The president was Margaret Gunn.

[125] Quoted in Jana Grekul, Harvey Krahn, and Dave Odynak "Sterilizing the 'Feeble-minded': Eugenics in Alberta, Canada 1929–1972," *Journal of Historical Sociology* 17, no. 4 (2004): 363.

[126] Jana Grekul, "A Well-Oiled Machine: Alberta's Eugenics Program, 1928–1972," *Alberta History* 59, no. 3 (2011): 20–1.

[127] Grekul et al., "Sterilizing the 'Feeble-minded,'" 363.

[128] Douglas Wahlsten, "Leilani Muir versus the Philosopher King: Eugenics on trial in Alberta," *Genetica* 99, nos. 2–3 (1997): 189.

[129] Grekul et al., "Sterilizing the 'Feeble-minded,'" 365–6. Sixty percent of these recommended sterilizations were carried out.

[130] Ibid., 367.

be in some sense immoral and veering from appropriate standards of feminine behavior.[131]

As one account underscores, the Board ran in classic, routinized bureaucratic style, in effect "a well-oiled machine."[132] Dr. MacEachran held the position and exercised a dominant influence throughout most of the Board's existence. There was also little turnover in Board members: over forty-four years, only nineteen people served on the Board.[133] Sociologist Jana Grekul measures routinization by noting the brief time each Board meeting gave to evaluating recommendations for eugenic sterilization. The evaluation included the Board meeting and holding an interview with the patient, an examination of the case file documents, reaching a unanimous agreement on the recommendation, and signing off on the necessary paperwork. This process took an average of fifteen minutes per patient in the 1930s, but, by the 1940s, the Board reduced the average time to eight minutes; the average increased again to sixteen and seventeen minutes per case in the 1960s and 1970s on account of new IQ tests that had to be administered to each patient.[134] The casual nature of these Board discussions, the failure to acquire confirmatory evidence of other mental illnesses in the patient's family, the failure to seek additional information about whether those recommended for sterilization were to be considered for discharge, and a disregard of some surprisingly high IQ test results imply a commitment to policy and procedure over scrutiny and individual case evaluation.[135]

An examination of the main mental institutions that sent recommendations to the Alberta Eugenics Board shows variation in the number of referrals submitted, which was itself a function of how active or inactive individual institution superintendents were. Superintendent activity was greatest in the Provincial Training School in Red Deer[136] and was also significant at Alberta Hospital (Oliver) in Edmonton and at Deerhome (also in Red Deer). The Board responded to recommendations for sterilization affirmatively and also delegated issues of liaising with patients to the province's network of institution superintendents. If patients were deemed to be mentally defective, their consent was not required as a condition of the sterilization treatment. Canadian Aboriginals were markedly overrepresented in the category of "mentally defective," with patient consent

[131] This analysis is developed in Grekul, "Sterilization in Alberta."

[132] Grekul, "Well-Oiled Machine." Grekul writes that "the conditions set out in the legislation got lost in the mire: prospect for discharge, 'negative' family history, and IQ tests faded into the background as the goal of sterilizing more people resulted in a sort of tunnel vision among key players." Ibid., 21. All records of the Board meetings, including patient case files, were deposited in the Alberta Provincial Archive in 1972, when the Eugenics Board was abandoned; in 1987, it was decided to destroy 80 percent of patient records, retaining only a 20-percent sample. See Grekul, "Sterilization in Alberta," 251–2.

[133] Grekul, "Well-Oiled Machine," 17. In the Board's final years from 1965 to 1972, the chairman was Dr. R. K. Thomson.

[134] Ibid., 18.

[135] Ibid., 19.

[136] Grekul et al., "Sterilizing the 'Feebleminded,'" 369–70.

sought from only 17 percent of Aboriginal inmates compared with 49 percent of Eastern European, 44 percent of Western European, and 38 percent of Anglo-Saxon patients. Three-quarters of those Aboriginal inmates recommended for the treatment were sterilized.[137]

The history of state-led sterilization in British Columbia shares much with the ideological beginnings and rhetorical campaign of that trend in other jurisdictions in North America. Fear of immigrants, as in Alberta, was a major motivation for firing up public support for surgical intervention, as immigrants were purported to be disproportionately feebleminded, epileptic, and criminal; Canada was at risk of being overrun by "lazy, degenerate, dissolute and mentally deficient" newcomers from such loathsome places as "sunny Italy."[138] The first call in the provincial Legislative Assembly for a sterilization law was made in 1925 by Mary Ellen Smith, the first female minister in the British Empire, who called for tougher immigration policy as well as a sterilization bill along the lines of the one passed by Washington state.[139] That Smith herself was an immigrant from England caused no logical dissonance because Canada was part of the British Empire.

Momentum for a sterilization measure picked up following an endorsement of such a law in a Royal Commission on Mental Hygiene report filed in 1928.[140] "Human thoroughbreds," argued activist Emily Murphy in 1932, should be the province's goal. Continuing her bald-faced barnyard metaphor, Murphy reasoned that the state "protect[s] the public against diseased and distempered cattle. We should similarly protect them [sic] against the offal of humanity."[141] The following year, the Legislative Assembly voted unanimously to consider a proposed sexual sterilization bill and passed it with "only cursory debate" just six days later.[142] As elsewhere, the new law established a provincial Board of Eugenics (to consist of a judge, a psychiatrist, and a social worker) and placed the authority to initiate a sterilization recommendation solely in the hands of the relevant institution's superintendent. That superintendent was empowered to do so if he felt the patient, if discharged "without being subjected to an operation for sexual sterilization would be likely to beget or bear children who by reason of inheritance would have a tendency to serious mental disease or mental deficiency."[143] The law was nominally voluntary – compared with the involuntary

---

[137] Ibid., 375.

[138] Angus McLaren, "The Creation of a Haven for 'Human Thoroughbreds': The Sterilization of the Feeble-Minded and the Mentally Ill in British Columbia," *Canadian Historical Review* 67, no. 2 (1986): 132, 134.

[139] McLaren, *Our Own Master Race*, 95. McLaren emphasizes that women were the "earliest and most vigorous proponents of sterilization" in British Columbia. "Human Thoroughbreds," 133.

[140] McLaren, "Human Thoroughbreds," 139.

[141] McLaren, *Our Own Master Race*, 101.

[142] Ibid., 102–3. The expedience of this passage might have been due at least in part to the deterioration of the government that was in power at the time; the bill and others were "rushed through three readings in the last week of the government's life." Ibid.

[143] McLaren, "Human Thoroughbreds," 144; *Our Own Master Race*, 104–5.

procedure adopted in Alberta – although consent could be supplied by a spouse, parent, guardian, or provincial secretary if the patient could not give it. But even this consensual basis was only loosely adhered to in routinized practice.[144]

Despite the apparent eugenic fervor when the law was passed, it "was infrequently and chaotically employed."[145] A lack of records makes the total number of persons sterilized by British Columbia hard to ascertain, but McLaren concludes that it was "not more than a few hundred."[146] Moreover, the Board of Eugenics "did not include a geneticist and did not show any great interest in the subject."[147] Sterilizations continued in the following decades, but the law was eventually repealed in 1973. In 2006, the Supreme Court of British Columbia awarded $450,000 to nine women, all of whom had been sterilized by the province between 1940 and 1968.[148]

DIVERGING ROADS

These state-by-state and province-by-province cases demonstrate the legal and constitutional barriers facing early advocates of sterilization laws. Although the practice of involuntary sterilization was to attain national legality later – at least in the United States – almost half of the laws (seven of sixteen statutes) were found unconstitutional in state or federal courts in the first half of the twentieth century. The grounds were unsurprising. When called to assess the sterilization laws passed by state legislatures, many U.S. judges found them inadequate in protecting basic rights such as due process and equal protection and in avoiding cruel and unusual punishment.[149] As historian Edward J. Larson underlines, the enthusiasm for eugenics was not a mass movement but, rather, part of the beliefs assimilated by elite and professional groups. Many, but by no means all, of these groups were keen on Progressivism and glad to lend their support to enacting laws for dealing rationally with the fiscal and social harms engendered by the "national breeding" of feeblemindedness. In the Southern United States, Progressivism functioned through a racist filter, which meant creating a policy for white feebleminded people only – but within that region, even this counted as progress. Throughout the states and provinces enacting eugenic sterilization laws, however, the embrace of eugenics was clear among those professionals and politicians involved in mental health issues, principally superintendents of institutions.

---

[144] Strange and Stephen, "Eugenics in Canada," 552: "when 'asexualization' was presented as a condition of release and when guardians of minors or of those diagnosed as incompetent so wished, sterilizations occurred, with and without individuals' consent."

[145] McLaren, "Human Thoroughbreds," 149.

[146] Ibid., 145.

[147] Ibid., 147.

[148] "BC Settles Sterilization Suit with Nine Women," *Catholic New Times* 30, no. 2 (January 29, 2006): 7.

[149] Larson, *Sex, Race, and Science*, 28.

Often too wedded to the particular cause of eugenic sterilization, prominent eugenicists failed to become public intellectuals themselves but nonetheless had profound influences on other prominent intellectuals and other opinion-framers who absorbed, often uncritically, the scientifically presented arguments of eugenicists and recycled them in newspaper columns, radio broadcasts, and other media. At the U.S. House of Representatives hearings on proposed immigration restriction, Harry Laughlin presented impressive statistical charts and predictions showing the unequivocal dangers of not limiting the influx of eugenically inferior migrants.

Without the dedication of such key individuals – in practice commonly physicians and superintendents allied with the ardent efforts of the American Eugenics Society and institutionally powered through the ERO on Long Island and the Human Betterment Foundation in California – sterilization laws would not have advanced. The indefatigable national propaganda effort by the likes of Davenport and Laughlin to affirm the justness of eugenic goals, including sterilization, did not translate into a mass popular movement.[150] But in the first couple of decades of the twentieth century, eugenic ideas about improving "human betterment" from scientific principles found a fervent audience among what we would now call public intellectuals and influential think-tank policy advisers, as they did in other advanced democracies.[151] Cultivating and defending the human "germplasm" was an obvious priority that drove the dual but linked aims of stopping the undesirable from procreating and encouraging the desirable to reproduce bountifully.

One historian of involuntary sterilization in the United States, Philip Reilly, concludes of the first two decades of the twentieth century that "when sterilization data are analyzed by institution, the influence of the superintendents is readily apparent."[152] This important proposition extends to subsequent decades, too. But for the superintendents' role to expand and to achieve legal certainty, assurance of the judicial and constitutional bases of their actions was necessary. In the United States, only a federal Supreme Court endorsement could furnish such a certainty. How this threshold was reached is the story of *Buck v. Bell*.

---

[150] Ibid., 32; Haller, *Eugenics*.
[151] Mottier, "Eugenics and the State."
[152] Reilly, *Surgical Solution*, 49.

6

Buck v. Bell and Beyond

In March 1924, seventeen-year-old Carrie Buck gave birth to a baby girl named Vivian in Charlottesville, Virginia.[1] On June 4, the young mother departed for Lynchburg with her social worker, leaving the baby with her foster parents, John T. and Alice Dobbs. After the ninety-minute train journey, the social worker, Caroline Wilhelm, signed Carrie over to the care of the State Colony for Epileptics and Feebleminded and its superintendent, Dr. Albert Priddy.[2] Carrie's committal to the institution was the recommendation of two court-appointed physicians assigned to her when her pregnancy became known. The doctors agreed with her foster parents' view that Carrie was feebleminded, as defined in state law, and concurred that she be moved to the State Colony after she gave birth. Despite this segregation from society, however, Carrie would not be completely alone: her mother, Emma, had already been committed to the Colony in 1920. Emma spent the remainder of her life in the institution until she died in 1944, aged seventy-one.[3]

Priddy used the law recently passed by the Virginian state legislature authorizing the involuntary sterilization of the feebleminded or the "socially inadequate" to recommend this treatment for Carrie. There was nothing surprising in this; it was effectively his law. Priddy had lobbied lawmakers and worked with the Colony's chief administrator, Aubrey Strode – a state legislator and the author of the bill that had established the State Colony – to draft the sterilization

---

[1] There is ample scholarly and research-based literature documenting the origins, circumstances, and significance of this major constitutional case; virtually all books dealing with eugenics have a discussion of it. We draw on this secondary literature and acknowledge scholars' major contributions, including Lombardo, *Three Generations*; Bruinius, *Better for All the World*; and Largent, *Breeding Contempt*.

[2] Lombardo, *Three Generations*, 105. Wilhelm also chaperoned a young man on the journey who was committed to the Colony.

[3] Carrie's own father had abandoned the family or died shortly after her birth. Bruinius, *Better for All the World*, 27.

bill and to see it through the legislature. The two men were close colleagues: Priddy had discharged the delicate task of conducting a confidential medical examination of Strode's fiancée, who was also an unreserved eugenicist, to determine if she could have children.[4]

Strode embraced the milieu and ideas of America's Progressive reformers. Senator Strode opposed corruption, was among the first sponsors of women's suffrage, wrote legislation in 1910 that established a women's college at the University of Virginia, advocated state assistance for the poor and the families of prisoners, was the author of the state's system for parole and probation, and was vocal in his support for Prohibition.[5] Strode threw his weight determinedly behind the new State Colony for Epileptics. It was built in Strode's district, just outside the town of Lynchburg.[6] The law stated that the institution should take in

> those indigent white persons who would be a greatest service to the Colony, who would in the judgment of the superintendent of the colony be most likely to receive benefit from colony care and training, and who are women of child-bearing age from twelve to forty-five years of age.[7]

Albert S. Priddy became the Colony's first superintendent. Within a few years of the Colony's opening in 1910, the first law to permit mandatory sterilization of the feebleminded was put before the Virginia legislature. The state's Board of Charities and Corrections, established in 1908, supported eugenic sterilization. Its chairman, Reverend Joseph Mastin, became a keen advocate for the feebleminded after undertaking a comprehensive study of the conditions in the state's existing institutions.[8] His report, published in 1916 for the General Assembly on Mental Defectives in Virginia, pronounced, "it is generally accepted that bad heredity is the cause of fully two thirds of the cases of feeblemindedness"; this view influenced Senator Strode's move to make addressing feeblemindedness a state priority.[9] Mastin's study included the standard eugenic family study charts showing the transmission of feeblemindedness across several generations and underlined the costs of this problem to the taxpayer. He concluded that, to control feeblemindedness, Virginia state law makers, must:

> [P]revent by segregation or sterilization the procreation of the feebleminded. Do this and most of them could be eliminated in two generations. Only by striking at

[4] Lombardo, *Three Generations*, 94–5.
[5] Dorr, *Segregation's Science*.
[6] Noll, "Public Face," 29.
[7] Quoted in Bruinius, *Better for All the World*, 40.
[8] Dorr, *Segregation's Science*, 122.
[9] Joseph T. Mastin, *Mental Defectives in Virginia: A Special Report of the State Board of Charities and Corrections to the General Assembly 1916, on Weak Mindedness in the State of Virginia; together with a Plan for the Training, Segregation and Prevention of the Procreation of the Feebleminded*, 20. Available at http://digitalarchive.gsu.edu/cgi/viewcontent.cgi?article=1001&context=col_facpub (accessed January 18, 2013).

the foundation head, i.e., by segregating or sterilizing all feebleminded child pro-
ducers, can we hope to dry up the springs of this evil.[10]

Strode welcomed the assumptions and recommendations for both segregation
and some form of sterilization to control the burgeoning problem, since feeblemindedness was hereditary and thus incurable. Reverend Mastin, who had
extensive experience with the state's institutions, threw his support behind the
measure, as did the Virginia Medical Society. Strode spoke in favor of the effort,
but to no avail; the bill went down to easy defeat in 1910. Public opinion
remained opposed, but success was not to be denied for long.

In most states, eugenicists mixed science with strategy: although eugenic
sterilization was justified on race-preservation grounds, it would also save
money. Not every legislator would agree with the former, but few, if any,
could disagree with the appeal of the latter. Priddy and Strode argued that
sterilizing the fecund feebleminded, like Carrie, would save Virginia unnecessary
future expenditures in two ways. First, there would be no costs to the state from
tending to offspring of the incompetent because they would no longer give birth
to government-dependent children. Second, such parental incompetents could
leave the institution after being treated and approximate self-sufficiency with
unskilled work. But Carrie's fate would have much wider significance than
enhancing mere fiscal rectitude.

Carrie Buck's proposed sterilization was nonetheless unexceptional on one
level. Superintendent Priddy sterilized almost one hundred Virginia women in
the 1910s. But a lawsuit against Priddy brought by the enraged husband of an
involuntarily sterilized woman (and their daughter), which added to doubts
about the legality of such operations, led to a pause in the practice. Priddy and
Strode wanted not only state law support for sterilizing the institutionalized but
also a declaration of the constitutionality of doing so. Under the 1924 state law,
sterilization was a treatment that the board of a state institution for the mentally
deficient or epileptic could order for a patient assumed to be "insane, idiotic,
imbecile, feeble-minded, or epileptic, and by the laws of heredity is the probably
potential parent of socially inadequate offspring likewise afflicted."[11] Thus,
when Carrie's routine sterilization became the subject of Supreme Court review
three years later, her name became forever bound to the American and wider
twentieth-century eugenic story.

Priddy sought to sterilize Carrie Buck as a "moral delinquent" and feebleminded under the definition of the new Virginia law. He recommended performing the operation to the board of governors of the Colony. The board, with a
guardian acting on Carrie's behalf, approved. Tested on the Stanford Revision of
the Binet-Simon test, which Goddard had introduced through Vineland in New
Jersey, Carrie was assessed to have the mental age of a nine-year-old and to be a

[10] Ibid., 110.
[11] Quoted in Lombardo, *Three Generations*, 99.

"middle grade moron." Priddy described her as "feebleminded of the lowest grade Moron class."[12] Priddy did not want this case to be just another routine sterilization but, rather, to be the test case that would establish a constitutional basis supporting a general policy of sterilization in the state. Carrie was particularly useful as a test case because her mother had also been deemed feebleminded, and the expectation was that Carrie's daughter, Vivian, would be too. Priddy therefore appointed an attorney, Irving Whitehead, to act as defender for Carrie in a court case he brought in the Circuit Court of Amherst County. The Circuit Court judge found in favor of sterilizing her following a presentation of experts and of evidence that characterized Carrie as feebleminded, a condition inherited from her mother. Furthermore, the condition was apparently already in evidence in the seven-month-old Vivian, who was described as "not quite a normal baby."[13] Sterilization would enable Carrie to leave the Colony, saving the state a great deal of money, and she would be incapable of future pregnancies. The judgment, issued in *Buck v. Priddy* (1924), approved the sterilization. Carrie's attorney, Whitehead, then appealed to the Virginia Supreme Court of Appeals, which upheld the original decision, and then to the U.S. Supreme Court in 1927. By this time, Priddy's death in 1925 meant the office of superintendent at the Colony had passed to Dr. J. H. Bell, whose appellation was attached to the Supreme Court decision, *Buck v. Bell*.[14] The law's constitutionality, the institution's lawyers argued, turned on two rights contained in the Fourteenth Amendment of the U.S. Constitution: individuals are entitled both to due process of law and to equal protection under the law. Agreement from the high court that the statute afforded Carrie due process and equal protection would validate Virginia's law and would also apply to almost identical eugenic laws in other states that were modeled on the same template.

Aside from confirming the constitutionality of state sterilization laws like the one in Virginia, the 1927 Supreme Court case is famous for the language used in the Court's decision, as it enshrined eugenic assumptions in American life. The decision received almost unanimous support (only one of the nine justices dissented), and one of America's great jurists, Justice Oliver Wendell Holmes Jr., penned a flamboyant embrace of eugenics in the majority opinion.

The Court was satisfied that the proper legal and administrative procedures had been followed in assessing Carrie's mental capacity and that Virginia's methods for undertaking sterilization (salpingectomy for women and vasectomy for men) were appropriate to the task and involved no "cruel or unusual punishment." Justice Holmes solemnly declaimed that if higher citizens could sacrifice themselves for the public welfare, such as through military or other public service, then the lower ones ought to be treated obversely. As he put it with characteristic eloquence,

---

[12] Quoted in ibid., 107.
[13] Ibid., 117.
[14] *Buck v. Bell*, 274 U.S. 200 (1927).

[W]e have seen more than once that the public welfare may call upon the best citizens for their lives. It would be strange if it could not call upon those who already sap the strength of the State for these lesser sacrifices, often not felt to be such by those concerned, in order to prevent our being swamped with incompetence. It is better for the world, if instead of waiting to execute degenerate offspring for crime, or to let them starve for their imbecility, society can prevent those who are manifestly unfit from continuing their kind.

The cerebral Holmes then added the phrase – soon to be an aphorism – with which both he and *Buck v. Bell* would be forever associated: "three generations of imbeciles are enough."[15] Carrie herself, her mother, and her now three-year-old daughter – notably, all females – were the three generations whose incompetence and imbecility elicited Holmes's favorable judgment. This conclusion was meant, the justices ruled, to indicate "a policy" with general application to states' sterilization laws.

LEGAL PRECEDENTS: HOW *BUCK V. BELL* CLARIFIED THE LEGALITY OF INVOLUNTARY STERILIZATION

Virginia was not the first state to draft and enact a law permitting the coerced sterilization of men and women assessed to be feebleminded. This supposed feeblemindedness meant – in the view of numerous Progressive social reformers, including eugenicist medical experts and institution superintendents – that the diagnosed individuals should be prevented from having children. Some of these early efforts, however, crashed against the rocks of constitutional opposition. In seven cases, courts struck down eugenic legislation passed at the state level because they violated constitutional rights.

These constraints and setbacks profoundly informed State Senator Strode's and Superintendent Priddy's work on the bill: they wanted air-tight legislation that could withstand legal and constitutional challenge. This goal was shaped not least by the fact that Priddy had suffered an injurious lawsuit, mentioned earlier, which had been brought by the husband of a woman sterilized unlawfully while confined in the State Colony. Above all, this aim meant ensuring that the legal precepts of due process and equal protection were afforded – or at least had the appearance of having been afforded – to those feebleminded individuals recommended for involuntary sterilization. A procedure including proper assessment by experts, the review of the recommendation by a notionally impartial board, and the opportunity for appeal all had to be included in the act. Several other states' eugenics-based sterilization bills had failed to comply with these legal standards and had been declared unconstitutional in various courts of appeal.

If Strode and Priddy could draft such a robust, legally unassailable bill and have it found constitutional, they would be achieving success for sterilization not

[15] Ibid. at 207.

merely in Virginia but nationwide. Like other reformers, Strode was able to draw on the model legislation drafted by Dr. Harry Laughlin, the Eugenics Record Office superintendent.[16] And, even before passage of the compulsory sterilization law in 1924, Strode drafted and guided the passage of two important measures in 1920 prompted by the grievous lawsuit that had been brought against Priddy.[17] The first prescribed that legal costs of any hospital superintendent sued by patients be covered by the hospital, and the second established that, by definition, confinement to an institution for the epileptic or feebleminded implied lawful commitment.[18]

The new Virginia law was passed with almost unanimous support in both chambers in 1924, having been introduced by a reliable colleague of Strode in concert with Governor E. Lee Trinkle, a long-standing associate of Strode.[19] The law's robustness and constitutionality, however, could only really be tested through review by the nation's highest court, the U.S. Supreme Court in Washington, DC.[20]

Once committed to the State Colony in Lynchburg and recommended for sterilization, Carrie's case was reviewed by the Colony's board of directors. This step was central to the claim that the Colony had provided due process. An attorney, R. G. Shelton, was appointed as Carrie's legal guardian, but only Superintendent Priddy spoke to document Carrie's supposed mental feebleness.[21] Her competence was judged high enough to permit making a living in society, hence relieving the state of fiscal responsibility for her well-being, but it was too low to permit Carrie to procreate. Indeed, prevention of her procreation was the principal eugenic rationale for the sterilization recommendation. Asked whether Carrie might return to society without the sterilization operation, Superintendent Priddy ruled the option out since Carrie was "congenitally and incurably defective."[22] The final question at the hearing was for Carrie, who had received written notification in advance of the hearing, as required by the Virginia law to ensure due process. Aubrey Strode asked Carrie whether she had a response to the recommended operation, to which she replied: "No sir, I have not. It is up to my people."[23] The board's approval for the sterilization

---

[16] Laughlin was the eugenicist engaged by the Committee on Immigration to help ensure that the bills winding through Congress in the 1920s included measures to exclude the eugenically inferior from entry to the United States.

[17] In the past, Priddy had relied on a "network of informants" comprised of various law enforcement officials, probation officers, and other bureaucrats to identify new inmates to be committed to the Colony. Lombardo, *Three Generations*, 64. Similar networks proved instrumental in the postwar sterilization efforts of other states, such as North Carolina, which is addressed in subsequent chapters.

[18] Lombardo, *Three Generations*, 92.

[19] Ibid., 95.

[20] Mastin, *Mental Defectives in Virginia*. The report included studies of criminality, almshouses, and charitable gifts finding in almost every cause of social concern a heredity source.

[21] Lombardo, *Three Generations*, 107.

[22] Quoted from transcripts, ibid.

[23] Quoted ibid.

procedure paved the way for the process of judicial review to state and federal courts of appeal. During the case, Superintendent Priddy's testimony stressed the benefits to both Carrie and society of her sterilization: she could nominally enjoy sexual relations and a full life without procreating and would not thereby burden society with the costs of her inevitably feebleminded offspring.

A star witness for the proponents of sterilization was Dr. Arthur H. Estabrook, a member of the national pro-eugenics elite and holder of a doctorate from Johns Hopkins University, one of the nation's most prestigious research universities.[24] Estabrook was an alumnus of the annual summer school run by Davenport and Laughlin at the ERO in Cold Spring Harbor, which trained more than 250 eugenic field workers between 1910 and 1924 to undertake multiple family and related eugenic studies.[25] Estabrook had also authored studies of degenerate families whose generations evinced the telltale signs of inherited imbecility or feeblemindedness, which, he claimed, were often manifest in such conditions as alcoholism or criminality, including prostitution. His published work included an updated version of Dugdale's widely cited and studied analysis of *The Jukes*, a study co-authored with Charles Davenport in 1912, and a study of miscegenation in Virginia published in 1926 – that is, concurrent with his trial work – titled *Mongrel Virginians*.[26] Estabrook traveled to the State Colony to examine Carrie and her mother, and he also traveled to Charlottesville to meet with former teachers and any other potential informants, such as social workers who had had contact with the family. This research allowed him to compile necessary information for a family analysis based on eugenic reasoning.

Dr. Estabrook's testimony aimed both to provide a general account of how eugenicists undertook their analysis of inheritance of particular traits and to tell the court how the Buck lineage fit such laws of heredity. Having examined both Carrie's baby and mother, Dr. Estabrook could offer a "three generation" pedigree analysis. His expert testimony in the Amherst County Circuit Court both explained general eugenic principles of inherited inadequate stock and the specific expression in the Buck lineage. Of Carrie Buck, he told the court, "I gave a sufficient examination so that I consider her feebleminded"; and of Carrie's baby, Vivian, he reported his application of the "regular mental test for a child of the age of six months, and judging from her reactions to the tests I gave her, I decided that she was below the average for a child of eight months of age."[27] He had concluded from a test that Carrie's sister, Doris, fit the feebleminded category too. Estabrook called a litany of "witnesses" from Charlottesville to testify at the trial, eliciting from each some view or opinion about Carrie or her

---

[24] Ibid., 1–4.
[25] Rafter, *White Trash*, 20.
[26] This made Estabook co-author of three of the fifteen family studies identified by Nicole Rafter. *White Trash*, 4.
[27] Quoted from the transcripts of the 1924 court hearing in Lombardo, *Three Generations*, 5.

parents' families. On some occasions, such witnesses merely reported second-hand rumors. Carrie's lawyer, Whitehead, failed to challenge any of the statements. Of three teachers called to the stand, none knew Carrie directly; the best one could do was cast a disparaging remark about her sister, Doris.[28] Strode finally introduced written testimony from Harry Laughlin, assistant director of the ERO. Laughlin's written remarks upheld the "three generations" view of imbecilic inheritance across Emma, Carrie, and Vivian. He also provided copious extracts from his widely cited book, *Eugenic Sterilization in the United States*, and proof copies of forthcoming articles including one entitled "Purging the Race." The theme of Laughlin's work introduced into the trial record was the fundamental importance of designing and implementing policy – at the state and national levels – to achieve what eugenicists called "race betterment."[29] Thus the murky line between negative and positive eugenics was inserted directly into the Virginian courthouse.

By the time the trial judge gave his verdict in February 1925, Superintendent Priddy was dead from Hodgkin's disease. Circuit Judge Bennett Gordon decided in favor of the sterilization recommendation. With Priddy's successor, Dr. John H. Bell, in the role of defendant, the appeals process continued on to the Virginia Supreme Court of Appeals and then to the U.S. Supreme Court. The appeals case that Carrie's lawyer – still Irving Whitehead – formulated was designed decisively to support the legality of sterilization of the feebleminded and to transcend the grounds on which seven other states had seen their sterilization laws struck down as unconstitutional. The appeal therefore was of momentous importance to the future of eugenics-justified sterilization. Defending the trial court's upholding of the sterilization recommendation before the Virginia Supreme Court, Strode explained that, without a policy of mandatory sterilization on its books, no state could exercise the police powers necessary to avoid the "multiplication of socially inadequate defectives."[30] Police powers meant that state policy prioritized the public good over individual rights.[31] The whole purpose of using Carrie Buck's recommendation in the courts was laid bare here.

The three grounds on which coerced sterilization was previously found wanting judicially were its violation of due process; its violation of the equal protection clause of the Fourteenth Amendment, because such sterilization would be applied only to those already institutionalized in such residences as the Virginia State Colony; and the charge that sterilization violated the Constitution's Eighth Amendment by inflicting cruel and unusual punishment on those treated. Eugenicist promoters of involuntary sterilization needed each of these arrows of rebuke to be deflected. Four years before the case reached the

---

[28] Ibid., 113–5.
[29] For a description of Laughlin's contribution to the trial see ibid., 133–5.
[30] Brief for Appellee, *Carrie Buck v. Dr. J. H. Bell*, Supreme Court of Appeals of Virginia, September 1925, quoted ibid., 151.
[31] Gerstle, "Resilient Power."

Supreme Court justices, Aubrey Strode had done his best to address these issues in the bill he drafted, which was passed by the state legislature in 1924. Design and implementation of the sterilization law was covered in six sections. These provisions set forth the rights of the patient, as well as the role of the superintendent in identifying for treatment those patients in institutions for the feebleminded. The latter included a stipulation that only those who possessed hereditary traits of the diagnosed condition would be suitable for sterilization.[32]

Happily for eugenicists, eight of the nine U.S. Supreme Court justices were content to conclude – in one of the Court's shortest written decisions – that these constitutional questions were satisfactorily addressed by the law and that sterilization in the circumstances of a woman like Carrie Buck was entirely constitutional, sensible, and imperative. The presiding chief justice, former president William Howard Taft, was himself a supporter of the national eugenics project and influenced Holmes's use of the "three generations" language. The analogy for the court's justices was vaccination: "the principle that sustains compulsory vaccination is broad enough to cover cutting the Fallopian tubes."[33]

Buck v. Bell was seized on by eugenic reformers throughout the states. Several states quickly enacted new sterilization laws in their legislatures, including some whose earlier laws had been found unconstitutional. Up until 1931, twenty-eight states had enacted sterilization laws modeled on the principles upheld by the Supreme Court case.[34] This rapid adoption shows how eagerly other states awaited the legal authority to pass their own eugenic sterilization laws as superintendents and reformers combined interests to persuade legislators to act.

CARRIE BUCK'S MISFORTUNE

Without question, *Buck v. Bell* was the most important legal decision in the history of coerced eugenic sterilization. One of the most famous justices in the history of the Court wrote the decision, and it is among the most quoted and quotable decisions ever.

And it rested on a fraudulent trial record.

The case itself was contrived from the beginning. Priddy and Strode had arranged for a suit to be taken in Carrie's name against the superintendent that challenged the recommendation for sterilization on grounds of feeblemindedness. Carrie's lawyer, Irving Whitehead, was a familiar figure at the State Colony: not only was he an advocate for sterilization, but he was also a former chairman of the Colony's board of directors *and* a boyhood friend of Aubrey

---

[32] Lombardo, *Three Generations*, 99–101.

[33] *Buck v. Bell* at 275. Eugenicists were keen to cite as precedent *Jacobson v. Massachusetts* 197 U.S. 11 (1905), in which the Supreme Court held that the state had a right to require an individual to be vaccinated (in this instance to prevent smallpox) against his or her will in the interest of public health.

[34] Bruinius, *Better for all the World*, 72; Largent counts thirty laws in the years up to and including 1931. *Breeding Contempt*, 72.

Strode.[35] On top of this, his fees for acting on Carrie's behalf were paid from the very institution in which she was assessed for sterilization and whose recommendation was now being challenged. It was Whitehead who made the case for Carrie against the superintendent in *Buck v. Priddy* in the circuit court in late 1924. This defense exercise was a sham: Whitehead's cross-examinations of witnesses, supposedly for the benefit of his client, purposefully revealed additional damaging information about Carrie's background and family. He neglected to cite Carrie's adequate record of school attendance (more on this below) and performance to the sixth grade, when her foster parents took her out of school. The spurious claim offered by a social worker that Carrie's baby, Vivian, seemed to be "not quite a normal baby" went unchallenged by Whitehead and was absorbed by the appeals judge.[36] At this stage, Harry Laughlin's praise of Strode and Priddy for selecting the middle of a three-generation family as a test case must have been ringing in their ears: in Laughlin's investigations of hundreds of intergenerational imbecility, he could "not recall a single instance in which feeblemindedness appeared in the grandmother, the mother, and the child (three generations), by environmental or accidental causes."[37] This result, therefore, must have been due to heredity and consequently made clear an urgent need to prevent further breeding.

There were still more problems with the case. So many of the facts and circumstances of Carrie's story prior to and even after 1927 have been exposed as partially inaccurate or entirely false that it is extraordinary how this woman's fate became elevated to the standard point of reference for proponents of compulsory sterilization legislation throughout the United States and often abroad.

Carrie had, in fact, performed well at school. She earned compliments for her neat handwriting and attendance. In her court appearances, this history was ignored, distorted, and deliberately maligned, such as when her writing notes to boys in class was cited as evidence of feeblemindedness. She was pregnant because the nephew of her adoptive parents – the Dobbses – had raped her, a fact that the couple wished to hush up. Prior to the pregnancy, the Dobbses had not expressed concern about Carrie's apparent mental state, but they petitioned for her admittance to the Colony as soon as it became known.[38] They described her as having "hallucinations" and "outbreaks of temper" to support their request, evidence of which behavior proved unforthcoming.[39] Carrie's mother, Emma, was frequently described as an unmarried mother, thus implying some

---

[35] Lombardo, *Three Generations*, 107.

[36] The description was given by the social worker Caroline Wilhelm, who had accompanied Carrie Buck on her trip to the State Colony after giving birth. Quoted from the trial records in Lombardo, *Three Generations*, 117.

[37] Laughlin to Strode, October 3, 1924, Colony Record No. 1692, quoted in Lombardo, *Three Generations*, 108.

[38] Ibid., 103.

[39] Ibid., 105.

degree of immorality, but the hospital records for the birth of each of her three children recorded her as a married woman.[40]

Carrie was selected by Priddy and Strode as an instrument of their eugenic ambition. The paper-thin defense offered by their friend, Irving Whitehead, confirmed a callous indifference and lack of genuine concern about Carrie as an individual. They were blinded by the assumption of Carrie's feeblemindedness and the rightness and urgency of eugenic treatment; they ignored the fact that the proposed operation was a life-changing one. Carrie's Colony-appointed guardian, Shelton, made no effort to defend her at the hearings before the board that approved her sterilization. Carrie's family background and the supposed mental ability of both her mother and daughter were sketched brutally and broadly in the trial court, appeals court, and Supreme Court proceedings. Ill-informed rumors and speculation based on second-hand knowledge stated by the witnesses called by Strode from Estabrook's investigations were permitted into the trial record without challenge. Witnesses with little or no direct acquaintance with Carrie or her family were allowed to inject prejudicial comments into the court transcripts. The expert testimony of Harry Laughlin was presented without him ever having met Carrie but simply by recycling – at points, word for word – the description provided to him by Superintendent Priddy. Doubt shrouded whether Frank Buck was in fact Carrie's father, and Priddy's efforts to establish reliable genealogies for either side of her parents' families resorted to rumor and hazy memories.[41] Laughlin's analysis of the hereditary stream in the Buck genealogy was nonetheless unequivocal:

> All this is a typical picture of a low grade moron. . . . The family history record and the individual case histories, if true, demonstrate the hereditary nature of the feeble-mindedness and moral delinquency described in Carrie Buck. She is therefore a potential parent of socially inadequate or defective offspring.[42]

This testimony was absorbed and accepted by eight U.S. Supreme Court justices.

Carrie's daughter Vivian died at the age of eight from pneumonia – but not before she had her name added to the honor roll at her school.[43] She showed no sign therefore of mental "imbecility" as ascribed to her speculatively by Superintendent Priddy and eugenicist Estabrook in the judicial hearings when she was a mere infant. In fact, the trial date at the Amherst Circuit Court in 1924 was delayed by Strode because evidence of Vivian's alleged mental deficiency was missing: even the social worker who took Vivian to the Dobbses after her birth noted the absence of any record of mental incompetence.[44] At the hearings

---

[40] Ibid., 105–6.
[41] Ibid., 109–10.
[42] Cited in Largent, *Breeding Contempt*, 100.
[43] Lombardo, *Three Generations*, xi.
[44] Ibid., 110.

to approve the sterilization, Priddy lied about the unwillingness of the Dobbses to receive Carrie back into the household if she underwent the treatment.[45]

The leading scholar of *Buck v. Bell*, Paul Lombardo, identifies further horrors in the judicial process. Whitehead did not even meet with Carrie, his client, before the trial.[46] In addition to the evident failure of Irving Whitehead to provide even a minimally competent defense, no effort was made to bring her foster parents to the 1924 circuit court hearing. Questioning of the Dobbses should have exposed their motive – ordinary embarrassment – for having Carrie removed. It also would have revealed the fact that her pregnancy arose from rape by the foster mother's nephew rather than from the young woman's alleged immoral behavior. Lombardo notes that Mrs. Dobbs's maiden name was Dudley, which made her a member of one of the families Priddy identified among a pool of the socially inadequate in Charlottesville. Lombardo concludes that the Dobbses feared prosecution of their nephew either for rape or, more likely, for the Virginian crime of seducing a woman "under the promise of marriage." Whitehead also made no effort to expose confusions in the evidence of the eugenic experts, notably Dr. Joseph DeJarnette's at times inaccurate explication of the hereditary theories cited by eugenicists. Casual use of terms like "shiftlessness" or "peculiar" to describe imbeciles went unremarked on. The trial was designed, in other words, to whitewash any doubts about either the evidentiary basis for sterilization as an effective means of ending the intergenerational distribution of feeblemindedness or the ideal suitability of Carrie Buck for such an operation. Such oversights and many others in Whitehead's role as defense lawyer were deliberate. His actions and questions were subordinate to the cause he shared with Priddy, Strode, and others: getting a mandatory sterilization law declared constitutional.

This spuriously formulated set of arguments and shadily mobilized evidence, crafted for the 1924 circuit court trial, served their ultimate purpose in the next stages of the process: in the appeals court in Virginia and then before the U.S. Supreme Court. Crucially, and consistent with his earlier stance, Whitehead offered no challenges to facts asserted in the trial, nor did he contest the validity of claims presented as "scientific" by eugenicists. His five-page brief was notably succinct – compare Strode's forty-page document defending Superintendent Bell. Moreover, as Lombardo points out, Whitehead failed to make use of the legal arguments that had led to successful appeals against compulsory sterilization laws in other states. These cases therefore did not come to the attention of the Virginia appeals justices who, on November 12, 1925, upheld the order for Carrie's sterilization.[47]

The indignities suffered by Carrie were only compounded as the appeals process progressed, as evinced by the extraordinarily pro-eugenic argument

[45] Bruinius, *Better for All the World*, 69.
[46] Lombardo, *Three Generations*, 139–40.
[47] Ibid., 153.

endorsed by an almost brazenly biased U.S. Supreme Court. Chief Justice Taft was a eugenicist: he was the author of pamphlets in support of the movement and was associated widely with key figures and organizations in it. Holmes, who wrote the judgment, was equally enamored of eugenic ideas. His private correspondence was punctuated liberally with eugenicist support and sentiment.[48] That Holmes authored the *Buck v. Bell* decision undoubtedly helped to fashion and deepen its eugenic flavor, something Taft anticipated when he assigned it to Holmes.

Again to Carrie Buck's misfortune (and many others subsequently), the Court's decision is remarkably brief for such a momentous intrusion into the individual liberty of one's right to procreate. The cursory nature of the opinion and the absence of the usual array of footnotes to previous decisions implied that Holmes and his seven concurring colleagues were positively predisposed to the eugenic argument for sterilization. Holmes was hardly alone, though, as evidenced by the wave of enthusiastic newspaper editorials published across the country in response to the decision.[49]

This roll call of injustices to Carrie Buck surpasses the alarming. It reveals the corrupted development of a national policy rooted in dishonesty and shaped by repeated malfeasance. So enamored of eugenic principles and so certain of the rightness of their cause were the eugenic advocates that they were fully willing to deploy deceit and half-truths to support their cause; they manipulated the judicial system not to serve justice but, rather, to produce a predetermined result. The "grossly negligent" legal counsel named on Carrie's behalf was not the least of the injuries inflicted on her.[50] Carrie was the center of a deliberate plot carefully orchestrated by powerful institutional actors bent on winning

---

[48] Ronald K. L. Collins, ed., *The Fundamental Holmes* (Cambridge: Cambridge University Press, 2010), 39–40. Holmes mentioned to one correspondent that his endorsement of eugenics in *Buck* gave him "pleasure." Ibid., 266–7. See also Robert M. Mennel and Christine L. Compston, eds., *Holmes and Frankfurter: Their Correspondence, 1912–1934* (Hanover, NH: University Press of New England, 1996), 125–6; Mark DeWolfe Howe, ed., *Holmes-Laski Letters: The Correspondence of Mr. Justice Holmes and Harold J. Laski, 1916–1935* (Cambridge: Harvard University Press, 1953), 1:207 [writing in 1919 that he believes in Malthus "in the broad – not bothering about details"], 1:761 [1925], and 2:942 [referencing his decision in *Buck v. Bell*]; M. DeWolfe Howe, ed., *Holmes-Pollock Letters: The Correspondence of Mr. Justice Holmes and Sir Frederick Pollock, 1874–1932* (Cambridge: Harvard University Press, 1941), 2:36: "I think that the sacredness of human life is a purely municipal ideal of no validity outside the jurisdiction." "[I] have said often enough, that it seems to me that every society rests on the death of men . . . I should be glad . . . if it could be arranged that the death should precede life by provisions for a selected race." Ibid.

[49] See Lombardo, *Three Generations*, 174–5.

[50] This is Lombardo's characterization of Irving Whitehead, ibid., 154. The author adds: "that description ['grossly negligent'] does not adequately portray the fraud he inflicted on his client and the court. Whitehead was not merely incompetent: his failure to represent Carrie Buck's interests was nothing less than betrayal. His poor showing in court was damning enough, but if the letter Aubrey Strode wrote recommending Whitehead for advancement to the position of general counsel to the Federal Land Bank at Baltimore the week before the *Buck* trial had become publicly

constitutional authority for compulsory sterilization at any cost. That cost, in the end, was paid by Carrie Buck and the thousands of other individuals whose sterilization was made legally justifiable through this disgrace to the American judicial system.

THE DIFFUSION OF INVOLUNTARY STERILIZATION

Buck v. Bell opened the way to further sterilization laws nationally and internationally. It vindicated pro-sterilization advocates' arguments and provided them an ironclad legal basis to carry out their agenda. This was true both in the United States and abroad. Former Lieutenant Governor of Ontario, Dr. Herbert A. Bruce, cited U.S. law as he explained Germany's 1933 law (An Act for the Prevention of Hereditarily Diseased Offspring) in his contribution to "The Future of the Race," a radio series sponsored by the Eugenics Society of Canada in 1938. Bruce approved of the twenty-nine American states that then had sterilization laws and explained to the radio audience that "although these laws demand voluntary application for sterilization, a Court decision of 1926 [sic] states that in certain cases compulsory sterilization has to be enforced ... [because the justices] think it would be better for everybody if those who are manifestly sub-normal and useless were prevented from propagating their kind."[51] Another speaker in the series, A. M. Harley, called for sterilization laws in Ontario. He praised the laws in British Columbia and Alberta but focused especially on the record in California, which had sterilized more than 12,000 individuals by 1938. In addition, Harley latched onto the fact that "the Supreme Court of the United States decided that the sterilization law of the State of Virginia was constitutionally valid." "Mr Justice Holmes," he concluded, "epitomized all the principles of sterilization when he said 'three generations of imbeciles are enough.'"[52]

These two speakers' separate references to the Court's decision in *Buck v. Bell* show that the decision had both domestic and international resonance. Domestically, Holmes's opinion empowered eugenicists already speaking the language of unfitness, mental defectives, bad breeding, dangerous differential rates of breeding between the best and the worst citizens, and so forth. The decades after *Buck v. Bell* saw the largest number of sterilizations across the country because eugenic-supporting superintendents and state boards could be assured of the legality of their actions.

This case study of Virginia, together with the studies of Indiana, Michigan, Alabama, Mississippi, California, and Oregon, confirm the institutional as the central causal factor in forced eugenic sterilization. Priddy's position – as a close

---

known at the time, it might well have raised serious suspicions about Whitehead's loyalties." Ibid., 154–5.

[51] The Eugenics Society of Canada, *The Future of the Race: A Series of Radio Addresses* (Toronto: Eugenics Board of Canada, 1938), issued as a pamphlet, 23–4.

[52] Ibid., 29.

contact of the state's power elite and, above all, as head of a state institution – allowed him to translate his eugenic fears into an effective legislative effort to make coerced sterilization legal.

It is also important to recognize that Priddy and Strode faced little opposition to that legislation. Decision makers in the state were, as noted, either indifferent or supportive. In addition, eugenic sterilization commanded even further enthusiastic support on the national level, drawing on former presidents, Supreme Court justices, and elite-university presidents. Those who could be counted on to otherwise oppose the bill – for example, Roman Catholics – were particularly weak in Virginia. Virginia had and has relatively few Catholics (at the time, less than 2 percent of the population).[53] In addition, they were scattered across the state and faced hostility in the 1920s that was unusual even by the standards of the day.[54] Reacting as shrewd religious minorities do, Virginia's Roman Catholics trod lightly and adapted themselves as much as they could to the dominant Protestant culture.[55] With average Catholics keeping their heads down, and with the diocese underfunded, understaffed, and reliant on missionaries to reach their scattered charges, Virginia was hardly fertile ground for organized Roman Catholic opposition to coerced sterilization.

Other states were.

---

[53] See Table 7.2.
[54] See Gerald P. Fogarty, *Commonwealth Catholicism: A History of the Catholic Church in Virginia* (Notre Dame: University of Notre Dame Press, 2001), chapters 20–21.
[55] Ibid.

7

Sterilization Thwarted

Buck v. Bell opened the floodgates. With the green light to eugenic sterilization provided by the U.S. Supreme Court in 1927, almost thirty American states either initiated or consolidated their own legislation on coerced sterilization. Most studies of sterilization, naturally enough, focus on these states. But a crucial consideration that they miss is that twenty states either lacked sterilization bill initiatives or had such initiatives that failed (Table 7.1).

For historians, this suggests that such studies overlook important source material. For political scientists, it means that they have selected on the dependent variable. That is, they have only chosen cases in which the effect – that which they wish to explain – occurs. The result is that when such studies develop explanations or theories, they are unable to ensure that these explanations or theories stand. For example, if we hold that x (e.g., high inflation) causes y (e.g., regime collapse), then we can only be confident of this if regimes never collapse in the absence of high inflation. This is obviously not true. A scholar must, therefore, compare both regimes that collapsed and those that survived under the same conditions. In the same way, one must consider states both with and without successful eugenic legislation in order to demonstrate with certainty the necessity of that legislation in producing sterilization as an outcome.

For this reason, we devote this chapter to further instances in which sterilization bills failed.

Historian Mark Largent draws attention to the importance of fiscal resources to the enactment and execution of compulsory sterilization laws. Financially weaker states, often Southern states such as Louisiana or Tennessee, lacked the network of mental institutions and therefore superintendents through which the sterilization campaign was achieved elsewhere.[1] In addition, any dissemination of Progressivist ideas in these states was racially skewed. This combination of

---

[1] Largent, *Breeding Contempt*, 81.

TABLE 7.1 *U.S. States without Sterilization Laws*[2]

Alaska	Missouri
Arkansas	Nevada
Colorado*	New Jersey
Florida	New Mexico
Hawaii	Ohio*
Illinois*	Pennsylvania*
Kentucky	Rhode Island
Louisiana	Tennessee
Maryland	Texas*
Massachusetts	Wyoming

* These states sterilized some residents (270 in the case of Pennsylvania) without any legal basis – that is, they did not pass laws permitting mandatory sterilization.[3]

material impoverishment and ideational recalcitrance was fortunate for African Americans, whom one might have otherwise expected to be targets of compulsory sterilization in this region; indeed, some were victims later in the twentieth century. In the more prosperous states that did sterilize energetically, it was usually poor, predominantly white, and, until the postwar years, always institutionalized individuals who were targeted.

It is notable how many false starts characterized the enactment of early eugenic laws and that some states sterilized without a corresponding law, such as in Colorado, because surgeon-superintendents preferred autonomy without a legal test of their powers (Tables 5.1 and 5.2).[4] The way state politics are shaped under federalism mattered. In several states, bills were passed by state legislatures and then vetoed by governors; further bills passed, were again vetoed, but were eventually accepted by the governor. Several laws were ruled unconstitutional by state and federal appeals courts. This pattern implies that adoption of eugenic measures was neither automatic nor uncontroversial. The same language, arguments, and type of evidence about the growing number of welfare dependents were advanced in those states where legislation failed as in the cases in which eugenics triumphed. In addition, anxieties expressed by opponents overlapped across state debates.

One predictor of success was institutional, reflecting a political culture already focused on the marginal and dependent: those states without laws proscribing marriage between individuals with some mental or physical defect were least likely to successfully put sterilization laws onto the books. In fact, in

---

[2] Alaska and Hawaii did not become states until 1959.

[3] Largent, *Breeding Contempt*, 72. See also Dorr, *Segregation's Science* and Larson, *Sex, Race, and Science*.

[4] Historian Mark Largent made this point to the authors; manuscript workshop, Munk School of Global Affairs, University of Toronto, December 10, 2011.

states that had enacted such marriage restrictions, state legislatures were twice as likely to embrace a sterilization measure.[5] An institutional trajectory had already been established.

Some of the policy makers recognized that laws of eugenic sterilization posed legal and ethical challenges and that their passage transgressed significant thresholds defining how states treated citizens eligible for coerced sterilization. Some vetoers resisted the implied Progressive and New Deal ideals. Passage in Georgia, for instance, came only when Eugene Talmadge (in office 1933–7) was succeeded in the Governor's Mansion by Eurith Rivers (in office 1937–41).[6] Likewise, a sterilization bill in Idaho passed by the state legislature in 1918 was vetoed by the governor; a subsequent bill was passed only seven years later. Initiatives in Florida to enact state sterilization laws were approved by legislative committees but were never allocated time on the chamber floors. Similarly well-mobilized efforts in Missouri to secure passage of bills to permit the sterilization of variously defined "undesirables" failed in 1929, 1931, 1933, 1935, and 1937; votes in the state house in 1929 and 1931 were against the bills by votes of 62–53 and 73–47, respectively.[7] New Hampshire's legislators rejected various bills until they finally passed one in 1929. Five efforts in Ohio between 1915 and 1963 to pass a eugenic sterilization law foundered. In Pennsylvania, a bill passed by the legislature in March 1905 was vetoed by the governor in the same month, and, despite subsequent initiatives to enact a statute in 1911, 1913, 1915, 1917, 1919, and 1921, none was signed into law.[8] Becoming a "eugenic nation" was an uneven and, in many ways, incomplete journey.[9]

To map this unevenness, we examine the case of Ohio, a state mirroring many common features of those states that did legislate for sterilization. Several bills to establish compulsory eugenic sterilization were introduced in its legislature but remained unenacted. We begin with the opponents of sterilization and how they mobilized.

OPPOSING STERILIZATION

It should be borne in mind that the absence of an enacted sterilization law in a given state does not indicate that such laws were never proposed. In other words, the fact that some states did not draw up sterilization bills at all should not overshadow the examples where states proposed such laws but failed to adopt them. This is an important reminder of situations in which some Americans not only organized into groups that opposed eugenic sterilization but, moreover,

---

[5] Largent, *Breeding Contempt*, 65.

[6] Kaelber, "Georgia," http://www.uvm.edu/~lkaelber/eugenics/GA/GA.html.

[7] Richard L. Lael, Barbara Brazos, and Margot Ford McMillen, *Evolution of a Missouri Asylum: Fulton State Hospital 1851–2006* (Columbia: University of Missouri Press, 2007) 97–8. See also Largent, *Breeding Contempt*, 72.

[8] Kaelber, "Pennsylvania," http://www.uvm.edu/~lkaelber/eugenics/PA/PA.html.

[9] This draws on the title of Alexandra Minna Stern's book of the same name.

were successful in their pursuit. Voices in the anti-sterilization lobby were often weak, but they were always present. Their ranks were comparatively smaller and their influence waxed and waned over time; the broad popularity and endorsement of eugenics meant that sterilization's opponents were mostly on the defensive.

Eugenicists enjoyed great support across partisan divides and in public intellectual circles, especially as the stamp of authentic scientific expertise was garnered by its key figures in the 1910s and 1920s. There were, to be sure, scientific skeptics, but they were very much in a minority during that period; the emerging field of genetics did not turn decisively against eugenics until the 1930s.[10] And the scientific opposition that did exist was qualified, with biologists and psychiatrists communicating cautiously with Davenport and other eugenicists rather than propagating their reservations. One eminent medical researcher, Abraham Myerson, published an attack in 1925 on the validity of eugenic arguments in a book entitled *The Inheritance of Mental Diseases*. His chapter on the menace of feeblemindedness was especially scathing of the prevailing eugenic logic. Myerson, a Boston psychiatrist prominent in national medical circles, concluded that no evidence existed for supporting a hereditary explanation of feeblemindedness. He maintained a steady public and professional critique of eugenics from 1917, when he published his first two critical papers in the *American Journal of Insanity*, and became more animated in voicing his concerns from the mid-1920s.[11] Despite his critique of hereditary eugenics, though, Myerson seemed to accept the suitability of eugenic sterilization in certain circumstances, a position he maintained into the 1930s.[12] He did, however, increasingly doubt the research agenda of popular eugenics and wrote pieces doubting its validity and, in particular, the weak methodology on which pedigree/family studies depended.[13]

Such weak scientific opposition competed on unequal footing against a Congress in which eugenicists and their adherents dominated. In the wholesale reform of immigration policy enacted in the 1920s, charts and data that showed the alleged dangers of straying from a "national origins" model of immigrant profiles structured the debate to a remarkable extent.[14]

---

[10] See Kevles, *In the Name of Eugenics*, chapter 13.

[11] James W. Trent, Jr., "'Who Shall Say Who Is a Useful Person?': Abraham Myerson's Opposition to the Eugenics Movement," *History of Psychiatry* 12, no. 45 (2001): 33–57. The papers were based on research conducted by Myerson in Taunton State Hospital and commenced with a critical assessment of the reliability of family-type studies of degeneracy beloved by many eugenicists.

[12] Largent, *Breeding Contempt*, 98–9. He notes that Myerson chaired the American Neurological Association's Committee for the Investigation of Sterilization, although he maintained his support of some forms of coerced sterilization.

[13] Trent, "'Who Shall Say Who Is a Useful Person?'" 40–3.

[14] Discussed in detail in Desmond King, *Making Americans: Immigration, Race, and the Origin of the Diverse Democracy* (Cambridge: Harvard University Press, 2000), chapters 5–7. See also Gary Gerstle, *American Crucible: Race and Nation in the Twentieth Century* (Princeton, NJ:

With the scientific community on its side, and with superintendents of homes for the feebleminded extolling the eugenic and fiscal benefits of coerced sterilization, opposition fell to an organization with a precarious existence in early twentieth-century America: the Roman Catholic Church. Catholic opposition to eugenic sterilization was founded on two pillars. The first was theological opposition to sterilization's perceived violation of natural law and the sanctity of human life. "Contrary to the secular worldview embraced by the eugenicists," writes historian Christine Rosen, "Catholics argued that natural, divine law – not the laws of biology – governed human behavior and protected, among other things, the indissolubility of marriage, the sanctity of procreation and human life (born and unborn), and the family. By interfering with these things, eugenicists violated natural law."[15] Although a tiny minority of Roman Catholics supported eugenics as a possible check on the social ills of poverty, alcoholism, and sexual deviance (suggesting an incorporation of eugenics into, respectively, Catholic social thought and Catholic morality[16]), this tentative support was soon overwhelmed by opposition to eugenic sterilization. Indeed, no eugenic proposal galvanized Roman Catholic opinion more successfully. "In the work of nearly every Catholic who engaged in a discussion of the eugenics movement, compulsory sterilization was the practical eugenic proposal that served as a flash point for debate."[17] The Church also used its campaign to develop an argument against eugenic sterilization that could only have been intended to resonate with the American public at large: involuntary sterilization was a violation of personal liberty. Carrie Buck, a priest wrote in the *Catholic Daily Tribune* in response to the Supreme Court decision, "has rights which the state has not given her, but which she possesses by the very fact that she is a human person." *Buck v. Bell*, the *Ecclesiastical Review* continued, was a "sustained legislative

Princeton University Press, 2001); Matthew Frye Jacobson, *Barbarian Virtues: The United States Encounters Peoples at Home and Abroad, 1876–1917* (New York: Hill and Wang, 2000); Philip A. Klinkner and Rogers M. Smith, *The Unsteady March: The Rise and Decline of Racial Equality in America* (Chicago: University of Chicago Press, 1999); and Mae M. Ngai, *Impossible Subjects: Illegal Aliens and the Making of Modern America* (Princeton, NJ: Princeton University Press, 2004).

[15] Rosen, *Preaching Eugenics*, 139.

[16] One such notable voice was that of social reformer John A. Ryan, who joined the American Eugenics Society in the 1920s. He stopped paying dues in the late 1920s and, in 1931, demanded that his name be taken off one of the standing committees on the grounds that the Society's "unblushing" advocacy of "compulsory legal sterilization," the "practice of birth control for certain classes," and an expansion in the legal grounds for divorce were "abhorrent for religious, moral and social reasons." John A. Ryan to Leon F. Witney, Executive Director, American Eugenics Society, April 9, 1931, 2/88, John A. Ryan Papers ("Ryan Papers"), Catholic University of America Archives (CUA), Washington, DC. For Witney's aggressive reply, see Leon F. Witney to John A. Ryan, April 17, 1931, 2/88, Ryan Papers. By contrast, Ryan exchanged very cordial letters with Margaret Sanger, who sought the priest's reaction to her book, *Motherhood in Bondage*. Margaret Sanger to John A. Ryan, February 6, 1929, and his reply on February 9, 1929, both at 18/33, Ryan Papers.

[17] Rosen, *Preaching Eugenics*, 149.

encroachment upon a purely personal right of liberty."[18] Eugenic sterilization thus violated both the sanctity of life and the sanctity of the person.

In places where the Roman Catholic Church could mobilize sufficiently against coerced eugenic sterilization, that legislation, often against the odds, failed. This dynamic is clearly illustrated in the case of Ohio.

OHIO: THE MIDWESTERN OUTLIER

On the face of it, Ohio seemed like hospitable territory for eugenic sterilization. Its geography and politics rendered Ohio comparable with many of the states that did enact eugenic sterilization laws, such as Michigan and Kansas. Ohio was one of the earliest to enact a law placing limits on marriage between legislatively defined unfit persons, which was in many states often a precursor or complement to the passage of a mandatory sterilization law.[19] It was a state in which doctors performed sterilization on mentally ill or developmentally disabled patients and prisoners at the end of the nineteenth century without legal authority.[20] Two of the fifteen eugenic family studies discussed earlier were conducted in Ohio: Mary Kostir's *The Family of Sam Sixty* (1916) and Mina Session's *The Feeble-Minded in a Rural County of Ohio* (1918). Both authors were graduates of the summer program at the Eugenics Record Office in Cold Spring Harbor.[21]

Furthermore, the Vineland eugenicist, Dr. Henry H. Goddard, whom we met earlier and who authored the widely distributed *Kallikak* family study, took up the cause of segregating "defective" young people in Ohio by promoting the use of IQ tests through the work of the Ohio Bureau of Juvenile Research.[22] The Bureau was established by law in 1913 and opened for inmates in July 1914, after Governor James Cox had sent a state delegation of welfare administration experts to consult with Goddard at Vineland about Ohio's feebleminded juveniles. The state at that point already had a long-established Institution for Feeble-Minded Youth.

The new Bureau was charged with identifying "morons" previously assigned to reformatories or other state institutions and bringing them to the State Institution for the Feeble Minded under Superintendent Edison Emerick, a keen backer of eugenics.[23] The category of "moron" was the great triumph of Goddard's adaptation of the Binet-Simon intelligence measurement system to

---

[18] Quotations from ibid., 151.

[19] Largent, *Breeding Contempt*, 32.

[20] Largent reports how Dr. C. A. Kirkly published reports of the surgeries he performed on women's reproductive systems in the Toledo Hospital for Insane in 1890 and 1891. Ibid., 19. But since this activity was twenty-five years ahead of the first eugenic sterilization bill, it is not clear how influential this earlier surgical work was. Kirkly did, however, focus on masturbation by women as a motive for his operations.

[21] Rafter, *White Trash*, 4, 20.

[22] Patrick J. Ryan, "Unnatural Selection: Intelligence Testing, Eugenics and American Political Cultures," *Journal of Social History* 30, no. 3 (1997): 669–85.

[23] Ibid., 672.

the U.S. world of institutions for the feebleminded.[24] Goddard introduced the test in the United States in 1908 – renamed as the Intelligence Quotient (IQ) – and during the first decade of its use wanted to incarcerate children who tested as morons; sterilization developed later. Goddard's influence in Ohio was marked.

What made Ohio different from its neighboring Midwestern states – except Illinois, which also did not pass sterilization laws – was its large Catholic population. According to the 1926 census of "Religious Bodies," Catholics in Ohio numbered just under one million, the sixth-highest Catholic population in the nation behind New York (3.1 million), Pennsylvania (2.1 million), Massachusetts (1.6 million), Illinois (1.3 million), and New Jersey (1 million).[25]

Ohio's Roman Catholics would play a decisive role in shaping the state's approach to eugenic sterilization.

Making a Start

As in most other states, pressure for coerced eugenic sterilization emerged from the institutions for the feebleminded and from those institutions' superintendents. The path was a familiar one: institutionalization came to be seen as the solution to feeblemindedness; institutions rapidly expanded; this expansion led to the usual problems of overcrowding; and officials responsible for patient populations, in this case Emerick, argued for sterilization as a solution to a problem that their own institutions had generated.

From the early 1900s, Ohio adopted a fitful policy of permanently incarcerating "moron" youth (although the subjects were not often systematically tested), leaving reform schools solely responsible for "normal" juveniles. The Ohio Board of Administration, concerned about the absence of systematic policy, embraced a tougher line with so-called moronic juveniles in the early 1910s.[26] Emerick's own institution, the Ohio Institution for Feeble-Minded Youth, expanded its numbers from 1,614 in 1911 to 2,614 in 1922. It also carried out testing of youth confined to other institutions between the years 1918 and 1922. The results indicated that, of this group, a mere 4.5 percent were normal, whereas 36.5 percent were feebleminded, 30.3 were psychopathic, and 17.4 percent could not be classified.[27] Goddard and Emerick wanted to respond to this challenging problem by expanding incarceration facilities. Efforts encouraging the legislature to expand both the state's responsibility for youth confined to public or state institutions and to build more confining facilities, however, made limited progress.

[24] Patrick J. Ryan, "'Six Blacks from Home,'" 256.
[25] *Religious Bodies, 1926,* "Vol. I: Summary and Detailed Tables," 44. United States Census Bureau, http://www.census.gov/prod/www/abs/decennial/special.html.
[26] Ryan quotes from the 1914 annual report of the Ohio Board of Administration the aim to confine permanently morons and release normals to reform schools. "Unnatural Selection," 683n13.
[27] Ibid., 672.

On at least ten occasions between 1915 and 1963, bills were introduced to the Ohio General Assembly and made various stages of progress.[28] Most withered quickly, such as those in 1929, 1931, and 1937. Those that progressed faced significant, and, above all Roman Catholic, opposition. The first sterilization bill in 1915 was drafted as Governor James Cox and Superintendent Emerick sought Henry Goddard's eugenic assistance in developing an expanded intelligence and confinement strategy for feebleminded youth through the new Bureau of Juvenile Research.[29] The measure was an obvious complement to hiring Goddard. In the two to three years before the bill was introduced, eugenic supporters, including Goddard, gave speeches, issued news reports, and used the opportunity presented by Board of Administration's annual reports to advocate for a greater role for the juvenile bureau in advancing a eugenic assessment framework for young people. For eugenicists, the apparently increasing size of the moron population marked an emerging crisis in need of urgent correction.

This early effort to use Binet-based intelligence testing as a stepping stone to eugenic sterilization ran into some turf-guarding opposition. The rival Ohio Boys' Industrial School resisted using new psychological testing. Its superintendent, Rupert U. Hastings, declared, "too much credence should not be given these results of rather hasty experiments ... [A] long journey of investigation and scientific research must be traveled to reach that standard of excellence the public has reason to demand." His views echoed the sort of objections to eugenic testing and policy developed by Abraham Myerson discussed earlier. In Hastings's view, the eugenic evaluations depended too "greatly ... at this time ... upon the personal equation of the investigator." And he did not want that investigator anywhere near his charges. Hastings wrote to Goddard, "to be frank with you Doctor, it will not be convenient at any time for any worker to come to the school to make such examination of our boys."[30]

The difference between the two institutions also concerned rival views of the end game: Goddard wanted inmates permanently confined, whereas the Boys' Industrial School planned on eventually discharging its residents and worried therefore about a system that tied release to a certain level of intelligence testing.

---

[28] Kaelber documents five bills based on Julius Paul, *Three Generations*, 587–603A: 1915, 1925, 1939 [two bills] and 1963. "Ohio," www.uvm.edu/~lkaelber/eugenics/OH/OH.html. A further two bills in 1927 are analyzed in Sharon M. Leon, "'A Human Being, and Not a Mere Social Factor': Catholic Strategies for Dealing with Sterilization Statutes in the 1920s," *Church History* 73, no. 2 (2004): 395–9. We are grateful to historians Mark Largent and Alexandra Minna Stern for guidance on interpreting these different sources and to Sharon Leon, who in correspondence (July 20, 2012) noted the bills introduced in 1929, 1931, and 1937. The bishops of Cleveland and Cincinnati were alerted to these bills by Catholic legislators.

[29] Cox served as governor in 1913–15 and 1917–21 and later had a prominent national career.

[30] Quoted in Ryan, "Unnatural Selection," 673. Hastings remained skeptical of the specialist claims of eugenics-influenced psychologists and believed the assessments developed by his own staff to be of equal value.

The argument was not without a certain irony: one of the justifications later dragged out in favor of coerced eugenic sterilization was that it would save funds by allowing the sterilized to be released into the community.

In the end, the Industrial School decisively won this dispute with the new Bureau of Juvenile Research about administrative jurisdictions, legal authority, and the appropriate physical location of affected juveniles. This victory was made clear after the state attorney general asserted the "importance of moral culpability" and individualism in preference to eugenic diagnosis of mental incompetence.[31] The Bureau lost its office space in the Industrial School and was relocated to the State Institution for Feeble-Minded Youth under the eugenics-supporting Emerick. The nationally known Goddard became director of the Bureau from May 1918, admitting children diagnosed as morons from 1919 to a state hospital in Columbus, Ohio. With legislative funding, he was also able to set up observation cottages in other institutions, but the scale of implementation was much more modest than the eugenicists had envisaged.

This temporary setback failed to dent the eugenicists' resolve. Goddard left the Bureau in 1922 to become a professor at Ohio State University, thereby retaining and widening his influence in the state. Emerick, brought back out of retirement, filled Goddard's position at the Bureau. Emerick dedicated his efforts throughout the 1920s and 1930s to fostering public anxiety about the menace of a large, undiagnosed feebleminded population and the threat it posed to the state.

The Rocks of Catholic Opposition in Ohio

The opposition to eugenic classifications developed by Superintendent Hastings at the Boys Industrial School and other state administrators won support from religious opponents to sterilization and from some activist groups that defended the rights of children and adults with learning disabilities. In the vanguard was the Roman Catholic Church, whose national leadership would later orchestrate opposition to eugenic sterilization in response to the *Buck v. Bell* decision.[32] Catholics mobilized in Ohio and helped to thwart the bill for eugenic sterilization of the hereditarily defective and criminals (and an accompanying bill to sterilize so-called incompetents before they were permitted to marry) in the late 1920s. The bill would have legislated powers to impose sterilization on the "mentally deficient" or "feeble-minded."[33]

When the bills were introduced in early 1927, Catholic clergy and lay people almost immediately began organizing to block passage of these laws through a coordinated program of public relations and legislative lobbying. The work of Ohio Catholics – specifically, their efforts to produce a body of literature

[31] Ibid., 674.
[32] Leon, "'Human Being.'"
[33] Paul, *Three Generations*, 592–3.

detailing theological, scientific, and political opposition to sterilization – was also instrumental in laying the groundwork for Catholics nationwide to combat sterilization legislation. Key figures in the mobilization efforts included Charles Dolle, Executive Secretary of the National Council of Catholic Men (NCCM), and John Burke of the National Catholic Welfare Council (NCWC).

In March 1927, Dolle informed the archbishops for Cleveland, Columbus, and Cincinnati of the need to send some "high medical authority" and other representatives to address the House Public Health Committee. The brief that Dolle and Burke prepared was later published as an "Information Bulletin" and was distributed to lay organizations in order to provide the public with "a greater realization of the evil consequences which will result from [the proposed sterilization law] and to assist them in opposing such measures."[34] The Information Bulletin provided a blueprint for subsequent materials to educate and inform Catholics and the general public. It therefore warrants examination in some detail.

The Information Bulletin provides a detailed description of both pieces of legislation, including the consent and appeal processes. It then goes on to discuss emergent medical opposition to sterilization and cites the work of recognized medical physicians. Rather than moving straight on to theology, the Bulletin gives an implicit nod to expert scientific authority by providing statistics on the inheritability of mental defects and ineffectiveness of sterilization to halt the spread of "psycho-sexual" diseases. Last, the Bulletin summarizes moral objections to the legislation, covering Roman Catholic doctrine on the inviolate self, natural law, and the lines between civil and moral authority. It was distributed to newspapers and other media nationally and in Ohio, as well as to politicians and Catholic priests; some dioceses distributed it to their parishioners.

The Bulletin had both normative and practical effects. In the former, it linked family and reproductive issues with demands for social and economic justice by providing a theological basis for the human rights of those deemed "degenerate" or feebleminded and thereby became a theme of defense against sterilization.[35] Second, the hybridized utilitarian-theological grounding (sterilization will not work anyway, and it also violates Catholic doctrine) was designed to appeal to the widest possible audience. The ordering here of the utilitarian before the theological was not a coincidence. Finally, and most importantly, the drafting, production, and dissemination of the pamphlet created linkages – vertically up and down the Roman Catholic hierarchy and horizontally across the state – that provided precisely the networks needed to organize a campaign against the sterilization bills. Correspondence between Cleveland's bishop, Joseph Schrembs, and Dolle about the Ohio legislation and the subsequent production

---

[34] AVS, "Special Bulletin on Ohio Sterilization and Marriage License Bills," issued by the National Council of Catholic Men, April 1927. See also Leon, "'Human Being,'" 395–6.
[35] Leon, "'Human Being,'" 395–6.

of the Information Bulletin ensured a linkage between state and local Catholic activism and the Church's national structures.

> Bishop Schrembs was Chairman of the Department of Lay Organizations, which oversaw the NCCM and monitored Dolle's administration of the day-to-day affairs of the organization. Through the Department of Lay Organizations, Schrembs had guided the foundation and development of both the National Council of Catholic Men and the National Council of Catholic Women to foster lay activism in local dioceses around the country. While the NCCW developed a vibrant national organization, the NCCM functioned more directly under Schrembs's control. The bishop envisioned the organization as being deeply involved with social issues, such as the fight against birth control, the campaign for parochial schools, the care of immigrants, and the education of Catholic men on social and moral teaching.[36]

The situation in Ohio allowed Schrembs to exercise his authority as spokesperson for the Catholics within the Diocese of Cleveland and to oversee the involvement of the NCCM in the hopes of providing materials to educate the laity about the variety of reasons for opposing eugenic policy initiatives.[37] He was a forceful opponent. Testifying before the Ohio Public Health Committee, the Bishop declared that mandatory sterilization of "so called feeble-mindedness in any of its degrees is a real menace to the Community and would lend itself in the hands of unscrupulous persons to the destruction of the most sacred human rights."[38]

The educational campaign mapped onto, reinforced, and was reinforced by an intense legislative lobbying campaign. Bishop James Hartley of Cleveland and Archbishop John McNicholas of Cincinnati wrote to the state legislature in protest of the proposed legislation. The NCCM's executive secretary arranged to meet the state governor, Democrat Alvin Victor Donahey, known as "veto Vic" for his record of vetoing 117 bills in the first two of his three gubernatorial terms (1923–9).[39] The Catholic District League of Cleveland condemned the law as "an outrage upon humanity ... a first step towards introducing into the civilization of the twentieth century the views and usages of semi-brutes."[40]

Throughout Ohio, Catholics and Catholic-based organizations lobbied intensely against the law, flooding state legislators and the governor with letters of protest. They sent in materials that synthesized disparate elements of anti-sterilization arguments. They presented legal, social, political, and economic objections, which cast a wide net and mobilized their vast preexisting institutional structures. And, above all, Catholics made it clear to election-sensitive

---

[36] Ibid., 396n30.
[37] Ibid., 396.
[38] Schrembs' message was contained in a telegram dated April 8, 1927, quoted in Leon, "'Human Being,'" 397–8.
[39] Ibid., 396.
[40] Ibid., 398.

legislators that there were votes to be lost among their large population in the state. This form of activism was far more effective than the isolated, disconnected efforts of academic biologists and social scientists airing doubts about the scientific validity of eugenics at rarefied professional meetings. Thanks in large part to the efforts of Ohio's Catholics, the prospective sterilization bills died in the state legislature in April 1927.

Pro-sterilization advocates were committed under all circumstances, and they became only more determined after suffering a defeat at the hands of this organization that infuriated them like no other in the United States. Following the immediate defeat of the 1926–1927 bills, multimillionaire and philanthropist Charles F. Brush founded the Brush Foundation for the Betterment of the Human Race in 1928 in Cleveland with an endowment of \$500,000.[41] He appointed T. Wingate Todd, a professor of forensic anthropology at Case Western Reserve University, as director. The Foundation provided support to the Race Betterment Conference convened by the Ohio Race Betterment Association at Dayton in October 1929. Brooke Sheppard, a trustee of the Foundation, noted that his organization was not a branch of the American Eugenics Society, thus acknowledging the limits of the eugenic label, and he urged more scientific research. As ever, superintendents at homes for the feebleminded threw their energies into the campaign. Dr. William Pritchard, superintendent of the Columbus State Hospital, strongly advocated for a sterilization law. He claimed the alternative – the physical segregation of the 21,000 residents of state hospitals and 11,000 prison inmates, half of whom he deemed beyond hope – was fiscally and socially impossible.

Over the next two years, Pritchard and the Ohio Race Betterment Foundation lobbied for new sterilization bills, but these efforts crashed once again against the rocks of Roman Catholic opposition.[42] The pattern continued for the next three decades, with the last bill being introduced in 1963, by which point civil libertarians had joined Roman Catholics in opposing coerced sterilization.

Ohio represents an important microcosm of the wider trends in American political history. Empirically, there is a rough correlation between the size of the Roman Catholic population and the presence or absence of a coerced eugenic sterilization law (Table 7.2).

That Catholics were firmly entrenched in the anti-sterilization camp was not a product solely of the Church's spiritual doctrine. The deep divide between the Roman Catholic Church and the self-styled progressive reformers who championed eugenics had been long in the making by the time the issue became a legislative topic in various states.

Sir Francis Galton himself had singled out the Church in the previous century for its apparent influence in exacerbating the condition of already declining

---

[41] Reilly, *Surgical Solution*, 81.
[42] Ibid., 81–3.

TABLE 7.2 *Percentage of Roman Catholic Population by State, 1916 and 1926*[43]

State	Catholics in 1916 as % of 1920 state population	Catholics in 1926 as % of 1930 state population	Eugenic Sterilization Law?
North Carolina	0.19	0.22	*
South Carolina	0.57	0.52	*
Georgia	0.63	0.61	*
Tennessee	0.98	0.95	–
Arkansas	1.21	1.33	–
Alabama	1.60	1.36	*
Virginia	1.59	1.59	*
Mississippi	1.80	1.63	*
Oklahoma	2.34	1.95	*
Florida	2.55	2.68	–
Utah	2.23	2.87	*
West Virginia	4.12	4.12	*
Idaho	4.16	5.20	*
Oregon	6.35	5.83	*
Kentucky	6.63	6.77	–
Washington	7.18	7.76	*
Wyoming	6.58	8.32	–
Kansas	7.29	9.10	*
Texas	8.64	9.54	–
Indiana	9.29	9.64	*
Nebraska	10.46	11.24	*
Iowa	10.92	11.62	*
Colorado	11.17	12.14	–
California	14.43	12.70	*
Montana	14.23	13.81	*
District of Columbia	11.75	13.83	–
South Dakota	11.33	14.01	*
Missouri	13.08	14.26	–
Maryland	15.14	14.34	–
Ohio	14.65	14.63	–
North Dakota	14.82	15.30	*

(continued)

[43] Data on religion were collected in 1916 and 1926 and are compared to full census data from 1920 and 1930, respectively. Alaska and Hawaii are omitted, as these states did not join the Union until 1959; neither state had a eugenic sterilization law. *Religious Bodies, 1916* Part I, Section 2, Table 62 [pp. 142–237] and Table 63 [pp. 238–329], United States Census Bureau; *Religious Bodies, 1926*, Vol. 1, Section 3, Table 4 [pp. 44–5] and Table 22 [pp. 142–275]. Information regarding laws here is drawn from Kaelber, "Eugenics."

TABLE 7.2 (*continued*)

State	Catholics in 1916 as % of 1920 state population	Catholics in 1926 as % of 1930 state population	Eugenic Sterilization Law?
Delaware	13.53	15.39	*
Michigan	15.60	17.43	*
Illinois	18.06	17.73	–
Minnesota	17.41	18.56	*
Nevada	11.29	21.71	Law passed, but no sterilizations took place before ruled unconstitutional
Maine	19.34	21.81	*
Pennsylvania	20.99	22.06	270 sterilizations reportedly carried out in public institutions despite lack of law
Arizona	25.36	22.15	*
Wisconsin	22.60	22.37	*
New York	26.44	24.75	*
Vermont	22.18	24.87	*
New Jersey	25.06	26.13	Law passed, but no sterilizations took place before ruled unconstitutional
Louisiana	21.38	27.98	–
New Hampshire	30.70	31.52	*
Connecticut	35.04	34.71	*
Massachusetts	36.61	38.34	–
New Mexico	49.32	41.17	–
Rhode Island	43.24	47.33	–

* sterilization law passed

"good stock" of society. Since the Middle Ages, Galton alleged, Roman Catholic monasteries "drained off the cream" of society and thus "brutalized the breed of our forefathers" through the prohibition on that "cream's" reproduction.[44] Galton nonetheless saw value in these Church institutions and prescribed a way in which they might be useful in the greater eugenic project. Instead of holding captive society's "cream," he suggested they serve as holding tanks for its dregs: the "weak" could be segregated from the world and confined in monasteries, "where, presumably, their degenerate traits would die with

[44] Rosen, *Preaching Eugenics*, 46, quoting Galton in 1907.

them."[45] Thus, as the first swell of eugenic and sterilization legislation gained energy in the early twentieth century, "eugenicists had already settled into the habit of singling out the Catholic Church for special criticism."[46]

By the time the Supreme Court issued its opinion in *Buck v. Bell*, then, Catholic opposition had had more than sufficient time to coalesce against the notion of sterilization. Christine Rosen points out that, "[a]lthough the spectrum of Catholic opinion on sterilization was never broad ... it shrunk considerably" after the Court's decision. This concentration of resolve, in turn, "served as a wake-up call for the lay and clerical Catholic community."[47] It is certainly true that the words of the unequivocal Justice Holmes must have landed like a bombshell among American Catholics. But that decision was hardly the first rallying point for the Church in its crusade against state-led sterilization. In fact, Catholics had in previous years already accumulated an impressive record of success in states where they were most prevalent.

"IMMOVABLE" CATHOLICS: OPPOSITION IN OTHER STATES

States with large Catholic populations were either slower to enact or simply never enacted eugenic sterilization laws.[48] Ohio is an important microcosm that represents the crucial role of the Church in frustrating eugenic ambitions. Even where Catholic-heavy states did pass sterilization laws, the statutes' lives were often cut short by court rulings of unconstitutionality, such as in New York, New Jersey, and Nevada. In states where laws stayed in effect despite a large Catholic presence, they were not always enforced with especial vigor (i.e., they did not produce large numbers of sterilized persons) and/or persistence (i.e., over a long period of time). Such examples include Connecticut, Vermont, and Arizona. Nonetheless, monographs of Catholic influence in several other states document its decisive role in preventing the passage of coercive sterilization measures elsewhere in the country.

In Massachusetts, for example, no eugenic sterilization laws were ever passed. The Roman Catholic Church enjoyed a particularly strong hold over its adherents in the state, which comprised almost 40 percent of Massachusetts's population at the time that *Buck v. Bell* drew the nation's attention. This broad presence was also a long-established one, which can help further in explaining the state's failure to produce a sterilization law. In addition to the direct influence of the pulpit, the Church further rooted its control through its long-insinuated

---

[45] Ibid.

[46] Ibid. For a general overview of religious and particularly Roman Catholic ideological approaches to eugenics and sterilization, see Rosen, *Preaching Eugenics*, chapter 1.

[47] Ibid., 150–1.

[48] For the purpose of these examples, we use a threshold of 20 percent (in 1926) to define a "large" Catholic population. See Table 7.2.

presence in Massachusetts's political institutions.[49] This power later began to diminish as the machinery of the federal New Deal supplanted some of the social welfare roles once fulfilled by the Church.[50] Nonetheless, the heavy concentration of Catholics in the Northeast – around half of the nation's total Catholics by the 1940s – meant that state legislators in this region "knew they faced political suicide by backing eugenic statutes."[51]

The Catholic influence also had decisive influence in preventing sterilization laws in the Rocky Mountain state of Colorado. Although the Church's presence was demographically lesser than in the Northeast – Colorado in 1926 was about 12 percent Catholic[52] – it nonetheless contributed to the defeat of four separate sterilization bills.[53] One attempt in 1927/1928 aroused particularly strong rancor among Catholic forces in the state. The local Knights of Columbus in Denver declared that the bill, if passed, "would deprive certain persons of a natural right and hence is fundamentally immoral." The city's Holy Name Diocesan Union also toed the fundamental line of Catholic objection to eugenics, stating that sterilization "is against the natural law" and could not be justified "no matter how much we lament the human deformities of society."[54] This battle was also fought out in print. *The Commonweal*, a lay Catholic magazine still published today, printed data related to sterilization and "urged its readers to lead the opposition to state laws." Periodicals on the other side of the debate took note of their adversaries: frustrated contributors to eugenic journals lamented that "the Roman Catholic Church 'furnished the main opposition' to bills in New York and Connecticut, and eugenic laws in general."[55] At risk, the Holy Name Union concluded in pithy aphoristic fashion, was a "flagrant injustice owing to 'human inability to pass judgment on human mentality.'"[56] Catholics prevailed, and Colorado's governor vetoed the sterilization bill in 1928.

One of the most robust examples of Catholic influence next to that of Ohio is that of Louisiana. Owing to its concentrated French influence, the state's large Catholic population makes it an outlier among Southern states. The role that this population played in combating efforts for eugenic sterilization in Louisiana is documented in thorough detail by historian Edward J. Larson.[57]

---

[49] Simone M. Caron, *Who Chooses? American Reproductive History since 1830* (Gainesville: University of Florida Press, 2008), 141.

[50] Ibid., 141.

[51] Dowbiggin, *Sterilization Movement*, 38

[52] *Religious Bodies, 1926.*

[53] Kaelber notes the failure of bills in 1908, 1913, 1925, and 1928. "Colorado," http://www. uvm.edu/~lkaelber/eugenics/CO/CO.html.

[54] Ibid., quoting Rosen, *Preaching Eugenics*, 151.

[55] Bruinius, *Better for All the World*, 230 citing *The Commonweal*, September 20, 1935 and Roswell H. Johnson, "Legislation," *Eugenics* 2, no. 4 (1927), 64.

[56] Reilly, *Surgical Solution*, 118.

[57] Larson, *Sex, Race, and Science.*

The only pro-eugenics bill to be passed into law in Louisiana was approved in 1918 and established a State Colony and Training School for the feebleminded. This legislation did not make any provision for compulsory sterilization, but it did establish an especially broad definition of "feeblemindedness." The definition included any mental defect "from birth or from an early age so pronounced that he is incapable of managing himself and his affairs, or being taught to do so, and requires supervision, control and care for his own welfare, or for the welfare of others, or for the welfare of the community."[58] The eugenic motivations behind the law – which was passed with a total of only four votes against it in both houses of the state legislature[59] – are evident in this short yet far-reaching passage. "The final phrase in this definition," Larson points out, "literally invited the introduction of eugenic considerations into any determination of 'feeble-mindedness.'"[60] This statute is particularly interesting not for what it accomplished but for the massive unrealized potential it contained with respect to coerced sterilization. In addition to the broad and flexible definition of what constituted "feeblemindedness," the law also provided for the usual "full and complete" authority of the Colony's superintendent.[61] It also laid out a fairly detailed legal procedure for having an individual declared feebleminded, but such a petition could be initiated by virtually anyone: by a relative, guardian, or "any reputable citizen" residing in the same area as the person targeted by the petition.[62] These legalities share some striking similarities with the conditions under which successful sterilization measures emerged in other states – not least of all with the example of Virginia, as seen in *Buck v. Bell.*

Sterilization laws never came to pass in Louisiana, but this was not due to an absence of support. Voices in favor of coerced sterilization were many in the state, especially in the 1910s and 1920s.[63] Joseph A. O'Hara, a prominent public health officer from New Orleans, exhorted the state in 1917 to establish an institution where sterilizations could be carried out and promoted a "campaign of education" to edify state legislators in order to reach this goal.[64] Although the State Colony was established without a provision for surgical intervention on its inmates, proponents continued their push toward that end for over a decade.

At the head of the movement was Jean Gordon, the youngest daughter of a prominent New Orleans family, who personally brought four sterilization bills to legislators throughout the 1920s. The "unstoppable" Gordon and other

---

[58] "Act No. 141 of 1918," in *Public School Laws of Louisiana and Sanitary Regulations of the State Board of Health*, ed. Chad. F. Trudeau, 11th compilation (Baton Rouge: Ramires-Jones Printing Company, 1919), 167–8.

[59] Larson, *Sex, Race, and Science*, 81.

[60] Ibid., 83.

[61] "Act No. 141," Section 9, 170.

[62] "Act No. 141," Section 11, 170.

[63] Consider, for example, the pleas of Superintendent John N. Thomas of the Louisiana Hospital for the Insane from 1916. Larson, *Sex, Race, and Science*, 44–5.

[64] Ibid., 53.

proponents, however, repeatedly clashed against the "immovable opposition of the locally powerful Roman Catholic Church."[65] That religious opposition ultimately led to the defeat, albeit usually a narrow one, of every single sterilization bill introduced in the Louisiana state legislature. A 1924 bill, for example, "stirred a prompt response from the local Catholic clergy." New Orleans Archbishop John W. Shaw lashed out at the bill and its proponents, condemning the bill as "unnatural legislation."[66] The statewide impact of the condemnation was decisive. As Larson notes, "following the archbishop's lead, priests throughout the state denounced compulsory sterilization as immoral mutilation and urged their parishioners to stand against the bill. This had a significant political impact in a state where half the voters were Catholic."[67]

Another bill in 1926 met a similar fate but not before its Catholic opponents eviscerated it on the floor of the state legislature. The bill's most notable opponent in the Senate, State Senator Grundy Cooper, declared, "my mind revolts at the whole theory of eugenic breeding of the human race."[68] Cooper was particularly disgusted by another senator's proposition that Louisiana should follow the example of California, despite the supposed mortality rate of 2 percent. "[D]o you realize that 120 persons have been killed in that state . . ., all of them operated on against their wills? What right had the state to take their lives?"[69]

Meanwhile, Roman Catholic opposition "waited to ambush the bill in the lower chamber," like a storm that "had gathered so quietly [that] no one noticed its forming."[70] The bill was killed by a two-vote margin, but not before State Rep. Julius P. Hebert reminded his colleagues, "God created these poor unfortunates just the same as he did legislators." Hebert was also a Grand Knight in the Knights of Columbus.[71]

Subsequent bills in 1928 and 1930, both spearheaded by the infamous Ms. Gordon, never made it out of the state legislature. Gordon died in 1931, never having seen her efforts bear fruit. "The institutional opposition of the Catholic Church," Larson concludes, "had outlived Jean Gordon's best efforts."[72]

Indeed, the Church's opposition to sterilization and other eugenic intrusions was not only adamant but also profoundly durable in Louisiana. As early as 1914, a eugenic marriage restriction bill was defeated thanks to "vehement" Catholic opposition.[73] This ideological influence persisted and was felt well into the postwar era. A commission charged with combating the perceived plague of illegitimate children in the state concluded in 1958 that coerced sterilization laws

[65] Ibid., 107. On Gordon, see ibid., 77–9.
[66] Ibid., 108–9.
[67] Ibid.
[68] Ibid., 111. Larson incorrectly identifies Cooper's first name as "Grandy."
[69] Ibid.
[70] Ibid., 111–2.
[71] Ibid., 112–3.
[72] Ibid., 115.
[73] Rosen, *Preaching Eugenics*, 69.

were "intrinsically evil, completely immoral and violative of all concepts of Christianity."[74] This fundamental opposition to eugenic measures, diffused through Roman Catholic churches and networks, thoroughly permeated Louisiana society and ultimately proved to be the crucial bulwark in preventing legalized coerced sterilization in the state.

The presence of the Roman Catholic Church did not provide an absolute guarantee of doom for sterilization laws – some states with sizable Catholic populations did pass and implement such legislation. But the mere absence of such legislation in other states cannot be written off simply as a lack of interest in sterilization measures and certainly not as a broad social consensus against them. An examination of the Church's role uncovers the other side of the history of sterilization legislation in the United States by revealing important reasons that prevented such laws from becoming viable. Especially where multiple bills for coerced sterilization were debated but failed, it is essential to acknowledge that Catholic activism was sufficiently intense and prevalent to kill them.

As time wore on, and as Catholics accumulated these victories, their opponents became increasingly exasperated. The "relentless opposition" of the Roman Catholic Church eventually "sapped the enthusiasm for state laws."[75] Following the Second World War, for example, initiatives for coerced sterilization in Wyoming and Pennsylvania were never approved because of "Catholic attacks," such as the successful campaign led by the bishop of Cheyenne that "easily defeated" a proposed bill.[76] And even where such bills did succeed, "Catholic prelates directed judges to stymie efforts to sterilize the handicapped."[77] In Wisconsin, even though a sterilization law was in effect, Roman Catholics' lobbying efforts scuttled an attempt to widen its scope.[78] None other than Marian Olden, president of Birthright, Inc., bemoaned an apparent onslaught of threats made by Catholic forces to boycott the businesses of the state's assemblymen if they were to vote in favor of the measure.[79] The Catholic Church proved the significance of its role, both as an institutional actor and as a network that could be readily mobilized, and it was decisive in many instances by preventing or curtailing measures that legalized coerced sterilization of the mentally handicapped.

---

[74] Mary Ziegler, "Reinventing Eugenics: Reproductive Choice and Law Reform after World War II," *Cardozo Journal of Law & Gender* 14, no. 2 (April 2008): 328.

[75] Dowbiggin, *Sterilization Movement*, 77.

[76] Ibid.; Reilly, *Surgical Solution*, 121.

[77] Dowbiggin, *Sterilization Movement*, 77.

[78] Catholic Social Welfare Bureau, Milwaukee, to the National Catholic Welfare Conference, July 1, 1939, NCWC 22–901, CUA.

[79] Reilly, *Surgical Solution*, 120–1.

THE DEVELOPMENT OF NATIONAL OPPOSITION

The decision in *Buck v. Bell* fanned the flames of sterilization legislation across a majority of U.S. states, but, in providing a central focus, it fostered the conditions for opponents of coerced sterilization to organize. The Supreme Court's tolerance – and, indeed, full-throated endorsement – of involuntary sterilization crystallized around a few key points set out by the Virginia law: respect of due process and equal protection (at least ostensibly) and the avoidance of a cruelly administered medical procedure. In making its decision, the Court clearly identified the issues on which opponents should concentrate their assault. It also energized a Roman Catholic Church that was fresh from its victory in Ohio only a month earlier. The NCWC's Family Bureau declared a "Holy War against Enemies of the Home and the Nation" in the aftermath of the decision, which it denounced as "totalitarianism."[80] The Church had by now become the principal energizer of nationally organized opposition to compulsory sterilization laws.[81] Any consideration of the potential value of eugenics aired by a small number of priests vanished.

The sole dissenting justice in *Buck v. Bell*, Justice Pierce Butler, was Roman Catholic. Whether this religious commitment influenced his dissent is unknown, and Justice Butler may have been equally prompted to dissent because of his commitment, expressed in other decisions, to protecting individual rights. Several scholars note that in asking Holmes to write the Court's opinion, Chief Justice Taft reminded his colleague in a note to recognize that "some of the brethren ... are troubled about the case, especially Butler." The decision, therefore, should stick with the core argument about imbecility, or, as Taft added, Holmes should focus on "the strength of the facts in three generations [of imbecility, which is] ... the strongest argument for such state action."[82] The caution seems to have had no impact on Justice Holmes, whose judgment conveyed neither a lack of certainty about the facts nor any doubt about the appropriate interpretation of those facts in determining the right eugenic road for the United States to follow.

But other Catholics stood against the eugenic sterilization doctrine and consolidated this objection. Indeed, two decades later, advocates of Planned Parenthood and Birthright, Inc. bemoaned the influence of Catholic opponents to sterilization:

---

[80] Rev. Edgar Schmiedeler, *A Holy War against Enemies of the Home and the Nation* (Washington, DC: Family Life Bureau, NCWC, n.d. [likely 1927]), 27, 3/62, CUA.

[81] Largent, *Breeding Contempt*, 96.

[82] Taft quoted in Matthew D. Martin, III, "The Dysfunctional Progeny of Eugenics: Autonomy Gone AWOL," *Cardozo Journal of International and Comparative Law* 15, no. 2 (2007): 385.

When attending a national convention, we were told by professional social workers that the limited application of sterilization laws was due in part to the difficulty in getting a case authorized in a court presided over by a Catholic judge. Michigan was given as the example where the courts in one county would authorize a proper case presented by the social worker while in the next county it was impossible to get a case authorized no matter how desperate the need, because of the judge being a Roman Catholic. This led to our question referred to in the opening sentence of the attached letter which we were requested to quote in its entirety or not at all.[83]

This concern, expressed in 1948, was more obviously anti-sterilization because the Catholic Church and its doctrine stood firmly against the measure as a violation of the sanctity of life. High-profile Roman Catholics displayed some uncertainty in the face of "expert" scientific opinion on eugenics in the 1910s and 1920s, but, after the issuance of the papal encyclical *Casti Connubii* (On Christian Marriage) by Pope Pius XI in late 1930, Catholic opposition to eugenic sterilization was categorical. The Pope declared:

[T]hat pernicious practice must be condemned which closely touches upon the natural right of man to enter matrimony but affects also in a real way the welfare of the offspring. For there are some who over solicitous for the cause of eugenics [and] ... put eugenics before aims of a higher order, and by public authority wish to prevent from marrying all those whom, even though naturally fit for marriage, they consider, according to the norms and conjectures of their investigations, would, through hereditary transmission, bring forth defective offspring. And more, they wish to legislate to deprive these of that natural faculty by medical action despite their unwillingness; and this they do not propose as an infliction of grave punishment under the authority of the state for a crime committed, not to prevent future crimes by guilty persons, but against every right and good they wish the civil authority to arrogate to itself a power over a faculty which it never had and can never legitimately possess.[84]

Such eugenic measures "are at fault in losing sight of the fact that the family is more sacred than the State and that men are begotten not for the earth and for time, but for Heaven and eternity."[85]

The Supreme Court decision and the papal encyclical created the conditions for a new battle between the Roman Catholic Church and supporters of sterilization. *Buck v. Bell* led to renewed confidence among eugenicists, new sterilization laws, and above all, a sharp uptick in the numbers subjected to coerced sterilizations. The Pope's intervention galvanized Roman Catholics, who lobbied against new sterilization laws and the implementation of old ones. The

[83] Birthright, Inc., *Report to Members, January 1 to April 29, 1948*, 2/14, AVS.

[84] *Casti Connubii, Encyclical of Pope Pius XI on Marriage*, paras. 68–9, December 31, 1930, http://www.vatican.va/holy_father/pius_xi/encyclicals/documents/hf_p-xi_enc_31121930_casti-connubii_en.html. See also Patrick J. Ward, "The Grave Issue of Sterilization," *NCWC Bulletin*, February 1, 1934, 16/2, CUA and Rev. Ignatius W. Cox and James J. Walsh, "What About Sterilization," *NCWC Bulletin*, 1934, 18/228, CUA.

[85] *Casti Connubii.*

figurative battle would continue into the postwar period, and it would extend to old fields of contraception and new ones of abortion. But before these developments occurred, a literal battle – against Nazi Germany – would take place. The war itself is not the subject of this book, but the regime that launched it is, for Nazi Germany's adoption and implementation of eugenic sterilization in the 1930s exceeded American eugenicists' wildest dreams.

PART B

8

Sterilization and Murder in Nazi Germany

Many countries had powerful eugenics movements, most eugenicists lobbied and argued for coerced sterilization, and politicians in many countries heeded the eugenicist call. Only one country, however, took eugenics to its murderous extreme. That it turned out so would not have been predicted in the early or even the late 1920s.

Germany was a latecomer to coerced sterilization: such policies had already been adopted in Canada, Denmark, Switzerland, and the United States. There had been discussion of eugenic sterilization since Alfred Ploetz, the key figure of German eugenics, published *Die Tüchtigkeit unserer Rasse* (*The Efficiency of Our Race*), but all legislative proposals had come to nothing. Sterilization bills died before the Prussian legislature in 1903, the national Reichstag in 1907, the Saxon legislature in 1923, and the Reichstag again in 1925. Interestingly, studies in the 1920s cited Germany as a country notable for its reluctance to adopt a sterilization law.[1] In this context, the fate of Gustav Boeters is instructive. Boeters was a sort of Saxon Hoyt Pilcher.[2] He worked as a district physician in Zwickau and lobbied for sterilization by instigating and advertising illegal sterilizations carried out by Heinrich Braun, a surgeon.[3] In May 1923, he sent a report to the government of Saxony in which he demanded compulsory sterilization for the hereditarily blind and deaf, the mentally handicapped, the mentally ill, "perverts," and fathers with two or more illegitimate children.[4] The "*Lex Zwickau*" (named after the Saxon town in which he lived) was his model law, but it received little support from the German government. On

---

[1] Sigwart Frank, *Praktische Erfahrungen mit Kastration und Sterilisation psychisch Defekter in der Schweiz* [inaugural dissertation for the Doctorate of Medicine, University of Zurich, 1925] (Berlin, 1925), 2.

[2] On Boeters, see Weindling, *Health, Race, and German Politics*, 392–3.

[3] Burleigh, *Death and Deliverance*, 36.

[4] Kühl, *Nazi Connection*, 23.

October 5, 1928, a frustrated Boeters penned a letter to the well-meaning but probably dim American philanthropist, E. S. Gosney.[5] In a plaintive and self-pitying letter (although, admittedly, one following an operation), Boeters complained to Gosney of his poor treatment at the hands of German officials:

> I am 59 years old and I do not expect to see any success of my lifes [sic] work devoted to the problem of sterilization for the purpose of 1) eugenic, 2) economy, 3) public safely, 4) philanthropy. In December 1925, I lost my employment as a medical officer in Saxony. I am living a sorrowful life. And the German Reichstag does not even think my "Lex Zwickau" worthy of consideration! May I ask you to publish my "Lex Zwickau" and "Outline" ... in the Journal of Hygiene? And may I ask you to publish – some weeks or months after – an explanation of the "Lex Zwickau" written by me?[6]

Other eugenicists – Fritz Lenz and Alfred Grotjahn in particular – concluded that a liberal democracy would be unable to overcome the opposition of the uninformed masses.[7]

There were nonetheless some pressures in Weimar Germany in favor of eugenic sterilization. In 1922, the German Society for Racial Hygiene called for legislation on sterilizing the mentally ill and those guilty of sexual crimes.[8] At a September 1925 meeting of the German Psychiatric Association, Robert Gaupp began a speech by extolling the efforts of Boeters in pursuing eugenic sterilization in Germany, American researchers such as Sharp and Ochsner in pioneering the application of vasectomies, and Swiss policy makers in adopting eugenic sterilization.[9] He ended by recommending sterilization as a German policy. The Protestant churches – which, along with their Roman Catholic counterparts, provided approximately 32,000 beds for the mentally and physically handicapped, the epileptic, and the mentally ill – reconciled themselves with eugenics in a way that the Roman Catholic Church never did.[10] Impressed by the "scientific" arguments, the Protestant Inner Mission (an umbrella organization representing Protestant churches in Germany) appointed a committee on eugenic questions led by Hans Harmsen. A year before the Nazis came to power, it recommended voluntary eugenic sterilization.[11] By that point, there had already

---

[5] Gosney's correspondence with Laughlin is riddled with errors of grammar and spelling. See Gosney Papers, Box 7.1.

[6] Dr. Gustav Boeters to Gosney, October 5, 1928, 11:8, Gosney Papers.

[7] Herlitzius, *Frauenbefreiung*, 86. It is worth noting that Grotjahn and Lenz were in many ways opponents. Grotjahn was a Social Democrat and sensitive to the role of environment in eugenic health, and he in any case died before the National Socialists seized power.

[8] A. Gütt, "Das Gesetz zur Verhütung erbkranken Nachwuchses," n.d. [1933–34], R 1501 126 249, BArch Berlin.

[9] Burleigh, *Death and Deliverance*, 36–7.

[10] Kurt Nowak, *Geschichte des Christentums in Deutschland* (Munich: C. H. Beck, 1995).

[11] Burleigh, *Death and Deliverance*, 41–2. After the war, Hans Harmsen, like American eugenicists, moved into the birth control movement and became a co-founder and long-term president of Pro-Familia, Germany's main birth-control organization. See Timm, *Politics of Fertility*, 250–2.

been three regional initiatives to introduce eugenic voluntary sterilization, with greatest support coming from a coalition of women and doctors in the Social Democratic Party.[12] On the eve of the Nazis' seizure of power, there was a coerced sterilization bill before the Prussian parliament.

Like the Kaiser's Reich before it, the Weimar Republic was caught in a struggle between Germany's liberal and illiberal trends, between democracy and imperialism, between anti-Semitism and integrationism. And, like the total ideological battle in which the debate over eugenic sterilization was nested, the outcome of that struggle was unclear. It is unclear whether a liberal democratic Germany would have adopted a moderate eugenic sterilization law along American or Scandinavian lines. But because some of the conditions conducive toward the adoption of such a law – massive economic, political, and social instability brought on by the war, reparations, and hyperinflation – were the same ones that destroyed the Republic, it is impossible to know.

THE NATIONAL SOCIALIST SEIZURE OF POWER

In May 1932, Heinrich Brüning resigned as German Chancellor. The aging President Hindenburg, a hero of the First World War, replaced him with a personal friend and landed aristocrat, Franz von Papen. A year later, Papen entered into a fatal pact with the leader of the National Socialist German Workers' Party, Adolf Hitler. With the support of the army, Hitler was appointed head of a new government in which all other posts but two were occupied by Papen's conservatives. Papen was himself Deputy Head. The conservatives hoped to use Hitler to establish a regime of aristocratic rule little checked by a rump parliament.[13] But their would-be pawn played them instead: within two months, the Communist Party was suppressed; within three, the Enabling Act abolished parliamentary democracy; and within six, Germany was a one-party state.

As early as 1925, when the National Socialists were still viewed as a motley collection of thugs, political outcasts, and foreigners that was not to be taken seriously, the party had publicly declared itself in favor of sterilization. The leader of the Nazis in the Reichstag, with the backing of the German People's Freedom Party, pleaded for the legal sterilization of "hereditarily burdened criminals" (*"erblich belasteter Verbrecher"*). The Federation of National Socialist doctors declared through their propaganda magazine, *Ziel und Weg*, its support for the sterilization of inferiors in 1929. This outspokenness in turn attracted more support. Keen supporters of sterilization policy joined the Nazi Party: Gustav Boeters, the Saxon doctor and sterilization advocate, and Herbert

---

[12] Michael Burleigh, *The Third Reich: A New History* (London: Macmillan, 2000), 353–4; Weindling, *Health, Race and German Politics*, 390–2, 419.

[13] Details from Richard J. Evans, *Coming of the Third Reich* (London: Allen Lane, 2003), chapter 4 and Evans, *The Third Reich in Power* (London: Allen Lane, 2005), prologue.

Linden, eventual co-author of Germany's Blood Protection Law, both joined the Party in the early 1920s.

AN "EXPERT COMMITTEE"

The high-profile German eugenicists we met in Chapter 2 – Alfred Ploetz, Fritz Lenz, and Eugen Fischer – kept their distance while the Nazis were out of power. Not all were instinctive National Socialists. In his 1895 treatise, Ploetz had ranked Jews with Aryans as one of the world's two premier races, and he derided anti-Semitism as a useless ploy ("*Schlag ins Wasser*").[14] They were rather late joiners of the National Socialist Party and probably did so under pressure: Ploetz and Lenz did not join until 1937, and Fischer held off until after Germany's victory over France in 1940.

When the National Socialists came to power, however, the eugenicists knew their policy day had come. They saw the Nazis as the opportunity to realize their eugenic utopia, and they showed no hesitation in seizing it.[15] The *Archiv für Rassen- und Gesellschaftsbiologie* welcomed the "new age of racial-biological revolution."[16] Lenz, its editor, had already lavished praise on Hitler as "the first politician of truly great import that takes racial hygiene seriously as a basic element in state policy."[17] Lenz's fawning was partly a strategic necessity, but there was more to it than this. Fritz Lenz's relationship to Nazism was complex and ambiguous: he welcomed them but did not join the party until 1937; he wrote much that could be construed as anti-Semitic yet insisted that much bound Germans and Jews together; he advised the regime on family policy and coerced sterilization, but he disagreed with Himmler about the racial value of illegitimate children.[18]

Like many in the Germany of 1933, Fischer was quick to sing from the new song sheet; like many in Germany after the war, he would do the same in 1945. But Fischer, too, set limits. He happily spouted in letters and from lecterns tirades against Jewish influence and racial incompatibility,[19] but he refused to concede that difference implied inferiority.[20] His scholarly stubbornness cost him National Socialist support in his bid for the rectorate of Berlin University

---

[14] Proctor, *Racial Hygiene*, 21. The translation is Proctor's.

[15] Fangerau, *Etablierung eines Rassenhygienischen Standardwerkes*.

[16] "Ansprache des Herrn Reichsministers des Innern Dr. Wilhelm Frick auf der ersten Sitzung des Sachverständigenbeirats für Bevölkerungs- und Rassenpolitik am 28. Juni 1933 in Berlin," *ARGB* Bd. 27, Heft 4 (1933): 419.

[17] Fritz Lenz, "Alfred Ploetz zum 70. Geburtstag am 22. Aug. 1930," *ARGB* Bd. 24, Heft 1 (1930): xiv.

[18] Lenz was possibly pressed by Gütt to join the Nazi Party. On Lenz's relationship to National Socialism, see Proctor, *Racial Hygiene*; Weiss, "Race and Class"; Müller-Hill, *Murderous Science*.

[19] Schneider, *Quality & Quantity*, 1–2.

[20] Müller-Hill, *Murderous Science*, 78.

(the professoriate elected him anyway) and his presidency of the German Eugenics Society.[21] Both he and Lenz viewed the Nazis with particular disdain in all matters academic and theoretical.[22]

But for all their social and intellectual vulgarity, the National Socialists offered two prizes that appealed to the eugenicists (and which, to this day, often appeal to academics). The first prize was access. Many academics like to think of themselves as people that matter, and they can, often in diffuse ways, shape intellectual and policy trends. But they rarely have political power, and they are as easily corrupted by it as anyone else. Within six months of the Nazis coming to power, Reich Interior Minister Wilhelm Frick announced the creation of an Expert Committee on Questions of Population and Racial Policy (*Sachverständigenbeirat für Bevölkerungs- und Rassenpolitik*) with the ambitious remit to chart the course of Nazi racial policy.[23] Its membership included Friedrich Burgdörfer (editor of *Politische Biologie* and director in the Reich's Statistics Office), Walther Darré (Reich Farmer Führer and champion of "blood and earth"), Hans Günther (professor of social anthropology at Jena), Charlotte von Hadeln (second leader of the short-lived German Women's Front), Ernst Rüdin (director of the German Institute for Psychiatry in Munich), Bodo Spiethoff (professor of medicine at Jena and expert in venereal diseases), Paul Schultze-Naumburg (Reichstag member and leader in the Nordic art movement), Gerhard Wagner (leader of the Nazi Physician's League), and Baldur von Schirach (head of the Hitler Youth). Arthur Gütt served as chairperson. They were later joined by Fritz Thyssen (an industrialist), SS chief Heinrich Himmler, and Fritz Lenz.

Whereas Lenz spent the Weimar years in Munich ostracized (particularly in the early postwar years) by the international community and driven to dependency on America for research funding, his fortunes were transformed by the Nazi seizure of power. He sat at the same table as senior officials from the Interior, Justice, Finance, and Defense Ministries, and he had their attention as he outlined Germany's need for eugenic population surveys, the training of doctors in the art and science of race hygiene, expanded sterilization, and fiscal transfers to eugenically sound large families.[24]

The second prize of National Socialism for eugenicists was the possibility of quick influence. Germany's educated middle and upper-middle classes, whose loyalty toward the Weimar Republic had been tenuous from the start, grew increasingly frustrated by the dithering of politicians, frequent elections, and their inability to cope with the social chaos that followed the economic crises of 1923 and 1929. The inflation of 1923 in particular, as Sebastian Haffner has

[21] Ibid.
[22] Ibid., 76–7.
[23] Proctor, *Racial Hygiene*, 95–6.
[24] See "Niederschrift über die Beratung des Sachverständigenbeirats für Bevölkerungs- und Rassenpolitk," June 25, 1934, R 1501 126 229, BArch Berlin.

written, perverted Germany's traditional value system: it rewarded the quick, reckless, and speculative, and it punished the conservative, cautious, and established.[25] Academics were a natural component of the latter group. The punishing effects of German postwar instability revealed themselves above all in funding. Throughout the 1920s, the major German research foundations – the Kaiser Wilhelm Institute (KWI) for Anthropology and Eugenics and Human Heredity in Berlin and the Institute for Psychiatry in Munich (later, the KWI for Psychiatry) – were desperately short of funds for equipment, buildings, and research. They were wholly dependent on support from the Rockefeller Foundation.

Scholars have taken this support as evidence of the tight links between American and German eugenics,[26] but many of the funds went to finance technical studies. Felix Plaut's work in 1933–4 on "the formation of spirochetes antibodies in syphilitic tissues and in pure cultures of syphilis spirochetes" was hardly likely to excite National Socialist fervor.[27] Indeed, the National Socialists ordered him sacked in 1935; Ernst Rüdin absorbed his research institute.[28] The more important point for German racial hygienists was that the National Socialists promised German eugenicists quick results. Intoxicated by this prospect, they let themselves be used.

The six months it took to form the committee coincided with the time it took to destroy what was left of German parliamentary democracy. The Nazis first had to deal with an old Prussian law coming up for consideration. On November 7, 1932, the German Doctors' Association (*Deutsche Ärztevereinsbund*) had sent a petition to the Reich Interior Minister in favor of a eugenic sterilization law. It contained, however, a number of conditions: (1) coerced sterilization would only be acceptable so long as it did not contradict the sensibilities of the public; (2) doctors must not be forced to perform the operation against their will; (3) economic considerations should not be taken into account, and the operation should not be performed on healthy couples; and (4) the law should protect the doctor performing the operation from legal uncertainty (*Rechtsunsicherheit*) and prosecution.[29] The third condition would have been especially objectionable to the National Socialists, as it would have protected healthy Jews and, for

---

[25] Sebastian Haffner, *Geschichte eines Deutschen: die Erinnerungen 1914–1933* (Stuttgart: Deutsche Verlags-Anstalt, 2000), chapter 10.

[26] See, in particular, Kühl, *Nazi Connection*, chapter 2.

[27] See "German Research Institute for Psychiatry: Department for Serology and Experimental Therapy (F. Plaut), 1933–1934," RF 1.1/717A/10/59 (2), RAC. Plaut was fired for being Jewish.

[28] Allan Gregg to Marie Kopp, November 25, 1938, RF 717A/10/58, RAC. Rockefeller's support for German research captures academic headlines but in its day was viewed as anodyne, not least because those living in the 1920s could not know what would happen in the 1930s. Moreover, Rockefeller and other U.S.-based organizations supported extensive scientific research activities of which eugenics research was only a part. For an overview of Rockefeller activities up to 1932, see International Health Division, n.d. [1932 or 1933], RF 3/906, Box 2/24, RAC.

[29] Bock, *Zwangssterilisation*, 80.

example, homeless people. But only condition four would have been readily acceptable to the Nazis. Their tyranny also faced opposition from institutional remnants of German democracy: the Reichstag, although humiliated and terminally weakened, stood in its way, and there was opposition from some state governments (*Länder*).[30] The National Socialists inveighed against the ineptitude of parliamentary democracy.[31]

The Nazis responded by delaying the deadline for presenting the law before the parliament for one month – from February 1, 1933 – and Hermann Göring used the March 5 sitting to transform the Prussian proposal by asking the Justice and Interior Ministries to present a paper on the topic. On May 17, a ministerial representative presented a paper in which a carefully selected set of petitions from organizations and municipalities demanded coerced sterilization and the protection of doctors.[32] Papen recommended to Reich Interior Minister Frick that he address the issue through a new criminal law,[33] which came on May 26, 1933. The Nazis' first sterilization measure was adopted by changing the language of the Prussian proposal.[34] The prohibitions on bodily harm were removed, and the opportunity for sterilization was created.[35] Just under a week later, on June 2, the committee was formed.

At its first meeting on June 28, Frick gave the members their marching orders: Germany was under attack from within and from without.[36] There were in Germany 500,000 seriously heredity defectives, mentally and bodily, and another 500,000 with milder illnesses; these figures were, if anything, conservative. At the same time, Germany was threatened by its eastern neighbors and by immigration, above all of Jews from Eastern Europe (*Ostjuden*): 4,000 had arrived in Berlin alone in 1930.[37]

The committee was either uncommonly unanimous in outlook, extraordinarily efficient, or irrelevant. Just over two weeks after its first meeting, on July 14, Frick and Minister of Justice Franz Gürtner presented the law to the Cabinet. Papen, increasingly isolated and superfluous, tried to moderate the bill. He argued that the hereditarian arguments were contestable, that some of those whom the bill wanted to sterilize were probably curable, that segregation and contraception were alternatives to coerced sterilization, and that Catholic

---

[30] Ibid., 84.
[31] A. Gütt, "Das Gesetz zur Verhütung erbkranken Nachwuchses," n.d. [1933–34], R 1501 126 249, BArch Berlin.
[32] Bock, *Zwangssterilisation*, 81.
[33] Ibid., 82.
[34] Proctor, *Racial Hygiene*, 101.
[35] Bock, *Zwangssterilisation*, 83.
[36] Falk Ruttke, "Abschrift. An die Deutsche Frauenfront," July 24, 1933, R 1501 126 227, BArch Berlin.
[37] W. Frick, "Bevölkerungspolitik und Rassenpolitik," speech from June 28, 1933 (printed as a pamphlet at Langensalza), 7, R 1501, BArch Berlin.

opposition had to be considered.[38] Papen also resorted to the tried-and-tested appeal for delay: sterilization might affect the concordat with the Vatican, and the law should be delayed until its signing. Papen, a relic of an old, aristocratic Germany that had sealed its fate when it opted to back the National Socialists, was silenced by Hitler. The dictator argued that the case for the law was "morally incontestable" (*"moralisch unanfechtbar"*): millions of healthy babies go unborn while the hereditarily ill are ever increasing in number. Hitler made it clear that he would tolerate no delay, and so the measure was adopted. In drafting the law, the National Socialists drew on international examples, above all from the United States. The Nazis cited California and Virginia's laws as precedents.[39]

The law outlined nine medical conditions as hereditary and likely to result in serious physical and/or mental defects in children: hereditary feeblemindedness, schizophrenia, manic-depressive insanity, Huntington's chorea, hereditary epilepsy, hereditary blindness, hereditary deafness, severe physical deformity, and serious alcoholism.[40] The Law for the Prevention of Hereditarily Diseased Offspring would take effect on January 1, 1934.[41]

On July 26, Goebbels ordered the mobilization of a nationwide propaganda campaign channeled through research institutes, health clinics, homes for the elderly, literary discussion groups, and public lectures.[42] Over the course of three months, from September to November 1933, the Nazis used their dense network of party organizations, along with infiltrated associations and groups, to disseminate millions of brochures.[43] The Nazi Party controlled the content of these brochures, of books and newspaper articles on race hygiene,[44] and of propaganda films.[45] Hitler himself took a direct interest in the campaign, and he issued instructions through his private secretary, Martin Bormann, to Interior Minister Frick.[46] The literature emphasized the race-suicidal effects of Germany's

---

[38] See Weindling, *Health, Race and German Politics*, 524.

[39] R. L. Dickinson to William Harris, Mount Sinai Hospital, April 2, 1948, 13/107, AVS; A. Gütt, "Das Gesetz zur Verhütung erbkranken Nachwuchses," n.d. [1933–34], R 1501 126 249, BArch Berlin. The latter also cites the Swiss canton of Vaud and applauds the canton and the state's use of compulsory measures.

[40] Weindling, *Health, Race and German Politics*, 525.

[41] Ibid.

[42] J. Goebbels to Arthur Gütt, Medical Officer, Interior Ministry, July 26, 1933, R 1501 126 227, BArch Berlin; "Das Reichsministerium für Volksaufklärung und Propaganda richtet an alle Organisationen, Verbände und Vereine im Deutschen Reich folgenden Aufruf," n.d. [1933], BArch Berlin.

[43] C. Thomalla, "Reichsverband Deutscher Schriften-Verlager e.V.," August 28, 1933, R 1501 126 227, BArch Berlin.

[44] R. Hess, "Der Stellvertreter des Führers. Anordnung. Betr.: Vereinheitlichung der Schulungs- und Propagandaarbeit auf dem Gebiet der Bevölkerungspolitik und Rassenpflege," November 27, 1933, R 1501 126 127, BArch Berlin.

[45] Burleigh, *Death and Deliverance*, chapter 6.

[46] M. Bormann to State Secretary Pfundtner, Interior Ministry, October 24, 1933, R 1501 126 249, BArch Berlin.

declining birthrate and the unchecked breeding of the physically and mentally unfit; it was a matter of "*Sieg oder Tod*": "victory or death."[47] Along with this Nazi bombast came the usual eugenic arguments. In a commentary on the law, Arthur Gütt outlined the same justifications that had circulated among eugenicists for two decades: differential class birth rates; the crushing costs of caring for the mentally ill, feebleminded, and asocial; the distinction between castration and sterilization; the safeguards built into the decision-making process; and the relative humanity of sterilization as compared to lifelong institutionalization.[48] In short, Gütt drew on the arguments that British, American, Canadian, and German eugenicists had been making almost since the turn of the century.

As the three-month propaganda blitz was drawing to a close, Frick began implementation. On December 11, he sent a circular to all state governments. It ordered them to set up genetic health courts and appellate genetic health courts (ruling on sterilization) and to specify the institutions in which the operations would take place.[49] Judges and doctors were selected according to their ideological compatibility. Frick decreed that only doctors with particular experience in the evaluation of mental illness and who were entirely committed to the law would be acceptable. Lest the state governments be in doubt about how to respect the second criterion, Frick issued a further circular on January 6, 1934, clarifying that membership in the National Socialist Doctors' Organization would suffice for eligibility.[50] Within a few years, successive amendments to the law broadened its reach to sanction the sterilization of children older than ten (with force over fourteen) years of age, compulsion to appear before a court of medical examiners, the withdrawal of legal representation for those appearing before the genetic health courts, the x-ray sterilization of women over the age of thirty-eight, and eugenic abortions up to the sixth month of pregnancy.[51]

The National Socialists set up 205 hereditary courts and twenty-six appellate courts in Germany.[52] The courts began rapidly processing cases (the generous payment of three Reichsmarks per hour attracted many interested candidates for the courts), and there was quickly a shortage of doctors. The entire machinery of the law was designed to ensure that sterilizations were performed as quickly and with as little bureaucratic fuss as possible. Frick expanded the definition of

---

[47] "Das Reichsministerium für Volksaufklärung und Propoganda richtet an alle Organisationen, Verbände und Vereine im Deutschen Reich folgenden Aufruf!" n.d. [autumn 1933], BArch Berlin. As part of its efforts to convince the German public, the propaganda hammered away at the antiindividualist claim that the common good trumps that of the individual good ("*Gemeinnutz geht vor Eigennutz*").

[48] A. Gütt, "Das Gesetz zur Verhütung erbkranken Nachwuchses," n.d. [1933–4], R 1501 126 249, BArch Berlin.

[49] "Niederschrift über die Beratung des Sachverständigenbeirats für Bevölkerungs- und Rassenpolitk," June 25, 1934, R 1501 126 229, BArch Berlin.

[50] Ibid.

[51] Burleigh, *The Third Reich*, 354.

[52] Some sources put these figures at 220 and 18, respectively. Ibid., 357.

examining doctors to include those working in institutions; he limited appeal rights; he altered courts' jurisdiction so that the patient's state (*Land*) of institutionalization, rather than his or her home state, made the decision; and he encouraged the courts to hold their hearings in the institution.[53] The result was that – as in the United States, Canada, and Scandinavia – institutionalized patients became the first easy target of sterilization; in the end, they constituted thirty to forty percent of those sterilized.[54] At the same time, Frick imposed a general duty on civil servants and teachers to report any individuals suspected of mental deficiency.[55] Once the state had this information, public health authorities would review the individuals' family background, school records, contact with welfare agencies, employment history, and even the opinions of neighbors and police officers.[56] A public health officer would then generally interview suspected individuals, and the doctor would file an application for sterilization with the hereditary health courts.[57] Because the health officer recommending sterilization often also sat on the court, and because all judges were to "always bear in mind Hitler's words that 'the right to personal freedom always gives way to the duty of preserving the race,'" it is not hard to imagine which way most decisions went.[58] Of the 64,500 sterilization rulings issued in 1934, 56,000 were in favor.[59] Estimates of the total number vary, but overall the hereditary courts ruled in favor of nearly 360,000 sterilizations, and approximately 5,000 people died as a result of surgical complications.[60] In addition, an unknown number of foreign workers, Roma, Jews, and people of mixed race were illegally sterilized.[61]

As illiberal as these measures were, committed National Socialists viewed them, and German eugenicists generally, as too timid. The Nazis wished to distinguish between a pure Germanic Aryan race, on the one hand, and racial inferiors, on the other. Eugenicist anthropologists in Germany, however, took a more complex view of the racial composition of the German population.[62] Anti-Semitism lay at the core of National Socialism; it was of secondary interest to the eugenicists.[63] Although eugenicists likely shared the anti-Semitic cultural prejudices that were common to the German middle classes and aristocracy of the time, they were interested in targeting hereditary defects and had no interest in

---

[53] Ibid.
[54] Ibid., 355.
[55] Ibid.
[56] Ibid., 354.
[57] Ibid.
[58] Ibid., 357–8.
[59] Evans, *The Third Reich in Power*, 508.
[60] Weindling, *Health, Race and German Politics*, 533 and Bock, *Zwangssterilisation*, 230–2 on the first statistic; Burleigh, *The Third Reich*, 358 on the second.
[61] Weindling, *Health, Race and German Politics*, 533.
[62] Ibid., 496.
[63] On this, see ibid., chapter 8.

an eliminationist program aimed at Jews. Above all, the essence of this particular brand of nationalism was war, murder, and death.[64]

Hitler himself drew the connection between euthanasia and war when he told the *Reichsärzteführer* (Reich Doctors' Leader) Gerhard Wagner around 1935, "if war should break out, he would take up the euthanasia question and implement it."[65] He was as good as his word. In May 1939, taking up the request of a Leipzig father who wanted a severely deformed child killed, Hitler ordered Karl Brandt to murder the child himself after confirming the diagnosis.[66] Brandt obliged, although Leipzig doctors did the dirty work.

Upon Brandt's return, Hitler ordered him to prepare the infrastructure for a major killing program.[67] Brandt dutifully complied and enlisted the support of Hitler's personal doctor, Theo Morell, and the head of the Chancellery, Philipp Bouhler. But the practical details of how the murders would be carried out remained undetermined. Hitler, responding to pressure from Bouhler, transferred responsibility for killing the handicapped to the Chancellery, and the killings began in the summer of 1939.[68] Hitler signed an order in October 1939 and backdated it to September 1, 1939, in order to highlight its connection to the war. The first "test run" began with the gassing of mentally handicapped patients in portions of Poland annexed by Germany. Bouhler, Brandt, and other officials oversaw the construction of gas chambers in Brandenburg, Württemberg, Hesse, and near Linz, Austria.[69] Centrally administered at Tiergartenstraße 4 in Berlin – one of the Nazi era's more infamous addresses, which generated the program's abbreviated name, *Aktion T4* – the program took over former hospitals and transformed them into killing centers. Buses and trains transported victims from hospitals and homes for the feebleminded to these facilities, where officials marked those with gold teeth and sometimes handed the victims soaps and scrubbing brushes.[70] They were then ordered into gas chambers; the doors were sealed and the gas turned on; and the victims died in the dark, often beating against the door in desperation.[71] By the time Hitler halted the program in August 1941, following protests by Catholic Bishop Clemens August Graf von Galen, doctors, nurses, and other personnel had murdered approximately 80,000 people at killing centers across the Reich. By then, Hitler's quota of 70,000 had been exceeded, and the SS began to think of other uses for the chambers.[72] In early 1941, Himmler had approached Bouhler and asked

---

[64] Timothy Snyder, *Bloodlands: Europe between Hitler and Stalin* (New York: Basic Books, 2010), preface; Richard Bessel, *Germany 1945: From War to Peace* (New York: Harper, 2009); Burleigh, *The Third Reich*; Evans, *The Third Reich at War*.

[65] Quoted in Burleigh, *The Third Reich*, 383.

[66] Evans, *The Third Reich in Power*, 78.

[67] Burleigh, *The Third Reich*, 383; Evans, *The Third Reich in Power*, 78.

[68] Evans, *The Third Reich in Power*, 79.

[69] Snyder, *Bloodlands*, 256; Evans, *The Third Reich in Power*, 84–5.

[70] Burleigh, *The Third Reich*, 395.

[71] Ibid.

[72] Evans, *The Third Reich in Power*, 100.

him about using the T4 facilities to murder "ballast existences" in the concentration camps. By November, Bouhler had sent the first T4 teams to the concentration camp at Belzec in southeast Poland.[73] "Medical killing," writes Paul Weindling, "became a pilot scheme" for the Holocaust.[74]

INTERNATIONAL REACTION

Eugenicists around the world reacted to the new National Socialist law with support and enthusiasm. One such person was Marie E. Kopp, an American author of a text on birth control and winner of an Oberländer Fellowship to visit Germany. The KWI liaised with the Interior Ministry and the Foreign Office to secure for Kopp a permit allowing her to interview officials and to visit any institution within the scope of her study.[75] Beginning in Bavaria, she drove across the breadth of a country that in ten years' time would be razed and truncated: from Munich she drove on to Berlin and Königsberg (now Russia's Kaliningrad), back across the Baltic Coast to Kiel, and back southward through Hamburg, Bremen, Frankfurt, and Freiburg.[76] She covered 1,500 miles by car, and if she did not begin her journey a true believer, she certainly ended it as one. A series of reports reflect the enthusiasm of the convert. "The legislation in Germany for the sterilization of those unfit for procreation," Kopp noted, "is simply a part of a very comprehensive program to improve the health as well as the mental and physical quality of its people."[77] She was fulsome in her praise of the legislation's scientific basis – thirty years of research by Ernst Rüdin at the Munich KWI for Psychiatry – and of the safeguards enacted by the German government – notably, provisions for prosecution if sterilization were carried out in the absence of eugenic or medical grounds.[78] Kopp was keen to make the point that the sterilization law could not be taken out of context: the legislation was part of a much broader, "comprehensive" program that included marriage loans for healthy couples, grants for third and fourth children, subsidies for children on farms, and laws to reduce unemployment by making loans contingent on women's willingness to return to the home.[79] She seemed irony-blind to the implications of this last provision for her own career.

One could spend pages quoting eyebrow-raising passages from Kopp's reports: that castration is a "therapeutic" measure appropriate for "dangerous

[73] Burleigh, *The Third Reich*, 403–4.
[74] Weindling, *Health, Race and German Politics*, 548.
[75] M. E. Kopp, "Untitled Report by Marie E. Kopp on her trip to Germany," n.d., 13/111, AVS.
[76] Ibid.
[77] M. E. Kopp, "The Application of the Sterilization Laws in Germany," n.d. [probably 1935], 13/111, AVS.
[78] M. E. Kopp, "The German Sterilization Legislation," n.d. [mid-1930s], 13/111, AVS.
[79] M. E. Kopp, "The Application of the Sterilization Laws in Germany," n.d. [probably 1935] and "The Nature of Operation of the German Eugenical Problem," n.d. [probably 1935], both from 13/111, AVS.

sex offenders" such as those twice convicted of sodomy, or that a shortcoming of the law is its failure to make provisions for sterilizing asocial groups such as "habitual paupers," a "serious handicap to any nation."[80] Kopp's statistics are even more striking. At one point, she blandly notes, "it is estimated that about 600,000 hereditary defectives in Germany come under the provision of the law. Some 225,000 persons so far have been adjudicated unfit for procreation and accordingly sterilized."[81] Kopp and anyone who would listen to her were thus fully aware that within a three- to four-year period, over 200,000 people had been sterilized in Germany – almost ten times the figure during the first twenty-five years of California's sterilization laws. Kopp viewed with absolute equanimity a total of 600,000 coerced sterilizations in the United States, a figure 50 percent higher than the most generous estimates of the total number of Germans forcibly sterilized by the Nazis. The results of Nazi policy, viewed with such retrospective horror by the American eugenicist community, were anticipated and endorsed by Kopp.[82] It can be said in her favor, perhaps, that her commitment to (German) eugenics was genuine and free of anti-Semitic undertones. When the Jewish former director of the Munich Institute for Psychiatry's Department of Serology, Felix Plaut, tried to escape Germany in the late 1930s, he appealed to Allen Gregg, the Director of Rockefeller's Medical Sciences Division who had arranged Rockefeller's support of Plaut's research. Kopp intervened on Plaut's behalf. She politely urged Gregg to secure an immigration visa (the Nazis had were stamping German Jews passports with a 'J' at this point) and a position (emigration meant forfeiting his pension).[83] Gregg told both Kopp and Plaut that he was sympathetic but, given the large number of German Jewish émigrés in America, he could do nothing.[84] As a final blow to Plaut's hopes, Gregg wrote that even "to advise you to take any decisive action is to imply a certain responsibility personally or officially. This I cannot assume."[85] No thanks to Gregg, Plaut eventually got out of Germany. He was able to secure a visa to the United Kingdom, where he died in 1940.

Kopp was not the most influential American eugenicist, but she was widely cited and not without connections.[86] She was also not alone. Marian Olden, founder of the Sterilization League of New Jersey and one of the most important

---

[80] Ibid.

[81] Kopp, "The Nature of Operation of the German Eugenical Problem."

[82] Kopp died before she was able to publish the study. Robert L. Dickinson expressed passing interest in seeing the material published, but nothing seems to have come of this. See Robert L. Dickinson to Norman E. Himes, October 15, 1947, 2/14, AVS.

[83] Marie E. Kopp to Alan Gregg, November 25, 1938, RF 7171/10/48, RAC.

[84] Allan Gregg to Marie E. Kopp, November 25, 1938 and Allan Gregg to Felix Plaut, November 15, 1938, both from RF 717A/10/58, RAC. Gregg drafted the letter to Plaut before the letter to Kopp but, fearing German censors, mailed it only sometime after November 29, 1938. See Gregg to Keith Merrill, November 29, 1938, RF 717A/10/58, RAC.

[85] Allan Gregg to Felix Plaut, November 15, 1938, RF 717A/10/58, RAC.

[86] Kopp's 1934 book, *Birth Control in Practice; Analysis of Ten Thousand Case Histories of the Birth Control Clinical Research Bureau*, was reissued by Arno Press in 1972.

coerced sterilization activists in the United States, visited Germany in 1938 and received something akin to a state welcome. Falk Ruttke, a member of the SS, participant in an Interior Ministry committee on population and race policies, and an admirer of Olden's diatribes against Roman Catholic opposition to eugenic sterilization, warmly welcomed her.[87]

Support for the German measures came from the American West Coast as well. In California, Gosney wrote to Norwegian Jon Alfred Mjoen, director of the Vindern Biological Laboratory in Oslo, stating that Germany's law was superior to California's.[88] Superintendents of mental health institutions acclaimed Germany's foresight.[89] Fellow Californian Charles C. Goethe, a successful businessman, philanthropist, environmentalist (he generously supported the Save-the-Redwoods League), anti-immigration activist, and eugenicist (he was president of the Eugenics Research Association and belonged to the American Eugenics Society and the Human Betterment Foundation), was equally supportive.[90] Thanks to the "characteristically thorough" work of German scientists, Goethe wrote at the brink of the Second World War,

> The Reich today has her social inadequates [sic] more thoroughly listed than any nation. . . . As a result, Germany believes that a few years will show the beginnings of a reduction in the costs of carrying defectives. This will continue until eventually her budget items for crime, insanity and dependency will almost be halved. To a land whose population approaches the saturation point, elimination by sterilization of those unfit, means room for high power.[91]

Goethe brought this piece of dreadful writing to a merciful close with a bit of credit-claiming: "It is well known . . . that [Germany's] leaders in the sterilization movement depended largely upon the material accumulated by the Human Betterment Foundation, using this as a California data foundation upon which to rear their present remarkable structure."[92]

In Canada, the Parents' Information Bureau, a pro–birth control organization,[93] applauded the German law for its legal safeguards and respect for Catholic sensitivities, and, like Kopp, noted approvingly the 180,000 to 200,000 sterilizations undertaken within two years.[94] The German Consulate in Montreal relayed with satisfaction the fact that Toronto's *Saturday Night* had

[87] Dowbiggin, *Sterilization Movement*, 37.

[88] E. S. Gosney to Jon Alfred Mjoen, Norway, January 2, 1936, 10.16, Gosney Papers.

[89] M. A. Tarumianz, "Sterilization as a National Problem," 1934, Reference Material, Medical Society of Delaware, AVS.

[90] On Goethe's life and activities, see Stern, *Eugenic Nation*, 134–49.

[91] "Report from C. M. Goethe," n.d.,. 5. 12. Gosney Papers.

[92] Ibid.

[93] The Bureau was set up by the philanthropist A. R. Kaufman to provide free contraception to the poor. See http://www.cpha.ca/en/programs/history/achievements/04-fp/history.aspx (accessed January 21, 2013).

[94] "Sterilization Notes," March 1, 1935, PBA, P. s. I. B., GR 0133, Box 6, 1.14.22, BCA.

compared the Nazi law favorably with Alberta's eugenic sterilization legislation.[95]

In 1935, more than a year after the law took effect, Nazi eugenic policies were showcased at the International Congress for Population Science in Berlin. Its president was Eugen Fischer, who extended an invitation to the American eugenicist Raymond Pearl to serve as vice president. Pearl, an early eugenicist who was by then moving into the field of population science, accepted this invitation as a "great honor." He however ultimately deputed Dr. Clarence G. Campbell, Honorary Chairman of the Eugenics Research Association (whose members included Harry Laughlin, Irving Fisher, and Robert L. Dickinson) and Harry Laughlin to serve in his place.[96] Campbell fawned over Hitler and effused about the possibilities offered by Nazi eugenics. Arthur Gütt, President of the German Academy of Public Health, later recalled his reaction to Campbell:

> At the international Congress of Eugenics ... attended by hundreds of foreign eugenicists Germany found one only who in the entire world of science could appreciate her eugenic and population-political procedures. The leading American eugenicist Campbell ... said[:] "In early times men had already recognized their eugenic responsibility: Cato, Seneca, Taitus, and others. Numbers of zoologists, paleontologists, and anthropologists in all countries have made their contributions to this science. Their work has inspired Adolf Hitler, the leader of the German people, to create, assisted by his co-workers, broad eugenic laws, which are sensational. All other countries must follow his example, if they do not wish to endanger themselves by lagging behind in their racial development and to endanger the continuity of their racial unity.[97]

In the same introduction, Gütt and his collaborators clarified that which they took to be the greatest threat to racial purity:

> Germany at the present time has only one race problem, the Jewish problem. Whoever has observed the increase of Jewish mixed marriages, whoever has seen the rape on German women, or the arrogance with which the Jews conduct themselves in all spheres of life, economies, civilization and morals, will understand that it is high time to erect a barrier between Germans and Jews.[98]

Sensitive to international criticism of prohibitions on marriage between Germans and Jews, the author cited a precedent: the thirty years of states forbidding intermarriage between whites and "Negroes" and the general institution of segregation in the Southern United States. Beyond the powerful American example, they could only come up with another: "a Spaniard" who

---

[95] German Consulate, Montreal to the Foreign Ministry, August 22, 1933, R 1501 126 249, BArch Berlin.

[96] Kühl, *Nazi Connection*, 33.

[97] Arthur Gütt, Herbert Linden, and Franz Messfeller, "Blutschutz – und Ehegesundheitsgesetz: Summary of the 'Einführung,'" 1936, 14/113, AVS.

[98] Ibid.

approved of German eugenic laws. However spotty it might have been, the Nazis expressed public gratification for international support.[99]

The support attracted in the United States by Germany's law and practices raises the question of the connection between American and National Socialist coerced sterilization. Were American, Canadian, and Scandinavian sterilization policies, as journalists like to argue, basically National Socialist?[100] Were the Allies implementing Hitler's ideas while fighting Hitler? In numerical terms, the claim is hard to credit: the 63,000 individuals sterilized in the United States pales in comparison to Germany's 360,000, particularly when viewed as a percentage of the countries' respective populations. Likewise, only Germany took eugenics to its murderous extreme: although there is strong anecdotal evidence that some mental health patients were murdered in the United States, only Germany enacted a systematic extermination policy.

Similarly, although there were close personal and intellectual links between National Socialist race hygienists and (some) American eugenicists, as historian Stefan Kühl has argued,[101] both movements were broad churches. They contained socialist/conservative, hardline/moderate, positive/negative, and (even in Germany) anti-Semitic/non–anti-Semitic strains. The American eugenicists who heralded Nazi foresight and courage – and there was no shortage of them – did so before German policy developed into mass murder.

There was nonetheless some clear common ground between German and North American eugenics, and that was the target population for eugenic sterilization. In February 1914, Harry Laughlin estimated the number of individuals who would have to be sterilized to rid the American population of defective genetic stock. He concluded that an effective program would require the sterilization, over two generations, of 15 million Americans, or 15 percent of the U.S. population; its achievement would be "consonant with modern humanitarian ideals and reasonable social endeavor."[102] The National Socialists, by contrast, sterilized less than half of 1 percent of the German population. Laughlin was, admittedly, among the most extreme of U.S. eugenicists, and he was writing during the prewar intellectual apogee of eugenics. Thirty-six years later, in 1950, however, the Human Betterment League of Iowa suggested a

---

[99] G. Herman Sieveking, "Sterilisierungsgesetze des Auslandes," *Deutsches Ärzteblatt* 35, no. 64 (August 25, 1934): 830.

[100] There is hardly a single newspaper article, radio broadcast, or television report on the subject of coerced sterilization in North America or Scandinavia that does not allege a close parallel or some misinformed causality between those cases and Nazi policies. Historical distortions aside, sensational assertions obviously provide a tried-and-true shortcut to exciting readers.

[101] Kühl, *Nazi Connection*.

[102] Harry H. Laughlin, "The Legal, Legislative and Administrative Aspects of Sterilization," *Report of the Committee to Study and to Report on the Best Practical Means of Cutting Off the Defective Germ-Plasm in the American Population*, Bulletin No. 10B (Cold Spring Harbor, NY: Eugenic Records Office, February 1914). This goal would be achieved by sterilizing 80,000 people per year at the start of the program and then raising the figure to 150,000 near the end.

figure almost identical in percentage terms to that achieved by the Nazis: 1.4 million.[103] The figure was based on the widely held assumption that "mental defectives" made up 1 percent of the population. In other words, the Nazis, in sterilizing 360,000 German citizens, only did what the American eugenicists would have done had they possessed the capabilities. In its policy implications, negative eugenics in Nazi Germany was, right up to the first murders of mentally handicapped Germans, indistinguishable from negative eugenics in democratic America.[104]

After that, the question is still more complicated: if, as scholars have argued, there is a basic relationship between the Nazi sterilization program, its euthanasia program, and, finally, the Final Solution, was there any connection between these horrors and American, Canadian, and Scandinavian eugenics? Such a relationship might be of two types: American eugenic sterilization might have made German eugenic sterilization (and, by extension, German mass murder) easier to implement, or American eugenic sterilization might have itself led to the murder of the mentally handicapped in the United States. The first suggestion is often luridly implied by Hitler's own citation of American sterilization law as an argument in favor of eugenic sterilization. It is one that need not be taken very seriously. The National Socialists gleefully murdered all manner of people – violence was integral to their creed – and they seldom felt the need to look abroad for reassurance or justification before doing so.[105] If anything, American legislation may have helped the Nazis in their curious need to achieve (often retrospective) legal precedence for their crimes. But even this is probably a stretch: the National Socialists hardly needed a precedent for the killing of 6 million Jews, 3 million Soviet prisoners of war, and millions of other people.

The Americans likely had little effect on the Germans, but the Germans had an effect, although an indirect one, on the Americans: Nazi mass sterilization, medical experimentation, and murder presented a major challenge for the proponents of eugenic sterilization. Americans, as one scholar notes, "had an exceedingly difficult time shaking the stigma of Hitlerian science."[106] However laudatory American advocates were in the early 1930s, by the end of that decade, eugenics "was well on its way to becoming ... a 'dirty word.'"[107] As

---

[103] Human Betterment Foundation of Iowa, "1,400,000 mental defectives: A problem for social workers," 1950, Pamphlets Misc. Box 34, AVS.

[104] Even the most moderate pro-sterilization measures, such as those suggested by the Brock Committee in 1934, called for mass sterilization. Depending on how one wishes to read the evidence, the British committee called for the voluntary sterilization of 250,000 people. A. Press, "British Survey Urges 250,000 Be Sterilized," January 18, 1934, 4/118, AVS. See also King and Hansen, "Experts at Work: State Autonomy, Social Learning and Eugenic Sterilisation in 1930s Britain," *British Journal of Political Science* 29, no. 1 (1999): 77–107.

[105] On violence and Nazism, see Bessel, *Germany 1945*; Burleigh, *The Third Reich*; and Evans, *The Third Reich at War*.

[106] Dowbiggin, *Keeping America Sane*, 233.

[107] Dowbiggin, *Sterilization Movement*, 30.

the war ended, it became imperative to differentiate their project from that of the Germans. American eugenicists would spend almost three decades, right up to the early 1970s, doing so.

SALVAGING SCIENCE: GERMAN EUGENICISTS AFTER THE WAR

For the German eugenicists themselves, having been deeply implicated in a project directly linked with mass coerced sterilization and mass murder, the task was even more difficult. At the end of the war, German eugenicists stood at the smoldering epicenter of the National Socialist disaster; their geographic and, by extension, moral proximity to Nazi eugenic atrocities produced an obstacle greater than that faced by eugenicists in other countries who sought to revamp their field in the postwar era. American eugenicists, on the other hand, could claim an ocean between them and the National Socialist project.

Whatever difficulties German eugenicists faced after the war, however, none of them was criminally charged. None of the leading eugenicists – Lenz, Fischer, Verschuer, or Ploetz – had been directly involved in the murderous T4 programs, although they may well have known about them.[108] Eugenicists, however, were neither easily nor uniformly reintegrated into postwar German academic life, despite claims to the contrary. As Hans-Peter Kröner demonstrates, of eleven German eugenicists who retained positions after the war, two went into psychiatry, one had held no position during the National Socialist era, one died within two years of taking up a position, and five, including Lenz, secured positions that were either unpaid or teaching-only.[109] None of those five secured a professorial chair, the only position in Germany for successful academics. The two who remained were Gerhard Heberer, an active National Socialist and member of the SS but not a eugenic researcher, and Otmar von Verschuer, a student of Eugen Fischer.

Although the National Socialists viewed von Verschuer as a "typical liberalist," he managed to retain his job at the KWI of Anthropology in Berlin throughout the war thanks to Fischer's diplomacy with the regime.[110] During this time, Verschuer's protégé, Josef Mengele, was installed as camp doctor at Auschwitz in May 1943. Mengele regularly sent body parts to Verschuer as research material, including the eyes of Roma and Sinti murdered in the camp[111]; such samples sustained the latter's work almost through the end of the war. He fled the capital, along with the KWI's files, in February 1945 and headed west to safety.

---

[108] Ploetz died in 1940.
[109] Kröner, *Von der Rassenhygiene zur Humangenetik*, 61.
[110] G. Brandt to the Ministry of Interior, June 12, 1933, R 1501 126 243, BArch Berlin.
[111] Miklos Nyiszli, *Auschwitz: A Doctor's Eyewitness Account* (New York: Fell, 1960). Nyiszli was Mengele's Jewish slave doctor at Auschwitz.

By early 1946, Verschuer had given up on reestablishing himself in Berlin, which was half-controlled by the Soviets, and sought instead to regain his old position at the University of Frankfurt, which he had left in 1942. He also sought to relocate and reconstitute the KWI to Frankfurt, to be led by himself.[112] Old adversaries soon surfaced, however. Hermann Muckermann, Lenz's Catholic predecessor at the KWI who had been sacked by the Nazis, and Hans Nachtsheim, the former head of hereditary pathology at the KWI, denounced Verschuer and protested his attempted academic coup.[113] A commission created to rebuild the overall Kaiser Wilhelm Gesellschaft, which reviewed Verschuer's publications, eventually came out with its ruling in January 1947:

> Verschuer should be considered not a collaborator but one of the most dangerous Nazi activists of the Third Reich. An objective judgment of the investigative committee must recognize this, and thereby take actions to guarantee that this man does not come into contact with German youth as a university teacher, or with the broader population as a scientist in the fields of genetics and anthropology.[114]

Three months later, the newspaper of the American occupied zone, *Die Neue Zeitung* (*The New Newspaper*), made the allegations public, accusing Verschuer of obtaining blood and eyeballs from Auschwitz through his collaboration with Mengele.[115]

Verschuer was convinced that Muckermann and Nachtsheim were behind the attack.[116] Publicly, he denied it all. Verschuer argued that his "Frankfurt Assistant," who went to Auschwitz against his will, gave him some blood samples from his "hospital" (*Lazarett*).[117] Although he had heard something about one "assistant" at the Institute doing research on eye pigments, Verschuer personally had nothing to do with it. Both statements were, of course, lies. Even more incredibly, he claimed to have been an open critic of the Nazis racist fanaticism.[118]

Kurt Gottschaldt, another KWI eugenicist who had enjoyed Eugen Fischer's support and funding under the Nazis,[119] saw which way the wind was blowing and began allying himself with Muckermann. Privately, Verschuer portrayed Gottschaldt's "poisonous" attack on him as a personal vendetta.[120] He then tried to smear his critics: he tried to convince a Miss Hupfer, probably a secretary

---

[112] Freiherr von Verschuer to Fritz Lenz, 20/9/45, Va 165, MPA.

[113] Verschuer to Lenz, 27/2/46, Va 165, MPA. See also: Lenz to Nachtsheim, 11/8/1946, N 18, MPA. Kühl suggests that Nachtsheim was in fact trying to protect his former boss (see Kühl, *Nazi Connection*), but this was contradicted by Verschuer at the time as well as by his son forty years later. For the latter, see Müller-Hill, *Murderous Science*, 116–9.

[114] Quoted in Proctor, *Racial Hygiene*, 307.

[115] "Vertriebene Wissenschaft," *Neue Zeitung* 35 (May 3, 1946).

[116] Kröner, *Von der Rassenhygiene zur Humangenetik*, 98.

[117] Ibid.

[118] Verschuer to the *Neue Zeitschrift*, n.d., A2-II-56, MPA.

[119] Proctor, *Racial Hygiene*.

[120] Otmar Freiherr von Verschuer to Fritz Lenz, 1/10/46, Va 165, MPA.

or an assistant, to go on record stating that she had paid for Gottschaldt's subscription to the SS.[121] Verschuer's talent for backstabbing was not one born of desperation: in August 1944, in a move worthy of the pettiest *Blockwart*, he denounced the janitor at his institute to the police because the tires of the Institute's bicycle were flat.[122]

Part of Verschuer's efforts at reconstitution involved shading, and then shifting, his intellectual efforts. Verschuer would eventually give up on eugenics altogether, but after the war, he hoped to reestablish both the discipline and his career in Germany along more acceptable moral and ideological lines. Language once used to describe the mentally handicapped and Jews was turned against the subject of eugenics itself:

> We have to lay bare the foundational sources of our discipline and cleanse it of all its pollutants. I am confident that eugenics can recover from the deep crises into which world events have thrown it. In the last few years I have, because of the horrifying abuses to which politicians have subjected eugenics under the now-fully corrupt name of "Race Hygiene," withdrawn more and more from the subject and concentrated on the apolitical study of genetics [*Erbpathologie*]. I now sense a stronger duty to serve the pure and true ideas of eugenics. Given the unspeakable suffering of mankind, and the disastrous situation of our people, I cannot withdraw into the isolated world of scholarship. Instead, I must do something in the hope that I can help people in their crisis.[123]

From a man who had welcomed Mengele into his home for Sunday lunches, the moral fears and human empathy were new, but the strategic motivation lying behind them had a long pedigree.

Verschuer's postwar pleading did not save his Frankfurt institute, but he did not suffer much personal adversity. The Americans' veto kept him unemployed for six years – the closest Verschuer came to justice – but the de-nazification tribunals, in the end, classified him merely as a "fellow traveler" (*"Mitläufer"*). In 1951, he was welcomed back into the academy. The University of Münster offered him a prestigious chair in human genetics, and he went on to build one of the largest centers for genetic research in West Germany.[124] He was shortly thereafter elected president of the Germany Society for Anthropology.[125] In his 1965 retirement speech at the University of Münster, Verschuer filled the room with lofty platitudes, only slightly less grating than they were in 1946, about human worth and scientific responsibility:

[121] Freiherr Otmar von Verschuer to Fritz Lenz, 8/11/46, MPA.

[122] Müller-Hill, *Murderous Science*, 76. The term *"Blockwart"* translates as "block warden" or "apartment superintendent," but it has far more negative connotations in German, referring to the National Socialist or East German official who would actively collaborate with the respective regime in spying on residents.

[123] Verschuer to Lenz, 27/2/46, Va 165, MPA; Verschuer to Lenz, 21/1/46, Va 165, MPA.

[124] Proctor, *Racial Hygiene*, 308.

[125] Kühl, *Nazi Connection*, 103.

Any attempt to select out or intervene in the heredity [*Erbsubstanz*] of man is absolutely utopian. Even on purely scientific grounds, such proposals have to be rejected because of insufficient evidence in their favor. But beyond this, man is spiritual and a social being defined by his belief in and responsibility [*Verantwortung*] to God. Eugenics can never neglect these basic truths: human worth, love of neighbor and responsibility before God! It is through research we best arm ourselves against every ideological manifestation of eugenics.[126]

Having bathed the room in the hypocrisy of his parting words, Verschuer retired. He was feted at his death – with, of course, no mention of his wartime activities – as one of the key postwar German scientists in the field.

The sorry Lenz-Verschuer saga was repeated time and time again across early postwar Germany. Those who had provided intellectual justification for mass sterilization and murder, or at least allowed themselves to be used in the justification of those ends, saw themselves reinstated. Lenz took up his position in Göttingen, although admittedly at a nontenured and badly paid one; the university rector clearly did not see him or his work as a priority.[127] Lenz nonetheless published well into the 1970s. Racial theorists Friedrich Burgdörfer, Egon von Eickstedt, and Günther Just were appointed to positions in Tübingen, Mainz, and Munich, respectively (although the last in the field of statistics and not as a professor).[128] Nachtsheim went on to a chair at Berlin's Free University, the buildings of which happened to include the old KWI facilities. His work there, according to one scholar, "on rabbit eye color, epilepsy and malformations paralleled Mengel's horrific human butchery."[129] His 1966 publication, *Kampf den Erbkrankheiten* (*The Battle against Heredity Diseases*), suggested a continuing concern with eugenics.

Although Verschuer's career flourished after the war, the old goal of social engineering through sterilization did not. He and other eugenicists changed their research topics, moved into related disciplines, or cleansed old research interests of now-discredited ideological trappings. Like eugenicists in the United States, they had their greatest success after they abandoned eugenics (see subsequent chapters), at least with respect to their rhetoric, and moved into related fields of population policy and family policy. Hans Harmsen, who did more than anyone else to reconcile forced eugenic sterilization with Christianity, had told the Protestant Inner Mission in 1931, "[t]he exaggerated protective measures for the anti-social and less valuable stemming from a misdirected humanitarianism have led to an ever stronger increase in the anti-social groups in the population."[130] Harmsen also co-founded and became the first President of Pro-Familia, Germany's branch of Planned Parenthood after the war.

---

[126] "Menschenwürde – Verantwortung" ("Human Dignity – Responsibility") *Münstersche Stadtanzeiger*, February 26, 1965.
[127] See Kröner, *Von der Rassenhygiene zur Humangenetik*, 65–70.
[128] Proctor, *Racial Hygiene*.
[129] Weindling, "Survival of Eugenics," 648.
[130] Quoted in Burleigh, *The Third Reich*, 362.

The move out of eugenics and into family and population policy was hardly unique to Germany – if anything, that shift was found to an even greater degree in the United States. There was, however, one important difference between the two countries: whereas Germany ended coerced eugenic sterilization in 1945, the United States did not. Indeed, in some states, it was just then reaching its peak.

9

Revival and Recovery

Eugenics in New Clothes

> Eugenic goals are most likely to be attained under a name other than eugenics.[1]
>
> Frederick Osborn, *The Future of Human Heredity*, 1968

By the late 1940s, eugenics' days seemed to be numbered. Discoveries in genetics in the 1920s and 1930s had challenged the hereditarian argument,[2] new psychoanalytic methods provided psychiatrists and psychologists with greater optimism about their ability to treat mental illness,[3] and Nazi atrocities cast eugenics and eugenic sterilization into ill repute.[4] One by one, the foundations of eugenics began to weaken, and its crowning achievement – coerced sterilization – faced its most serious challenge.

As it did, pro-sterilization advocates, who were still overwhelmingly eugenicists, did not quietly retreat from their project; for many of them, it amounted to a lifetime of work. In institutions, many superintendents and boards of eugenics carried on much as before, except that, in many cases, their level of activity expanded. But, for public intellectuals and advocates of sterilization, the challenges were more enduring. They were compelled to regroup both politically and intellectually in recognition of the severe threat to their project's credibility posed by the German experience.[5]

---

[1] Frederick Osborn, *The Future of Human Heredity; An Introduction to Eugenics in Modern Society* (New York: Weybright and Talley, 1968), 104.

[2] Kevles, *In the Name of Eugenics*, chapter 13; Connelly, *Fatal Misconception*, 55.

[3] Dowbiggin, *Keeping America Sane*, 110–16.

[4] Those atrocities, however, did not wholly delegitimize these topics, as segregation in the U.S. South into the postwar era makes clear.

[5] In April 1948, for instance, the executive director of Birthright, George Rundquist, wrote to Dr. Tage Kemp of the University Institute for Human Genetics in Copenhagen. Referring to the "abuse of sterilization" in Germany as a "stock argument against" eugenic sterilization, Rundquist requested any information on the "German attempt to improve the quality of the race." Rundquist to Tage Kemp, April 26, 1948, 2/15, AVS. See also Marian S. Olden to Sheldon Glueck, December 10, 1947, 2/14, AVS. Quoting Olden: "You ask if I am sure that the Nazis

Pro-sterilization advocates faced two challenges: (1) coping with scientific challenges to their hereditarian arguments beginning in the 1930s, and (2) reconciling themselves, tentatively from the 1940s and robustly from the 1960s, to a new, rights-based culture. In a way that they probably did not predict, some of the strongest supporters of eugenic sterilization addressed the hereditarian challenge in the interwar period in a manner that left them better prepared to cope with the rights culture in the postwar period.

THE SCIENTIFIC CHALLENGE

The scientific challenge was the less straightforward of the two because of the way in which scientific arguments filter down (or do not) from their originators to policy makers and the public. It is arguable that, as scholars of the various disciplines that lent intellectual support to eugenics (or even grew out of it) concluded that eugenics' scientific pillars were collapsing, they withdrew their support and left the once-universally admired movement to the cranks. This process began in the 1920s, leading many observers to conclude that eugenics had intellectually peaked in the 1910s. As a serious subject of academic study, eugenics was in freefall by the 1930s.[6]

Students of the eugenics movement have assumed, logically enough, that the project of coerced sterilization declined in tandem with the ideology that justified it.[7] "Enforcement of United States sterilization laws plummeted sharply in the early 1940s," writes historian Daniel Kevles, "and was minuscule by 1950."[8]

In making this empirical inference, however, scholars have assumed a tight correspondence between the development of scientific knowledge and its dissemination to and interpretation by politicians and law makers. None exists.

did not use sterilization as a political measure. All I can say is that I have yet to find a shred of evidence that they did so use it, though I am searching constantly for the facts. Being ruthless, they employed more brutal measures it seems." Either Olden was being disingenuous, or her constant search for the facts failed to extend to the Nuremburg Doctors' Trial, during which senior National Socialist physicians were tried and indicted for war crimes that included sterilization experiments (Count II). The trial had ended in late August 1947. In an earlier letter to Glueck, Olden similarly wrote, "The myth that sterilization was one of the Nazi atrocity measures has received impetus from many writers who are usually more careful and accurate." See Marian S. Olden to Sheldon Glueck, November 6, 1947, 2/14, AVS. In the 1930s, Glueck and his wife, now famous for their studies on delinquency in Boston, conducted a study on female delinquency. Using the family history method pioneered in the Jukes and Kallikaks studies, they recommended coerced sterilization. Sheldon Glueck and Eleanor T. Glueck, "Five Hundred Delinquent Women," 1934, 4/11, AVS.

[6] Kevles, *In the Name of Eugenics*, chapter 11; Barker, "The Biology of Stupidity"; Dowbiggin, *Keeping America Sane*; Sean A. Valles, "Lionel Penrose and the Concept of Normal Variation in Human Intelligence," *Studies in History and Philosophy of Biological and Biomedical Sciences* 43, no. 1 (2012): 281–9.

[7] Loren R. Graham, "Science and Values: The Eugenics Movement in Germany and Russia in the 1920s," *American Historical Review* 82, no. 5 (1977): 1158; Gould, *Mismeasure of Man*; Degler, *In Search of Human Nature*.

[8] Kevles, *In the Name of Eugenics*, 169.

Simply put, although geneticists may have been convinced that feeblemindedness was not hereditary, it took a long time before everyone else also reached that conclusion. Policy makers, journalists, and activists do not always keep abreast of the latest intellectual developments. As political scientists have demonstrated, once institutional arrangements and routines are established for a policy, they prove resilient.[9] There is almost always lag time between the making of a scholarly discovery and its dissemination to the public and policy-making community. This lag was, if anything, longer than usual in the case of eugenics. Eugenics resonated so loudly because it granted legitimacy, through nothing less than science itself, to people's fears and prejudices. Furthermore, it had the endorsement of the U.S. Supreme Court. These fears did not vanish because geneticists developed more sophisticated explanations of heredity's essential complexity or because new techniques in psychiatry turned the practice from its previous therapeutic pessimism. On the contrary, the more sophisticated and nuanced the arguments became, the less likely they were to filter out of the academy. Part of the appeal of eugenics arose because, once stripped down to its hereditarian and apocalyptic cores, the ideas it expressed were easy to understand.[10] From the 1930s to the 1960s, lay people were casually working with assumptions that, for many scientists, were long disproven.

Case in point: in 1940, the state of New Jersey's "Special Committee for the Study of Eugenic Legislation" published its report. The committee's minority report was twice as long as its majority counterpart, and it made what are now familiar arguments about the complicated and unpredictable nature of heredity. The majority report, conversely, confidently concluded that feeblemindedness is "a proven inheritable characteristic [and] all the children of a feebleminded couple are doomed to be feebleminded."[11] Two years later, a bill introduced in March 1942 into the New Jersey legislature enshrined eugenic and hereditarian language:

> Since human experience has demonstrated that mental deficiency, familial mental disease, familial epilepsy, familial blindness, familial deaf-mutism, familial gross deformity and familial neurological diseases are genetically transmitted; and since the State of New Jersey has, both at large and in custodial care, in various private and State institutions many defective persons and those suffering from mental deficiency, familial mental disease, familial epilepsy, familial blindness, familial deaf-mutism, familial gross deformity and familial neurological disease who would become by the propagation of their kind a menace to society but who if incapable of

---

[9] Steinmo, Thelen, and Longstreth, eds., *Structuring Politics*; Paul Pierson, "Increasing Returns, Path Dependence, and the Study of Politics," *The American Political Science Review* 94, no. 2 (2000): 251–67; and Thelen's case study of German training and vocational education systems in *How Institutions Evolve: the Political Economy of Skills in Germany, Britain, the United States, and Japan* (Cambridge: Cambridge University Press, 2004).

[10] See the valuable discussions in Béland and Cox, eds., *Ideas and Politics*.

[11] W. MacMillan, *Report of the Special Committee for the Study of Eugenic Sterilization*, 1940, Special Collection, Medical Society of New Jersey, AVS.

procreation might become self-supporting with benefit to themselves and society; and since sexual sterilization may be effected in males by the operation of vasectomy and in females by the operation of salpingectomy ... it is hereby declared to be the policy of the Legislature to ameliorate the living conditions of afflicted individuals and to promote the social welfare by sexual sterilization of persons suffering from mental deficiency, familial mental disease, familial epilepsy, familial blindness, familial deaf-mutism, familial gross deformity and familial neurological disease under careful safeguard and by competent conscientious authority.[12]

The bill was referred to the Ways and Means Committee. The General Assembly's Roman Catholic speaker ensured that it died there.[13] The language and concepts in the bill retained the standard expectations for sterilization: it would benefit society by preventing procreation among those liable to produce offspring with serious and incurable illnesses. Furthermore, by addressing the "living standards of afflicted persons," it would permit those sterilized to live independently in society without fear of acquiring dependents.

Policy makers had paid at best passing attention to intellectual trends in the hard sciences, even when the results had been laid out for them.[14]

THE RIGHTS CHALLENGE: DEVELOPING NEW ARGUMENTS TO JUSTIFY OLD ENDS

Supporters of developing a strong state policy of sterilization were neither blind nor indifferent to the weakening hereditarian case or to the enduring taint of National Socialism.[15] As Ezra Gosney, founder of the Human Betterment Foundation, wrote in 1940 with characteristic opacity, "we have little in this country to consider in *racial integrity*. Germany is pushing that. We should steer clear of it lest we be misunderstood."[16] With scientific support for eugenics declining, the chief lobby groups in the United States developed and propagated new arguments defending coerced sterilization. These arguments, in turn, attracted new supporters and new funding. Equally importantly, they provided intellectual cover for continuing sterilization under prewar laws. Emboldened by new arguments, mental health superintendents continued to recommend patients for sterilization. Indeed, Paul Popenoe felt sufficiently emboldened in

---

[12] State of New Jersey, Draft Sterilization Bill, Introduced March, 1942 by Mr. Shepard, GRO133/6, BCA.

[13] Irene Headley Armes to Mattie DeV. Ford, March 29, 1949, 2/16, AVS; Dowbiggin, *Sterilization Movement*, 38.

[14] Thus, in 1929–30, a New York State Committee casually remarked: "Not all mental deviates owe their condition to heredity, but it is reasonably certain the children of insane or feebleminded will resemble their parents." New York State Committee on Birth Control and Sterilization (1928–30), "Sterilization," 13/104, AVS.

[15] See Rundquist to Kempe, April 26, 1948, 2/15, AVS; Marian S. Olden to Sheldon Glueck, October 20, 1947, 2/14, AVS; Mabel Law to the Editor, *Time Magazine*, July 1, 1948, 2/15, AVS.

[16] E. S. Gosney to Frank Reid, September 9, 1940, 1.2, Gosney Papers.

early 1949 to confide to a birth control activist his belief that "[t]he public is more ready now to talk about sterilization, than it was during Hitler's heyday."[17]

All of this meant that the political turn against eugenics was more fitful, more drawn out, and ultimately less complete than the scientific shift that had preceded it.

These new justifications for coerced sterilizations took two approaches. The first, anticipating and then embracing the postwar rights culture, was individualistic: it centered on the right of the unborn child to be born to adequate parents (the subject of this chapter). The second justification was social: it concerned the consequences of unfettered population growth (the subject of the next chapter). All of this was wrapped in a new rhetorical packaging: "eugenic" was discreetly dropped and replaced with "selective."

THE RIGHTS OF THE CHILD

The main interwar lobby groups played the most important role in recasting the argument in favor of eugenics. In the 1930s, there were three main American eugenic organizations: the American Eugenics Society, the Human Betterment Foundation (HBF), and the Sterilization League of New Jersey. The American Eugenics Society had about six hundred members, and its interests were as varied as those of its director, Irving Fisher: temperance, nutrition, and human breeding, including sponsoring the famous "Fitter Families" fairs, at which families' IQs and eugenic-genealogical histories were tested and the best given a prize.[18] It was an ineffective organization, and a lack of funds almost forced it to disband in 1932.[19]

The HBF, which was funded by the wealthy pro-eugenicist Ezra Gosney, claimed to have a broader mandate,[20] but it and other groups like it were essentially pro-sterilization advocacy groups. Paul Popenoe directed the HBF

---

[17] Paul Popenoe to Dr. Dickinson, January 2, 1949, 30, Private Papers of Robert L. Dickinson, Center for the History of Medicine, Francis A. Countway Library of Medicine, Harvard University, Cambridge, MA.

[18] Kevles, *In the Name of Eugenics*, 172. On "Fitter Families," see Robert W. Rydell, *World of Fairs: The Century-of-Progress Expositions* (Chicago: University of Chicago Press, 1993), chapter 2; and Katherine Swift, "Sinister Science: Eugenics, Nazism, and the Technocratic Rhetoric of the Human Betterment Foundation," *Lore* 6, no. 2 (May 2008): 1–11. On the role of genealogical studies as a tool of eugenic fieldwork, see Patrick J. Ryan, "'Six Blacks from Home,': Childhood, Motherhood, and Eugenics in America," *Journal of Policy History* 19, no. 3 (2007): 256.

[19] "People's Foundation Invitation to Colonel Woods to become a Member of the Board of Governors," April 11, 1932, RF 176–189, 3/2 Box 8, RAC. Irving Fisher was one of the organization's two chief contributors.

[20] The organization also claimed "to foster and aid constructive and educational forces for the protection and betterment of the human family in body, mind, character, and citizenship." David A. Valone, *"Eugenic Science in California: Guide to E. S. Gosney Papers and Records of the Human Betterment Foundation* (Pasadena: California Institute of Technology, 1996).

and sponsored a series of studies that, without fail, concluded that eugenic sterilization was safe, cheap, and beneficial to the "race." It was a cheerleader for California's eugenic sterilization legislation, a law that Gosney himself believed to be exceeded only by Nazi Germany.[21] The HBF would later be folded into the Sterilization League of New Jersey.

The Sterilization League of New Jersey was the brainchild of its founder, Marian S. Olden. She was one of the most extraordinary figures in a movement exhibiting no shortage of them. Olden was, in the words of one historian, "[i]ntelligent, attractive, determined, opinionated, yet at the same time touchingly vulnerable."[22] Like many others in the eugenics movement, early experience informed her commitment: witnessing the deplorable conditions in state mental hospitals and the destructive effect of poverty on families converted her to the cause of coerced sterilization.[23] Her marriage to a Princeton professor took her to New Jersey in the 1930s. It was during this time that superintendents of homes for the feebleminded convinced her, as they convinced many others, of the need for eugenic sterilization. They told her that the hereditarily defective were "scattering" throughout New Jersey and "menac[ing] our better stock."[24] She became a convert, and would until her last days display all the zeal of one. She joined the progressivist League of Women Voters, spearheading its efforts to collect signatures from physicians supporting a eugenic sterilization bill.[25] Through it, she received an invitation to set up a social hygiene chapter.[26] Also under the aegis of the League, Olden gave a class on heredity and social problems, lectured at state institutions, and authored five booklets.[27] With the support of the director of the New York School of Social Work, Olden founded the Sterilization League of New Jersey in Trenton, in January 1937.

Olden led efforts by the League of Women Voters to draft a sterilization bill,[28] collected signatures from physicians who supported it, and helped draft the 1942 bill that died in committee.[29] Olden quickly became convinced that the cause of eugenic sterilization was too important to be subsumed under the League of Women Voters. "Wherever sterilization was but one of several objectives in a platform," she wrote, "it was sure to be abandoned in favor of more popular measures when pressure was applied."[30] The Sterilization League's views were

[21] Gosney to Jon Alfred Mjoen, Norway, January 2, 1936, 10.16, Gosney Papers.
[22] Dowbiggin, *Sterilization Movement*, 34.
[23] Ian Dowbiggin, "'A Rational Coalition': Euthanasia, Eugenics, and Birth Control in America, 1940–1970," *Journal of Social Policy History* 14, no. 3 (2002): 227.
[24] Quoted in Dowbiggin, *Sterilization Movement*, 36.
[25] Marian S. Olden, *Birthright, Inc. – Its Roots, Fruits, and Objectives*, 3, 1945, 9/9 AVS; Reilly, *Surgical Solution*.
[26] Ibid.
[27] Olden, *Birthright, Inc. – Its Roots, Fruits, and Objectives*, 3.
[28] See *Points to Incorporate in a Model Sterilization Bill*, 1942, 9/9, AVS.
[29] Reilly, *Surgical Solution*, 131.
[30] Olden, *Birthright, Inc. – Its Roots, Fruits, and Objectives*, 4.

hateful. Its 1937 publication, *Selective Sterilization in Primer Form*, included grotesque photographs of the severely physically and mentally handicapped and bears great resemblance to the National Socialist *Ich klage an* (*I accuse*). The foreword introducing the photographs assures the reader that

> if every voter saw the squatting, slobber-soaked idiots tied to benches, those that rock themselves back and forth all day or those that endlessly box with the air ("shadow boxing"), and the women so depraved that at the sight of a visitor they pull their breasts forth and try further exhibitionism. In our institutions hundreds and hundreds of caricatures of humanity are being spoon-fed soft pap to keep them alive, even when they are only what the resident physician termed "hunks of protoplasm."[31]

For the League, the solution to these ills – indeed, a prerequisite for solving any social ill – was plain: "the better type in every social class of every race should be encouraged to increase; the worst type should be helped to die out."[32] For Olden, as for so many other eugenicists, the chief obstacle to implementation was the Roman Catholic Church. Olden bitterly attacked the Church for standing in the way of eugenics, birth control, and sterilization but also, she believed, for plotting to use Roman Catholics' high fertility rate to make them a majority in America capable of dictating government policy.[33] That Catholics seemed to contribute a disproportionate number of "defectives" made a counterattack against Catholicism all the more urgent.[34] The time had come, she declared in 1947, to "expose the methods of that church which acts as an obstacle to progress in every form."[35] The most recent examples of such obstructionism were then to be found in Wyoming and Pennsylvania, where Catholic Church leaders stymied early postwar efforts to introduce coerced sterilization bills into the legislatures.[36] And, as we saw in Chapter 7, well-organized Catholic opponents of eugenic sterilization played a role in defeating attempts to pass laws in Ohio and elsewhere.

Olden's Sterilization League began its work too late to see the adoption of new sterilization laws; the high tide of interwar eugenic legislation had already ebbed. The League did not secure the adoption of eugenic sterilization in New Jersey, with its large Roman Catholic population, or anywhere else. Georgia passed the last eugenic sterilization law in the United States in 1937. Although Olden's bill made no progress in the New Jersey legislature, her organization was not too late to play an important role in providing arguments for the

---

[31] *Selective Sterilization in Primer Form*, 1, 9/9, AVS.
[32] Ibid., 5.
[33] Dowbiggin, "'Rational Coalition,'" 227. This old canard has recently been cited to explain Roman Catholic support for the legalization of illegal Mexican migrants in the United States.
[34] Dowbiggin, "'Rational Coalition,'" 227.
[35] Quoted ibid.
[36] Dowbiggin, *Sterilization Movement*, 77.

maintenance and rigorous application of those laws that were already on the books in other states.

To do so, the organization, later renamed Birthright, Inc.,[37] had to get rid of Olden. By the late 1940s, a move was under way to push the founder out of the organization. The move was partly motivated by the fact that she was a trying colleague.[38] Her desire to control all aspects of the organization (one colleague compared her to a "dictator") made administration chaotic and demoralizing.[39] She insisted that she had the right to read all correspondence coming to the organization and to vet all replies going out.[40] Two female members of staff broke down in tears after one confrontation with her in late 1947 or early 1948,[41] and an exasperated H. Curtis Wood, president of the organization from 1945, threatened mass resignation. He wrote to Olden in January 1948,

> If you feel you must continue this battle for the sake of your ego I can assure you that most of us will drop out of the picture entirely, as we are just not willing to give our time to these fruitless and upsetting personality problems. It is Birthright we are concerned about and not the emotional mal-adjustments of our staff, regardless of the causes and justification for them.[42]

Men such as Wood and later Robert L. Dickinson, Hugh Moore, and Alan Guttmacher, were ill inclined to take orders from women and showed little respect for the female revolutionaries who had founded the birth control movement.[43]

But the conflict was more than personal. Olden was an unreconstructed supporter of eugenic sterilization, and she was not one to mince words when making the case in favor of a cause. This put her in conflict with younger

---

[37] In 1943, the organization changed its name, briefly, to the "Sterilization League for Human Betterment" and then to "Birthright, Inc." (which should not be confused with the present-day pro-life organization, Birthright International). Although intellectual shifts dating to the 1930s were behind the rebranding, the trigger was Ezra Gosney's death on September 14, 1942. His daughter, Lois Castle, agreed to transfer the Human Betterment Foundation's records to the East Coast. As a condition, she insisted that Olden's organization drop "Human Betterment" from the title to avoid confusion with her father's organization. Marian S. Olden to Mrs. Castle, February 4, 1947, 2/14, AVS. Olden resented the change (see Birthright, Inc., *Report to Members, January 1 to April 29, 1948*, 2/14, AVS), but the name was better insofar as it captured the organization's new justification for sterilizing people with developmental disabilities.

[38] According to the new executive director of Birthright, appointed on February 24, 1948, "it seems that Mrs. Olden is both Birthright, Inc. and Mrs. Olden; therefore everything is personal. I listened to her problems and difficulties, mostly her relations with other people." George E. Rundquist to Dr. H. Curtis Wood, Jr., March 9, 1948, 2/15, AVS. See also the delicately worded letter from H. G. Wood to Mrs. Olden, April 18, 1948, 2/15, AVS and Lovett Dees to Marian S. Olden, December 30, 1947, 2/14, AVS.

[39] H. Curtis Wood Jr. to Clarence Gamble, November 5, 1949, 2/16, AVS.

[40] H. Curtis Wood Jr. to Olden, January 20, 1948, 2/16, AVS.

[41] Ibid.

[42] Ibid.

[43] On this, see Dowbiggin, *Sterilization Movement*, 39.

members of the organization who wished to adopt softer language and new arguments. Above all, they wished to make the rights of the child, rather than the threat of the feebleminded, the basis of their case in favor of sterilization. In a pamphlet entitled *If the Unborn Could Choose Their Parents!*, the organization repackaged the argument for coerced sterilization:

> Should we not protect the Birthright of the unborn from violation by defective and irresponsible parents?
>
> If the unborn were given the opportunity of choosing their parents would they elect a mother whose feeble wits indulge themselves in the careless drowning of her baby in the bath tub, a schizophrenic father that in recurrent fits of insanity hurls a child against a wall, or a family where already are to be seen neglected fragments of intelligence clothed in pitiful little bodies? Would the unborn feel that parenthood is an *inherent* right to every person regardless of his lack of intelligence and emotional stability? Would unborn children willingly sacrifice a happy, intelligent home and a good heredity for feeblemindedness, neglect, and institutional care so that mentally defective persons might have the *right* to parenthood?[44]

These rhetorical arguments were followed by a raft of statistics and arguments, unchanged since the 1920s, on feeblemindedness, heredity, and differential fertility. An unaltered argument had been repackaged. In its platform,[45] and wherever else possible, Birthright disingenuously quoted a 1930 statement supposedly made by then-President Herbert Hoover at the 1930 White House Conference on Child Health and Protection: "There should be no child in America that has not the complete birthright of a sound mind in a sound body and that has not been born under proper conditions."[46] Others gladly joined the chorus. Smaller organizations across the United States made the same argument.[47] In 1936, the president of the American Association for Mental

---

[44] *If the Unborn Could Choose Their Parents!*, n.d. [likely 1950], 9/9, AVS.

[45] *Platform of Birthright, Inc.*, 1943, 9/9, AVS.

[46] Throughout the 1940s, the quotation was on Birthright's letterhead. See 2/14 and 2/15, AVS. The quotation in fact did not originate at the 1930 White House Conference on Child Health and Protection. Hoover had been president of the American Child Health Association (ACHA) from 1922, at which time he formulated a simple bill of rights for children. This was expanded to include the "sound mind" and "sound body" references in May 1928, six months before his November election as president. This exact quotation was not part of the 1930 conference, and, more importantly, Article XIII of the 1930 Conference charter states: "For every child who is blind, deaf, crippled, or otherwise physically handicapped, and for the child who is mentally handicapped, such measures as will early discover and diagnose his handicap, provide care and treatment, and so train him that he may become an asset to society rather than a liability. Expenses of these services should be borne publicly where they cannot be privately met." There is no evidence that Hoover would have condoned Birthright's use of his words. See Dominique Marshall, "Children's Rights and Children's Action in International Relief and Domestic Welfare: The Work of Herbert Hoover between 1914 and 1950," *Journal of the History of Childhood and Youth* 1, no. 3 (2008): 351–88.

[47] Human Betterment Foundation, Des Moines, Iowa, "1,400,000 mental defectives: a Problem for Social Workers," 1950, and Human League Foundation of North Carolina: "Selective Sterilization," n.d. [likely 1950], both found in AVS Pamphlets, Misc. Box 34, AVS.

Deficiency declared, "the most powerful argument for sterilization today is that which urges that no feeble-minded person is fit to be a parent, whether or not his condition is heredity."[48] Birthright had, it seemed, backed a winner.

CLARENCE G. GAMBLE

Neither Birthright nor its institutional predecessor was a wealthy organization, and they relied, as the birth control movement did generally, on wealthy bene-factors. This reliance increased as well-endowed foundations began withdraw-ing their support. The Carnegie Institute, increasingly embarrassed by Laughlin's extremism and worried about the growing scientific challenge to eugenicists' methods, had shut down the Eugenics Record Office in 1939, and, six years later, it rejected Birthright's request for support.[49] At this critical juncture, Clarence J. Gamble, heir to the Procter & Gamble Ivory Soap fortune, made an appearance. Following Harvard and a failed medical career, in the 1930s, Gamble became a supporter of birth control as a tool of social engineer-ing.[50] The timing was more than coincidental: Gamble opposed Roosevelt's New Deal and argued that the President's policies merely exacerbated a problem that only birth control could solve. As one historian put it, for Gamble "federal relief offered only to ameliorate the system; birth control went to the heart of the problem – the poor were having too many children, and the better sorts of people were not."[51]

Gamble would become a key figure both in the postwar American birth control movement and in Birthright's efforts to refound its pro-sterilization argument. In the 1940s and 1950s, Gamble fought a series of pitched battles with Birthright and its successor, the Human Betterment Association of America (HBAA). These arguments had nothing to do with ends; Gamble, like Birthright generally, sup-ported eugenic sterilization. In 1949, he bemoaned the low number of steriliza-tions relative to the size of the insane and feebleminded populations:

> By a conservative estimate, feebleminded persons constitute at least 1 percent of the population. The assumption that they have an average life span of 50 years indicates that each year there are at least 20 additional feebleminded persons per 100,000 population. *This is nine times the 901 sterilizations of the feebleminded reported for the twenty-seven states and nearly twice those in the most active state.*[52]

The arguments were, rather, about means. There were two steps in Gamble's campaign for sterilization. The first involved taking the campaign away from the

[48] Dowbiggin, *Sterilization Movement*, 30.

[49] Ibid., 49.

[50] Donald T. Critchlow, *Intended Consequences: Birth Control, Abortion, and the Federal Government in Modern America* (New York: Oxford University Press, 1999), 34.

[51] Ibid., 34–5.

[52] Clarence Gamble, "Preventative Sterilization in 1948," November 12, 1949, 141, AVS Pamphlets Misc. Box 34, 1–4, 2–3. Emphasis added.

East Coast and into the South. To avoid Southern suspicions that eugenic sterilization was some sort of Yankee conspiracy, Gamble had successfully encouraged the foundation of Human Betterment Associations throughout the South and refused to integrate the local organizations into Birthright.[53] Gamble established and funded the first birth control programs in West Virginia, North Carolina, and Florida.[54] These programs created nothing like the opposition they would have in the north because (a) there were few Roman Catholics (Louisiana excepted) in the South, and (b) it was no secret that the controlled births would chiefly be black.[55]

The second step involved dragging Birthright into the postwar era by urging it to take seriously and to advertise the importance of parental responsibility and the rights of the child as justifications for coerced sterilization. In June 1947, facing severe resistance from Olden, Gamble resigned as director of Birthright but continued his campaign.[56] He argued at length with Irene Headley Armes, a member of Birthright/HBAA's Board of Directors, over the name of the organization, its aims, its methods, and whether he would integrate his federations. In response to Gamble's argument that one of the advantages of sterilization is that *"most important of all, potential children are protected,"* Armes stated,

> These words in italics bring us from a practical publicity emphasis into the imaginary world of never-never land – how can you protect children who are not born and who, if the feebleminded are sterilized, will never be born? The only people being protected are the generations of the future as to the level of their mentality and, possibly, the taxpayer's pocket-book.[57]

Although Armes was speaking the truth, Gamble won the argument. A month later, Armes offered an argument similar to Gamble's in a letter to Douglas Arant, an Alabama lawyer drafting a sterilization bill for presentation to the state legislature:

> Our organization believes that even more important than the hereditary transmission of mental defect is the fact that no feebleminded person can be or should be expected to bring up children, of whatever ability, and for this reason we strongly feel that from this point of view sterilization is even more unassailable.[58]

---

[53] H. Curtis Wood to Irene Headley Armes, October 5, 1950, 4/113, AVS; H. Curtis Wood to Clarence Gamble, September 1, 1950, 4/33, AVS.

[54] Critchlow, *Intended Consequences*, 35.

[55] Ibid., 36 for point (b).

[56] On the resignation, see Clarence J. Gamble to the Board Directors of Birthright, June 14, 1947, 2/14, AVS. A month before the resignation, Robert Dickinson roundly condemned Birthright (which, in a clumsy effort at humor, he suggested be rechristened "Birthwrong"): "It seems to me a grievous error on the part of our organization to unfriend one who has worked for, and can do much in the cause of sterilization."

[57] Irene Headley Armes to Gamble, October 5, 1950, 4/33, AVS.

[58] Irene Headley Armes to Douglas Arant, November 9, 1950, 4/33, AVS; *To Protect the Retarded Adolescent*, n.d. AVS, Pamphlets Misc. Box 34, AVS.

Armes continued in her role as executive director of HBAA until her death in 1955. By the time Gamble and Birthright parted ways in 1957, the organization, by then with yet another name, had buried its overtly hereditarian rhetoric.[59]

But it had not rid itself of its hereditarian intent. Birthright's commitment to the new rights culture was rhetorical and instrumental. Birthright officials were glad to make the hereditarian case, which they continued to believe, in select circles. A Birthright pamphlet from 1949 or 1950, entitled *Wanted – Increased Action*, said it all. The pamphlet bemoaned the derisory number of sterilizations occurring in the United States – about 2 per 100,000 of the population, or 3,000 per year.[60] It compared this to an average annual increase in the number of feebleminded of 45.5 per 100,000 and noted that Virginia came closest to "protection of the feebleminded" with 13 sterilizations per 100,000.[61] Assuming – as we must – that Birthright wanted to reach the figure of 45.5, the organization was, in effect, calling for almost 70,000 sterilizations per year. The implied sterilization of 1,300 people per week across the United States would have approached National Socialist ambitions.

MENTAL HEALTH SUPERINTENDENTS: THE CASE OF FRED BUTLER

These arguments may have meant nothing if people with power had not been prepared to listen. Such people – and, specifically, the superintendents of mental health institutions – were ready to do so. As the argument shifted within pro-sterilization foundations and lobby groups, the end results still lay within the institutions for the feebleminded. One of those who grounded the argument for sterilization in the language of parental responsibility was an unlikely convert to rights-based arguments: Fred Butler.

For thirty-six years, from 1913 until 1949, Butler had served as the superintendent of Sonoma State Home in California. In the prewar years, California led the country in sterilizations. Butler was the architect of this policy at Sonoma and a zealous defender of coerced sterilization. During his tenure, he personally sterilized 5,400 men and women,[62] making him one of the most productive sterilizers in the democratic world.

In 1938, Butler advanced the argument that coerced sterilization was not about protecting the normal from the mentally defective; it was, rather, about protecting the defective and their children. Sterilization allowed the mentally defective to marry (something Butler would likely have viewed with horror in the 1920s):

---

[59] Dowbiggin, *Sterilization Movement*, 54.
[60] The U.S. population at the time was 150,697,361. *Census of Population: 1950* (Washington, DC: U.S. Census Bureau, 1952), http://www.census.gov/prod/www/abs/decennial/1950.html.
[61] Birthright, Inc., *Wanted – Increased Action*, Publication no. 5, 7/62, AVS.
[62] Fred O. Butler to Karl Bowman, January 11, 1951, 4/34, AVS.

[The problem is that] the average defective individual who is sterilized is really incapable of having the responsibility of rearing children. Of this I am more convinced as time goes on. *Regardless of whether their condition can be traced to hereditary or secondary causes,* they should not be given an opportunity to have children on account of their inability mentally, financially and otherwise to properly care for them. One of my saddest experiences is seeing a normal child come to the Institution to see their defective mother. Therefore, I firmly believe that unless a child can be born of normal parents, it is better not be born at all.[63]

The admission that there could be a "normal" child visiting its "defective" mother quietly removed one of the most important pegs from Butler's earlier support for coerced sterilization, but it did not affect his positive attitude toward or his implementation of sterilization policy. A decade after this fulsome embrace of childhood entitlement, Butler threw his efforts into blocking a proposed California law providing judicial oversight of sterilization. California's law, Butler argued, "is not perfect, yet it has been functioning apparently better than any law in the United States ... and thereby has permitted us to sterilize around thirty-five percent of all the cases sterilized in the United States."[64] In 1949 or 1950, Butler, who by then was Birthright's medical director, organized a tour of the states. He congratulated state institutions in which sterilizations had increased, politely chided those in which they had decreased, and sought to judge the scope for new sterilization laws in states lacking them. In organizing his trip, he drafted two slightly different letters: one for institutions showing an increase in sterilizations during 1948 over 1947 and one for those that did not. Both letters contained, however, the same argument in favor of sterilization: "if enough sterilizations are performed, we can certainly look forward to removing some of the hereditary drain on the quality of our population."[65] The main impediment to doing so was the limited nature of the institutions; they could, by definition, only house a fraction of the target population. With this in mind, he wrote in the autumn of 1950 to an Alabama legislator and argued the case in favor of coerced sterilizations of the "mentally deficient" who existed outside mental health institutions and prisons.[66] By focusing only on the institutions, Butler argued, American policy targeted a mere 10 percent of the total population.[67] This would be disastrous for the

---

[63] F. O. Butler, "Sterilization," July 10, 1938, 11.7, Gosney Papers. See also Paul Popenoe to Hardwig-O'Reile, February 11, 1936, Folder 5.5, Gosney Papers.

[64] F. O. Butler to Marian S. Olden, May 6, 1947, 2/14, AVS.

[65] F. O. Butler, letter for institutions showing decrease in sterilizations, n.d., 7/65, AVS; Butler, letter for institutions showing increase in sterilizations, n.d., 7/65, AVS. The words "some of" were added during editing of the second letter.

[66] F.O. Butler to Douglas Arant, November 10, 1950, SWO15.1/72, Alabama, AVS.

[67] Ibid.

United States because, as he casually observed, "we know that the mental defective can never, *so called* recover – once defective always defective."[68]

Public intellectuals sympathetic to sterilization also fell into line with the new rights-sensitive language. One of the most important of these was Butler's fellow Californian at the Human Betterment Foundation, Paul Popenoe. Philanthropist Ezra Gosney, it will be recalled, hired Popenoe to direct Pasadena's Human Betterment Foundation in 1928. Throughout the 1930s, Popenoe was "America's sterilization guru," and his organization produced dozens of papers, articles, and presentations making the case in favor of eugenic sterilization.[69] In 1936, as he was confidently making hereditarian arguments in public, he expressed doubts in private. And these doubts in turn led him to a new concern with parental responsibility. Referring to British geneticist J. S. Haldane's anti-eugenic arguments, Popenoe, argued that "even if mental deficiency and mental disease were not inherited at all, but simply due to [a bad] environment and training, surely it is not to the advantage of society that normal children be brought up by insane and feeble-minded parents."[70]

The shift to nascent rights-based arguments could also be seen among scholarly supporters of coerced sterilization. Gladys C. Schwesinger was a prominent psychologist and co-author, with anti–population-growth lobbyist Frederick Osborn, of a text on heredity and the environment. She gave a talk to the New Jersey Health and Sanitary Association, an organization formed in 1875 and driven by Progressivist concerns – sanitation and hygiene – that shifted easily into eugenic concerns.[71] Schwesinger argued that the eugenicist cause would be better served if its apostles avoided favored phrases such as "pollution of the race," "scum of the earth," "tainted streams," understandable though they were, and made an appeal to eugenicists to refocus their arguments around children. As the whole personality of a child can be marred by the parent, "for this reason the twentieth century insists that the child's right to be protected from the wrong kind of parents is even more 'inalienable' than the adult's right to bring forth children."[72]

The move out of hereditarian and into rights-based arguments was never complete, which is hardly surprising given that the conversion was a matter of presentation rather than a change of heart or mind. Hereditarian arguments continued to be made when eugenicists thought such views would resonate. Both arguments were often made at the same time. Thus, in the early 1950s, the Human Betterment League of North Carolina distributed pamphlets in favor of

---

[68] Ruth Proskauer Smith, Executive Director, HBF, to Miriam T. Tannhauser, Maryland Board of Education, April 11, 1956, SWO15.1/73, Maryland, AVS. Emphasis in original.

[69] Stern, *Eugenic Nation*, 107.

[70] Paul Popenoe to Hardwig-O'Reile, February 11, 1936, Folder 5.5, Gosney Papers.

[71] At its twenty-first annual meeting, the Association sponsored a paper on the "hygienic supervision of school children." *New York Times*, "New-Jersey Sanitary Association; Twenty-First Annual Meeting to be Held in Atlantic City," November 29, 1896.

[72] G. C. Schwesinger, "Sterilization and the Child," December 10, 1937, Folder 11.7, AVS.

"selective sterilizations." The League quoted worthy authorities to make the case. One was Paul Popenoe: "The amount of feeblemindedness in the population probably could be reduced by more than one-third in a single generation if all the frank cases of feeblemindedness were prevented from reproducing." Another was Robert L. Dickinson, later president of the HBAA: "Parenthood is too important an undertaking to be entrusted to men and women who have not the mental, physical, spiritual, and economic capacity for its successful fulfillment."[73] A variety of arguments was invoked to defend sterilization, including protecting the race, saving money, and the avoidance of overcrowding. This newfound concern with the child and the family ensured that long-standing hereditarian arguments carried through to the 1940s, 1950s, and into the 1960s.[74] Even those scholars who show how eugenics dodged the genetic assault of the 1930s suggest that the move into family policy marked a more definitive break with hereditarianism than it did.[75] Rights-based justifications for sterilizations were layered onto older, hereditarian arguments and only gradually came to supplant them.[76] At the same time, eugenicists gladly employed other pro-sterilization arguments when they suited the context – old ones about saving money[77] and new ones about sterilization as a way of protecting the sterilized from procreative burdens.[78] Writing in 1949, Butler

---

[73] Quotations from Human League Foundation of North Carolina: "Selective Sterilization," n.d. [likely 1950], AVS Pamphlets, Misc. Box 34, AVS. The third authority was Oliver Wendell Holmes ("Three generations ... is enough"); the fourth, as befits the Cold War context, was Harvard's Ernest A. Hooton, who informed readers, "The fifth column which offers the greatest threat to democracy is not that of the totalitarian agents but of mental and constitutional inferiors."

[74] Thus, a showcase exhibit organized by Birthright at the University of Pennsylvania from April 14–16 showed a film entitled, "The fatal chain of hereditary disease." Birthright, Inc., *Report to Members, January 1 to April 29, 1948*, 2/14, AVS. In the same document, Sheldon C. Reed, Director of the Dight Institute in Minnesota, says publicly, "those of us who work with heredity in man are a miserably small group but we know that our numbers will grow because the public is beginning to wake-up to the fact that heredity works in man as well as in wheat and oats."

[75] See Kline, *Building a Better Race*.

[76] There were other arguments as well. Throughout the period, eugenicists and other supporters of coerced sterilization would make the argument – a fairly specious one given how often patients were promised and then denied release – that it was in the patients' interests because it ensured their liberty. Bond, "Prophylactic Mental Hygiene in Iowa," n.d. [probably early 1940s], 14/113, AVS.

[77] Clarence Gamble, n.d. [likely 1949], 2/16, AVS; press release from the Human Betterment League of Iowa, April 28, 1949; memo from Irene Headley Armes to Members of Birthright, 1949.

[78] Clarence Gamble and William P. Richardson, "The Sterilization of the Insane and Mentally Deficient in North Carolina," *North Carolina Medical Journal* 8, no. 1 (January 1947): 19–21. The ability of actors to tailor arguments to fit the context was nothing new: after 1900, the same superintendents of mental institutions who wrote in private correspondence of the "sweet defective" capable of learning and progress warned legislators of the menace posed by the incurable feebleminded. See Trent, *Inventing the Feeble Mind*. Even National Socialists argued that sterilization would ease the misery of the afflicted and their families and that the state had a duty to help these unhappy people. Arthur Gütt, "Niederschrift über die Besprechung am 3. November

argued, "[i]n the 27 states having sterilization laws, 24,257 feebleminded persons have been protected from parenthood." He added, for good measure, that doing so had saved more than $1 million.[79]

BIRTH CONTROL AND COERCED EUGENIC STERILIZATION

This rebranding of eugenic sterilization by its most strident proponents induced tighter links between them and their counterparts in the birth control movement. Before the war, all the major advocates of birth control – Clarence Gamble, Margaret Sanger (founder of the American Birth Control League, later called the Planned Parenthood Federation of America [PPFA]), Marie Stopes (the leading figure in the British birth control movement), and Robert L. Dickinson (an American gynecologist, birth control advocate, and eventual president of both the Euthanasia Society of America and the HBAA) – enthusiastically supported eugenics. Gamble's views have been discussed. For her part, Sanger outlined in April 1932 a "Plan for Peace" in the *Birth Control Review*. Sanger advocated the following:

- First, put into action President Wilson's fourteen points, upon which terms Germany and Austria surrendered to the Allies in 1918.
- Second, have Congress set up a special department for the study of population problems and appoint a Parliament of Population, the directors representing the various branches of science: this body to direct and control the population through birth rates and immigration, and to direct its distribution over the country according to national needs consistent with taste, fitness and interest of individuals. The main objects of the Population Congress would be:
  a. to raise the level and increase the general intelligence of population
  b. to increase the population slowly by keeping the birth rate at its present level of fifteen per thousand, decreasing the death rate below its present mark of 11 per thousand
  c. to keep the doors of immigration closed to the entrance of certain aliens whose condition is known to be detrimental to the stamina of the race, such as feebleminded, idiots, morons, insane, syphilitic, epileptic, criminal, professional prostitutes, and others in this class barred by the immigration laws of 1924
  d. *to apply a stern and rigid policy of sterilization and segregation to that grade of population whose progeny is already tainted, or whose inheritance is such that objectionable traits may be transmitted to offspring*
  e. to insure [sic] the country against future burdens of maintenance for numerous offspring as may be born of feebleminded parents, by pensioning all persons with transmissible disease who voluntarily consent to sterilization

---

1933, betreffend Gesetz zur Verhütung erbkranken Nachwuches," November 3, 1933, R 1501 126 250, BArch Berlin.
[79] Fred O. Butler, "The Mental Defective and His Future," *American Journal of Mental Deficiency* 54 (October 1949): 163–5, AVS Pamphlets Misc. Box 34, AVS.

   f. to give certain dysgenic groups in our population their choice of segregation
     or sterilization
   g. to apportion farm lands and homesteads for these segregated persons where
     they would be taught to work under competent instructors for the period of
     their entire lives.[80]

In Britain, Marie Stopes was similarly unequivocal. In *Radiant Motherhood* (1905), one of her most famous and best-selling books, she states, "I would like to see the sterilization of those totally unfit for parenthood made an immediate possibility, indeed made compulsory."[81]

    Another figure in the U.S. birth control movement was Robert Dickinson, second in prominence only to Margaret Sanger.[82] Dickinson, who urged Gamble to fund Birthright,[83] was an influential gynecologist and a more acceptable face of American birth control advocacy, one that appealed to doctors and eugenicists put off by Sanger's radicalism.[84] His commitment to eugenics was, however, every bit as entrenched as hers.[85] In 1923, Dickinson had joined Sanger in founding the National Committee for Maternal Health (NCMH).[86] Fortified with Gamble's cash, NCMH officials fanned out across the United States and enlisted the aid of local nurses and doctors to test and apply birth control methods. As pharmaceutical companies entered the consumer market with contraceptives in the 1930s, the NMCH entered into agreements with them, purchasing large quantities of contraceptives at a low price in exchange for testing their products on local populations. Schoen writes,

> Such trials tested foam powder in North Carolina and Puerto Rico, condoms in the Appalachian Mountains, contraceptive jelly in Logan County, West Virginia, and, in the 1950s, the birth control pill in Kentucky and Puerto Rico. Health professionals tested a wide range of products, followed each other's progress, exchanged formulas, and recommended or discouraged the use of one product over another. Researchers commented on each other's tests and negotiated with doctors, nurses, and women over the policies and practices of contraceptive trials.[87]

---

[80] Margaret Sanger, "A Plan for Peace," *Birth Control Review* 16, no. 4 (April 1932): 107–8. Emphasis added. The article is a summary of an address given by Sanger to the New History Society on January 17, 1932 in New York.

[81] Marie C. Stopes, *Radiant Motherhood: A Book for Those Who Are Creating the Future* (London: G. P. Putnam's Sons), 221.

[82] Dowbiggin, *Sterilization Movement*, 55.

[83] Clarence J. Gamble to H. Curtis Wood, April 8, 1947, 2/14, AVS.

[84] Kline, *Building a Better Race*.

[85] Robert Latou Dickinson, M.D. to the Editor of the *Mirror*, April 19, 1947, 2/14, AVS.

[86] Schoen, *Choice and Coercion*, 29.

[87] Ibid., 30.

These tests involved often unreliable and sometimes dangerous contraceptives, and the women involved were rarely informed of the risks.[88] The tests were plainly undertaken in areas, such as Puerto Rico, that were considered quasi-isolated from mainstream political pressures.

It is, of course, entirely possible that those who signed up for these tests, lacking other alternatives, did so voluntarily. But the motives for the tests – and the work of Gamble, Sanger, and Dickinson more broadly – could not be in the remotest sense understood as driven by a concern with women's rights to contraception. Dickinson was a member of the American Eugenics Society and had regular and friendly contact with Gosney, Popenoe, and Laughlin. He corresponded with the latter about the need for a more efficient way of sterilizing women. In 1927, Laughlin wrote to Dickinson:

> In answering one of Mr. Gosney's recent communications I have taken the occasion to explain to him that I felt that one of the most important practical services which could be rendered by eugenics just now would consist in developing a simplified method for sterilizing the human female.[89]

Both Laughlin's and Dickinson's concerns were less with the pain and suffering associated with the operation than with the time and cost it entailed. Dickinson wrote:

> I wish I had the chance at 100 women out here in California who could be held or brought back at the fourth or six months for tests of tubal patency. One could also do what has not yet been done. In a few instances of women whose abdomens were to be opened anyway I could, after the abdomen was opened, burn, with my platinum tipped uterine sound, straight through the uterine cornu and determine just what the danger time was. My experiments in the morgue are of course on uteri the subjects of rigor mortis and without the wet succulence of life, or the easy bleeding from certain mucus membranes in this region.[90]

Dickinson got his opportunity: in February 1928, Gosney invited him to California.[91] For three weeks, Dickinson was given free access to eight hospitals in the state, to interview the staff and to practice his sterilization techniques on the patients.[92] Dickinson was impressed with California's program. Twenty years later, in 1948, he wrote in his notes of the success of California's experience, evinced in the admiration it had inspired among the Germans in 1935.[93]

Like many of his counterparts, Dickinson's commitment to eugenics survived the end of the Second World War. In 1947, when pro-sterilization lobbyists were

---

[88] Ibid.
[89] H. H. Laughlin to Robert L Dickinson, Committee on Maternal Health, August 9, 1927, 13/106, AVS.
[90] R. L. Dickinson to H. H. Laughlin, August 9, 1927, 13/106, AVS.
[91] L. S. Bryant, letter from Dickinson's Executive Secretary to H. H. Laughlin, February 29, 1928, 13/106, AVS.
[92] Kline, *Building a Better Race.*
[93] R. L. Dickinson to William Harris, Mount Sinai Hospital, April 2, 1948, 13/107, AVS.

adjusting to the language of rights-based thought, Dickinson fired off a letter to the editor of the *Mirror*, a Memphis-based publication. "The average intelligence in the U.S.," he informed the editor,

> is probably falling about 2 or 3 percent each generation ... [Therefore,] [t]he world may not be all it should be because the people in it are not all they should be. Who shall inherit the earth and all its problems? The 1940 U.S. Census shows that two thirds of all children of this nation are growing up in the "least fortunate" third of our families. Who are these "least fortunate" of our people? They fill our charity wards, our mental institutions, and our prisons; they absorb huge appropriations levied on taxpayers; they are the tools of dishonest political leaders. Their progeny fill our juvenile courts and our children's homes. Burlingame [author of *Heredity and Social Problems*] states, "The low class of social inadequates is probably increasing in the U.S and England by 15 to 20 percent per generation."
>
> Obviously, if our civilization and our American democracy is [sic] to be preserved, our most pressing obligation is to safeguard our biological heritage. As one measure, and before it is too late, we must have a nation-wide program of selective sterilization for those who are unfit for parenthood. The number of the unfit are [sic] increasing so rapidly that lifelong segregation has become impractical. If sterilized, half of the adults in custody could safely be released to lead self-supporting lives. Can we not learn from the experience of progressive countries like Sweden and Denmark what great benefit to a nation is derived from a sound eugenic program?[94]

In this missive, Dickinson recycles uncritically the prewar assumption that the numbers of the "unfit are increasing so rapidly" that urgent action was necessary, with sterilization being the most attractive option for state policy. Three years later, Dickinson became President of the HBAA, the last position he held before he died.

After Dickinson's death in 1950, there were plenty of other figures, old and new, prepared to carry HBAA's torch. Throughout the 1950s, HBAA's sponsors included Margaret Sanger, Lois Gosney-Castle, and Paul Popenoe.[95] Gosney-Castle was the daughter of Ezra Gosney and, like her father and Popenoe, was an exponent of coerced eugenic sterilization.[96] Other figures would also play a central role in the subsequent pro-choice movement. Most notably, Alan F. Guttmacher chaired the Association's Medical and Scientific Committee.[97] Guttmacher had earlier served as the Maryland Director for Birthright, Inc.[98] He

---

[94] Robert Latou Dickinson, M.D. to the Editor of the *Mirror*, April 19, 1947, 2/14, AVS.

[95] Invitation from Sophia M. Ribison, Chairman, Conference Committee, Human Betterment Association of America, February 3, 1961, 4.39, AVS.

[96] See, for example, letter from Lois Gosney Castle, February 9, 1947 [following her return from China, where she attempted to persuade members of Chiang Kai-shek's circle of eugenic sterilization's merits], 12/4, AVS.

[97] "First Public Meeting on Surgical Birth Control," March 21, 1961, 4.39, AVS.

[98] H. C. Wood, "The Physician's Responsibility in the Decline of National Intelligence," 1940, RF 632–44, 3, 59, 635, RAC.

was active in the HBAA throughout the 1950s, resigning from the Association for Voluntary Sterilization (AVS), HBAA's successor, in 1966, by which point he had moved on to the PPFA.[99] Guttmacher's interests in sterilization were social: there were, he said, "millions of children being brought into being and reared in an atmosphere of moral and economic irresponsibility. . . . In New York nearly half the babies in the metropolitan area are born to families on welfare."[100] Only by offering voluntary sterilization to the men and women who were a "burden" on society could the cycle be broken.[101]

Guttmacher would later become national director of PPFA. In 1962, he succeeded William Vogt, author of the *Road to Survival* (1948), one of the first postwar books to argue that runaway population growth would lead to mass starvation.[102] Vogt was a neo-Malthusian and a misanthrope who recycled crude, turn-of-the-century arguments about the dangers of medical intervention in preserving lives that nature would have consigned to death. "Why," he asked the World Health Organization, "are you trying to save the lives of children when you'll just doom them to starvation?"[103] Vogt sat on HBAA's medical and scientific committees in the 1950s and was national director of PPFA from 1951 until 1962. Vogt was decisive in creating a PPFA agenda that urged making food aid conditional on population control.[104] But even that was not enough: many poor people would not make it and should simply be left to die.[105]

Although there were tactical differences between lobbying organizations throughout the 1950s and into the 1960s – Planned Parenthood at this point was uneasy about publicly supporting sterilization and instead put contraception and abortion ahead of sterilization – advocates of birth control and previous and current advocates of coerced sterilization sat side by side.[106] At the November 14, 1961 annual business meeting of the HBAA, delegates considered a motion to amalgamate Planned Parenthood with the HBAA as a "great step forward for the whole movement."[107] And, in April 1964, the International Planned Parenthood Federation (IPPF) members participated in their first

---

[99] See, for example, "Minutes. Meeting of the Board of Directors, Human Betterment Association of America," October 15, 1958, 3/24, AVS.

[100] Quoted in Dowbiggin, *Sterilization Movement*, 61.

[101] Ibid.

[102] Ibid., 111.

[103] Quoted in ibid., 112. See also Connelly, *Fatal Misconception*, 129.

[104] Connelly, *Fatal Misconception*, 130.

[105] Ibid.

[106] See remarks by H. Curtis Wood, President of Human Betterment Association at its annual 1958 business meeting: "Annual Business Meeting. Remarks by the President," October 15, 1958, 3/24, AVS. Planned Parenthood had, however, supported sterilization since at least 1948. See *Planned Parenthood Federation of America, Inc: National Policies*, January 1948, 2/15, AVS.

[107] "Minutes of the Annual Business Meeting," November 14, 1961, 3/24, AVS. Opposition came from the Association on the grounds that the latter's clinics were too reluctant to push "surgical birth control" – that is, sterilization. Planned Parenthood, for its part, rather ironically feared that, as birth control became more acceptable in the 1960s, the radical taint of sterilization would

AVS-sponsored International Conference on Voluntary Sterilization in New York, with Alan Guttmacher chairing the medical and science committee.[108] The battle for reproductive rights was waged by individuals whose previous aim had been large-scale coerced eugenic sterilization. By the 1960s, their aim was simply large-scale sterilization. At the 1964 conference, observes Matthew Connelly, "[w]hat was contemplated was not ... just targeting poor countries, but rather the sterilization of poor people worldwide, including in the United States."[109] In the end, however, C. P. Blacker convinced the conference that India presented the greatest need, and the conference gave birth to the International Association for Voluntary Sterilization, with India's Minister of Health and Family Planning as chair.[110]

FROM THE SOCIAL TO THE INDIVIDUAL AND BACK AGAIN

By the end of the 1950s, the HBAA quietly abandoned the argument about the "rights of the child" because it was so tenuous a claim that it could provide only a very weak foundation for expanded sterilization. Moreover, acknowledging the premise that unborn children have rights anchored the Association in the Roman Catholic Church's argument against contraception and abortion, which was hardly an appealing prospect for the organization. Instead, the HBAA struggled to find other ways of reconciling itself to the emergent rights culture. H. Curtis Wood, longtime advocate for birth control and president of the HBAA from 1945 to 1960, played a central role in this initiative.

Like many in the sterilization movement, Wood was influenced by empirical data and by grand theory. He treated and was deeply moved by a woman who, unable to feed her five children, risked and suffered horribly from a botched abortion.[111] From his readings in social theory, he was convinced that the high fertility of the mentally defective lowered America's intelligence year after year.[112] Unlike many in the sterilization movement, Wood early on learned about tactics and framing. He helped convince Birthright that Marian Olden had to go in 1948, urged it to ignore rather than provoke the Roman Catholic Church, and sought throughout the 1950s to reframe the argument for sterilization in a manner that drew in the largest number of supporters while

---

offend its mainstream members. See Rebecca M. Kluchin, *Fit to Be Tied: Sterilization and Reproductive Rights in America, 1950–1980* (New Brunswick, NJ: Rutgers University Press, 2009), 35. By the mid-1960s, as the population "crisis" made sterilization more acceptable, Planned Parenthood and the HBAA/AVS moved closer together both ideologically and physically, occupying adjacent offices in New York by 1963. In 1969, Planned Parenthood endorsed sterilization as an acceptable form of birth control. Dowbiggin, *Sterilization Movement*, 101.

[108] Connelly, *Fatal Misconception*, 207.
[109] Quoted ibid., 208.
[110] Ibid.
[111] Dowbiggin, *Sterilization Movement*, 57.
[112] Ibid., 62–3.

providing its opponents with as few openings for attack as possible.[113] By the end of the 1950s, he found such arguments in the defense of liberty (there being no more powerful frame in America) and in the defense of medical autonomy. As he put it in his address to the organization's annual business meeting:

> *I like to think of our organization as one that is not only interested in sterilization but also a crusader for human rights and individual liberty. There is no question but that, as we become more and more organized and perhaps civilized, there is a slow and insidious decline in the freedom of the individual.* ... In Philadelphia there is a campaign at the present time to make jay walking a legal offense, punishable by fine or imprisonment. We have reached such a point in our "Welfare State" in this country that a man may not cross the street in the middle of the block if he wants to, without running the risk of being arrested for a misdemeanor. ... With the ever increasing security that Americans seem to want almost more than anything, come more and more rules, regulations and the restriction of individual liberty. ...
>
> The same trend is apparent in the practice of medicine. There is not much room for the rugged individualist and the average doctor is being more and more tied up in red tape and regulations of one kind or another. It seems as if the day may come when a surgeon will not be allowed to perform any major surgery without a written consultation and approval by some other so called "qualified" physician. The public is being "protected" on all sides from doctors who appear to be so poorly trained or else immoral and irresponsible that they must not be given a free hand with their patients and allowed to decide medical issues by themselves.[114]

The speech is noteworthy for two features. First, behind this embrace of freedom lay a thinly veiled frustration with the new procedural safeguards, still inadequate in some ways, designed to protect individuals from coerced sterilization. Six months earlier, almost to the day, HBAA minutes had optimistically reported the results of a meeting with a representative of the Philadelphia Youth Services Board, responsible for dealing with juvenile delinquents. To HBBA's delight, the representative believed that "mental retardation and excessively large families for which parents could not provide either physically or economically are two of the largest contributors to delinquency."[115] Unfortunately for the HBAA, on the same day, it received a response to its enquiries to a New York law firm. The law, the firm opined, afforded a great deal of protection to the feebleminded or retarded adult, and that for any sterilization, the consent of their caregivers and/or spouses would be needed.[116]

This opinion and other setbacks greatly frustrated HBAA activists, a frustration that would only be partially reduced by the post-1960s liberalization of restrictions of contraception. For, as the executive director of the HBAA put it in

---

[113] On Wood's efforts, see ibid., 64–6.
[114] "Annual Business Meeting. Remarks by the President," October 15, 1958, 3/24, AVS.
[115] "Minutes. Meeting of the Board of Directors, Human Betterment Association," April 17, 1957, 3/24, AVS. Emphasis added.
[116] Barry H. Siner, Greenbaum, Wolff & Ernst to Ruth Proskauer Smith, April 17, 1957, 3/24, AVS.

1959, "contraception does not and never will solve the problem for those who need fertility control the most – the morons, the ignorant, the irresponsible."[117]

Second, the defense of doctors' autonomy reflected more than a professional's annoyance with bureaucracy. As the postwar net closed on sterilization, doctors remained the last hope for any serious effort to secure large-scale sterilization outside institutions. They enjoyed immense prestige and a high, although declining, degree of patient deference, and their vocabulary and expertise enabled them to invoke arguments powerful enough to persuade even the most skeptical patients of sterilization's merits. Doctors would do so in increasingly large numbers in the late 1960s and early 1970s.

In the end, the effort on the part of coerced sterilizers to accommodate themselves to the rights culture proved difficult and eventually (although it took some time) insurmountable, for the obvious reason that the individual rights most affected by sterilization belong to the sterilized. To buttress the argument for large-scale sterilization further, those in favor of it needed to ground their argument in social, rather than individual, concerns. They found a rich opportunity in the core idea of the emergent postwar movement for global population control.

---

[117] Ruth Proskauer Smith to Harry Emerson Fosdick, April 28, 1959, SW15.1/23, AVS, quoted in Dowbiggin, "'Rational Coalition,'" 254n31.

Eugenics and World Population Growth

The eugenicists' espousal of the rights of the child was rhetorical. Invoking an unborn child's rights – to anything but being born, that is – was a one-liner that allowed Gamble, Popenoe, Butler, and others to reframe their argument in favor of sterilization. This reframing based on rights, however, ultimately challenged and constrained eugenicist projects. It challenged them because conceding that an unborn baby has rights rests, at best, uneasily with abortion: the same premise is, of course, a foundation of the anti-choice/pro-life position. It constrained them because it suffered from an age-old problem affecting eugenic sterilization in North America: it would only ever encompass a small number of people. With the exception of North Carolina, only people in institutions were targets for sterilization in North America, and those in institutions were only a fraction of the targeted population of feebleminded – a fact supporters of eugenic sterilization frequently bemoaned.[1] If the pro-sterilization movement were to survive the onslaught of Hitlerian science *and* increase its impact, it needed another ideological basis.

World population growth supplied part of the new basis. In the pre-First World War period, a wide range of thinkers sympathetic to eugenics were chiefly motivated by population decline, above all among the genetically fit. Indeed, the differential birthrate between the upper and lower classes was an anchor of the argument that eugenic sterilization was needed to ward off race suicide. As

---

[1] F. O. Butler to Douglas Arant, November 10, 1950, SWO15.1/72, Alabama, AVS; M. Morton, "Sterilization Law: Health Officer Says Provisions May Broaden," n.d., SWO15.1/93. South Dakota represented something of a hybrid case. State institutions were responsible for identifying and caring for the feebleminded; the institutions sterilized those individuals if they were judged likely to have children but not be able to care for them. Institutions would, however, sterilize individuals as an alternative to institutionalizing them. Because the institution made the decision, however, it remained the central actor. See State of South Dakota, *Seventeenth Biennial Report of the State Commission for the Control of the Feeble-Minded*, June 30, 1958, SWo15/73, AVS.

historian Richard A. Soloway comments on the British case, "one of the most compelling factors in the rise of eugenic analyses was the declining birthrate in general and its class differential characteristics in particular."[2] During the interwar period, a number of eugenicists switched horses dramatically: population growth, not population decline, was the greatest threat to racial health. This emphasis intensified after 1945, with the conventional concern about class differentials in birth rates (too many poor, not enough births in the educated classes) becoming an implicit rather than explicit focus.

Moderate, or "reform," eugenicists moved into population policy early in the interwar period. The immediate cause was their disillusionment with what came to be seen as the crude hereditarianism of mainline eugenics. In 1927, Raymond Pearl, a Johns Hopkins University biologist, prolific writer, and long-time supporter of eugenics, published an article in *American Mercury* for his friend and fellow bon vivant, H. L. Mencken. Entitled "The Biology of Superiority," it was a strident attack on eugenicists' failure to incorporate recent genetic discoveries. This included, for example, Wilhelm Johannsen's argument that the genotype of an organism (its genetic makeup) could not be determined with any certainty from its phenotype (the physical expression, in intelligence or stupidity, of those genes).[3] Mocking Galtonian work on ancestry, Pearl wrote that "by superior people, whether individuals, classes, or races, seems always to be meant either: a. '*My* kind of people,' or b. 'People whom *I* happen to like.'"[4] Eugenics, he argued, had "largely become a mingled mess of ill-grounded and uncritical sociology, economics, anthropology, and politics, full of emotional appeals to class and race prejudice, solemnly put forth as science, and unfortunately accepted as such by the general public."[5] The journal had a wide readership, and local newspapers gave Pearl's attack further coverage. His critique may have cost him a job offer at Harvard, but it dealt a serious blow to hereditarian eugenics as a respectable scientific principle.[6] Mencken had softened the target up by publishing an attack of his own on eugenics, which was printed in a Baltimore newspaper prior to the appearance of Pearl's piece:

> In none of their books have I ever found a clear definition of the superiority they talk about so copiously. At one time they seem to identify it with high intelligence. At another time with character, i.e., moral stability, and yet another time with mere fame, i.e., luck. . . . But their vagueness about the exact nature of superiority is not the only thing that corrupts the fine fury of the eugenists. Even more dismaying is their gratuitous assumption that all of the socially useful and laudable qualities (whatever they may be) are the exclusive possession of one class of men, and that

---

[2] Soloway, *Demography and Degeneration*, xiii.
[3] Raymond Pearl, "The Biology of Superiority," *The American Mercury* 12, no. 47 (1927): 257–66.
[4] Ibid., 261.
[5] Ibid., 260.
[6] Melissa Hendricks, "Raymond Pearl's 'Mingled Mess,'" *Johns Hopkins Magazine* 58, no. 2 (April 2006): 50–6, http://www.jhu.edu/jhumag/0406web/pearl.html.

the other classes lack them altogether. This is plainly not true. All that may be truthfully said of such qualities is that they appear rather more frequently in one class than in another. But they are rare in all classes, and the difference in the frequency of their occurrence between this class and that one is not very great, and of little genuine importance.

If all the biologists in the United States were hanged tomorrow (as has been proposed by the Mississippi clergy) and their children with them, we'd probably still have a sufficiency of biologists in the next generation. There might not be as many as we have today, but there would be enough. They would come out of the families of bricklayers and politicians, bootleggers and bond salesmen. Some of them, indeed, might even come out of the families of Mississippi ecclesiastics. For the supply of such men, like the supply of synthetic gin, always tends to run with the demand. Whenever the supply is short the demand almost automatically augments it.[7]

Pearl's widely read piece signaled his shift out of old-style eugenics and into the field of population policy. The next year, he founded the International Union for the Scientific Study of Population Problems and served as its first president. In his book, *The Biology of Population Growth*, Pearl rejected the claim that high fertility among the lower classes is in itself an indication of feeblemindedness or, indeed, of anything at all:

> Some early eugenic studies, especially those of Professor Karl Pearson and his students, demonstrated the fact that various sorts of obviously undesirable persons had high fertility rates, while some equally obviously desirable persons had low fertility rates. This situation he rightly regarded as racially unsound in principle. But a curiously inverted deduction from these results seems to have become current. The doctrine appears to have gained wide acceptance that somehow high fertility in a group is *in itself* an indication of probable racial unfitness. In actual fact nearly every group of persons that I can think of which does show high fertility seems to be quite generally regarded by eugenists as socially and biologically undesirable. They are alleged to be *genetically* unfit. It has even been suggested that there is a gene, or combination of genes, for poverty – that a man is poor only because he inherited stupidity, lack of ambition, or some other similar traits of character. Now, while I am certainly no violent environmentalist, I gravely doubt if such a position is tenable.[8]

A few lines later, however, Pearl trades in one correlation – between poverty and feeblemindedness – for another – between high fertility and poverty:

> If the facts presented in this chapter ... are in any degree generally true ... the thought at once suggests itself that it is not only desirable in the eugenic interest of the race to cut down, indeed completely extinguish, the high birth rate of the unfit and defective portions of mankind, but it is also equally desirable because of the

[7] H. L. Mencken, "On Eugenics," *Baltimore Sun*, May 15, 1927.
[8] Raymond Pearl, *The Biology of Population Growth*. rev. ed. (New York: Knopf, 1930), 169. Emphasis in original.

menacing pressure of world population, to reduce the birth rate of the poor, even though that unfortunate moiety of humanity be in every way biologically sound and fit.[9]

If the eugenic echoes of the argument were not yet clear enough, he later clarified them: "if it is not possible," he asks, "to make desirable people have more babies, why not try teaching other people how to have fewer?"[10]

Many other eugenicists were asking the same question. In 1922, the first research institution devoted to population studies, the Scripps Foundation for Research in Population Problems, was established in Miami. Its first director was Warren S. Thompson, a sociologist who rejected simple hereditarian arguments ("human nature as we know it in everyday life is by no means a pure product of heredity") and who viewed IQ testers' claim to measure "native intelligence" as absurd, given the obvious role of culture and environment in ensuring success on the tests.[11] Worse still, the use of these tests in immigration policy meant that eugenics was "in process of becoming a propaganda to urge increase of the Nordics because they seem to meet these tests adequately and because of race bigotry. . . . It is exceedingly unfortunate that a study as important as eugenics is degenerating from its high estate and in this country is becoming a mere class propaganda."[12] He viewed differential birth rates as a judgment on the upper classes, not the lower ones:

> I am doubtful of the value to the race of people who having had the opportunity to view life in a large way are unwilling to raise fair-sized families or are unwilling to jeopardize their economic security by undertaking to so change the present social order that there will be opportunity for a healthy family life and a reasonable satisfaction of ambition for the majority of mankind.
>
> I cannot say more here. The future belongs to people who raise children. But at present the upper economic classes are unwilling to participate in the future in this way (the only sure way) and yet are bewailing the fact that they cannot control the future. One cannot eat one's cake and have it. If we prize a book written, a trip to Europe, expensive dinners, a fortune accumulated, prominent position socially or economically, etc., etc., more than children, we should be willing to let the people who do raise the children determine the future. We should do it gracefully as it ill becomes us to rail at new immigrants and "the lower orders" who instinctively understand nature's requirements better than we of the so-called upper classes.[13]

Although distinctly moderate in its framing and expression, there is in these words nonetheless a concern over upper-class "race suicide" and a belief that

[9] Ibid., 171.
[10] Ibid., 176.
[11] Warren S. Thompson, "Eugenics and the Social Good," *The Journal of Social Forces* 3, no. 3 (March 1925): 414–9.[
[12] Ibid., 418.
[13] Ibid., 419.

eugenics, properly conceived, could serve the social good. There is also skepticism about the failure of those most concerned about fertility trends to act themselves to address the issue. Thompson joined the American Eugenics Society Board in 1935.[14]

This concern and this belief would inform future developments in demography and population policy. In 1925, the Sixth International Neo-Malthusian and Birth Control Conference took place in New York. It was organized by Margaret Sanger, heartily endorsed by Raymond Pearl, and attended by Irving Fisher and two dozen members of the American Eugenic Society's Advisory Council.[15] "Birth Control," Pearl wrote, "seems to offer the only method at once humane and intelligent of meeting the menace of population growth without fundamentally altering our civilization and standards of living."[16] Two years later, in 1927, Sanger organized, with heavy steering from Raymond Pearl, the International Population Conference in Geneva, the first World Population Conference.[17] Eugenicists were once again prominent on the participants' list.[18] At the close of the meeting, Pearl founded the International Union for the Scientific Investigation of Population Problems (later, the International Union for the Scientific Study of Population, or IUSSP).[19] IUSSP, like many population organizations, was partly a research institution and partly a lobbyist organization for population control.

The organization also retained a eugenic accent. Its American affiliate, founded in 1931, was the Population Association of America (PAA), with Henry Pratt Fairchild as its first president.[20] Fairchild's biography was a reflection of the close interconnections of eugenics, birth control, and population control: he was a one-time president of the American Eugenics Society, founding member of the Planned Parenthood Federation of America, and long-time anti-immigration activist.[21] In December 1930, Margaret Sanger and Harry H. Laughlin took part in PAA's organizational meeting, drawing up a list of invitees for the organization's first conference.[22] The list included "a full range of eugenicists ... from the unrestrained to the temperate": Guy Irving Birch, Charles Davenport, Clarence Gamble, Frederick Osborn, Raymond Pearl, and

[14] Ordover, *American Eugenics*, 272.
[15] Connelly, *Fatal Misconception*, 63–4.
[16] "Up to the Doctors: A Review of the International Birth Control Conference," "The Savants Say," *The Survey* 54, no. 2 (April 15, 1925): 73.
[17] "History of the International Union for the Scientific Study of Population," IUSSP, http://www.iussp.org/en/about/history (accessed January 22, 2013); Connelly, *Fatal Misconception*, 69–70.
[18] Connelly, *Fatal Misconception*, 69–70.
[19] Ibid., 70.
[20] *Constitution of the Population Association of America*, 1931, RF 190–200, 3,2,9, RAC.
[21] Connelly, *Fatal Misconception*, 84.
[22] Ibid.; "Preliminary Conference on a Population Association for the United States, Held with the Cooperation of the Milbank Memorial Fund," New York University, December 15, 1930, RF 190–200, 3,2,9, RAC.

Leon Whitney (Executive Secretary of the American Eugenics Society).[23] When the chairman, Henry Pratt Fairchild, opened the conference, he stated that "[p]opulation presents two aspects – qualitative and quantitative. What kind of people comprise a society and how many people are there in society? Particularly, the word population has been shifting away from its proper meaning as a term that should cover the whole field, to a restrictive quantitative meaning. ... So it seems to me it is the crucial time to step in and check this degradation of a very significant word."[24] In its mission statement, the organization stressed that its interest in population policy included "both its quantitative and qualitative accents." Language on the "qualitative and quantitative aspects" of population growth was retained through to a 1974 revision of the organization's mission statement.[25] The PAA's very first research project examined the fertility "of socially inadequate classes."[26] "No one," observes Matthew Connelly, "disagreed with the idea that the PAA should study eugenics"; indeed, if anything, it was the study of population control that was pushing boundaries.[27]

The Milbank Memorial Fund provided most of the funding for the Association.[28] Set up in 1921 with large donations from philanthropist and public health advocate Elizabeth Milbank Anderson and thereafter directed by Albert G. Milbank, the Fund's interest in public policy was essentially eugenic. "[I]f public health activities," its 1929 report stated, "are to be guided intelligently, it is highly desirable to have some knowledge of the changes in the rate at which various social groups reproduce themselves, of changes in the extent to which specific social groups are recruited from other social classes, and of the general constitutional or physical characteristics of these groups or classes."[29] One of its key areas of support was research into differential birth rate by social class.[30]

In 1928, as the IUSSP was being formed, the Milbank Memorial Fund decided to set up a research division staffed with demographers to study population issues. According to Dorothy G. Wiehl, one of the first demographers working for the fund, the chief interest was eugenic:

---

[23] Dennis Hodgson, "The Ideological Origins of the Population Association of America" *Population and Development Review* 17, no. 1 (1991): 2, 28. The invitation list for the second was much the same. "Persons Requested to Attend Second Conference on Population Association," n.d., RF 190–200, 3,2,9, RAC.

[24] "Preliminary Conference on a Population Association for the United States, Held with the Cooperation of the Milbank Memorial Fund," New York University, December 15, 1930, RF 190–200, 3,2,9, RAC

[25] "Population Problems," n.d., RF 190–200, 3, 2, 9, RAC; Angela Franks, *Margaret Sanger's Eugenic Legacy: The Control of Female Fertility* (Jefferson, NC: McFarland & Co., 2005), 135.

[26] Connelly, *Fatal Misconception*, 84.

[27] Ibid.

[28] *Constitution of the Population Association of America*, 1931, RF 190–200, 3,2,9, RAC.

[29] "Interest of Milbank Memorial Fund in Population," March 20, 1931, RF 190–200, 3, 2, 9, RAC.

[30] Ibid.

In the mid-1920s, the interest in population centered more in the eugenic implications of differential birth rates and the rapid decrease in the mortality of infants and
children. ... The question asked by many of us was: "Are the unfit being kept
alive?" Again, [the report on] Health and Environment ... [MMF staff demogra-
pher Edgar] Sydenstricker referred to a then current argument as to whether "the
increase in death rate of persons over 50 years of age is due wholly or in part to a
deterioration of vitality"' which, he said, raised the question "Are the American
people breeding a stock with a lower inherited capacity to survive?" It was in this
climate of opinion and speculative philosophizing that the Fund began its activities
of differential fertility and became involved in studying the effectiveness of birth
control methods.[31]

FROM INTERWAR TO POSTWAR

The interwar period was a period of rock-bottom birth rates across the West.
After the Second World War, birth rates increased sharply, and, across the West,
there was a twenty-year baby boom. In Japan and parts of the developing world,
notably India and China, birth rates also shot up. As they did, the old menace of
population decline was replaced by a new one: overpopulation driven by the
baby boom. Individuals who had decried low and especially differentially low
birthrates lined up to decry high fertility. Two fears drove the white, middle-,
and upper-middle class activists who would make up the well-funded, high-
profile, and influential population lobby: poor Americans' – and often particularly poor black Americans' – high birth rates in the United States. In addition,
all high birth rates in the so-called Third World were a cause of concern. Similar
to the dynamic observed in the Sterilization League of New Jersey's uneasy
adoption of rights-based language, activists' concerns with population growth
compounded continuing eugenic concerns. Thus, Birthright, Inc.'s 1948 Report
to Members discussed a Birthright film entitled "The Fatal Chain of Heredity
Disease," which was shown on university campuses. The Report noted with
approval the disappointment expressed by viewers that Pennsylvania had not
been "progressive enough to enact sterilization legislation";[32] Pennsylvania was
one of the states where a large Catholic population thwarted such efforts in the
1920s and 1930s. If the first half of the report looked backward, the second
looked forward: recommended reading for members included a *Harper* article
entitled "Too Many People" and an *Atlantic Monthly* article called "Crowded
off the Earth."

The battle against population growth, although it would eventually draw in
many volunteers, was at first fought by eugenicist veterans. In 1949, Robert
Dickinson became first chairman of the Human Betterment Association of
America's new medical and scientific committee and then director of the organization itself a year later. He had for several years been taken by the

[31] Quoted in Franks, *Margaret Sanger's Eugenic Legacy*, 142.
[32] Birthright, Inc., *Report to Members, January 1 to April 29, 1948*, 2/14, AVS.

Scandinavian example, in which compulsory sterilization was bound up with a comprehensive progressive program in population, family, and social policy.[33] The link with population policy seemed particularly appealing, and it provided Dickinson an opportunity to help sterilization escape the "paralysis" it suffered within institutions.[34] Dickinson linked the rechristened organization's mission with population policy. He wrote on April 22:

> I have been asked to attempt an outline of a program of possible basic research of a comprehensive character looking toward discoveries of methods of conception control suited to the poorest peasant folk of countries of such countries [sic] as India, China and Puerto Rico.
>
> The National Committee on Maternal Health has not failed to keep this goal in mind. Our limited resources have handicapped this most difficult of all our endeavors. *The new stress on world-wide overbalance of population should expand our opportunities and secure personnel on an extended scale backed by large funds.*[35]

It did, at least indirectly. Dickinson died on November 29, 1950, but the irrepressible Clarence Gamble took up the cause.

By 1957, in a sort of rerun of his fights with Birthright, Gamble had been drummed out of the International Planned Parenthood Federation (IPPF). This came in reaction to his insistence on promoting, as he had done in North Carolina, cheap, untested, and unreliable birth control methods in Asia.[36] In response, Gamble refounded the National Committee on Maternal Health and created the Pathfinder Fund, the institution through which he would take the practices he pioneered in North Carolina to the developing world. One of the Fund's early official documents claimed, "since its formation the National Committee on Maternal Health has considered that an important element in maternal health is the freedom to choose the number and time of arrival of children."[37] This was, to put it no more strongly, a revisionist reading of its organization's history. Probably conscious of the organization's eugenicist past, the document noted that findings of a five-year study (funded with $150,000, probably given by Gamble) of the "effectiveness, acceptability, and safety of contraceptive methods" would be more "impartial" if they came from a separate organization, namely the Pathfinder Fund. At the same time, the work of "organizing birth control groups and clinics" in Asia, Africa, and the Near East, as well as in the United States, were transferred to the Fund.[38] Gamble also helped to pay the salaries of staff at Puerto Rico's Family Planning

---

[33] Dowbiggin, *Sterilization Movement*, 84.
[34] Ibid.
[35] Robert L. Dickinson to Frederick Osborn, April 22, 1949, 2/16, AVS. Emphasis added.
[36] Connelly, *Fatal Misconception*, 179.
[37] "The Pathfinder Fund," October 1957, 4, Clarence James Gamble Papers ("Gamble Papers"), Center for the History of Medicine, Francis A. Countway Library of Medicine, Harvard University.
[38] "Tentative Draft, October 3, 1947: the Pioneer Fund," October 10, 1957, 42, Gamble Papers.

Association, which was an advocate of sterilization rather than of contraception.[39] The National Committee was meant to restrict itself to research, but, as evidence of the overlap between research and advocacy in the new organization, it was closed down two years later.

One of those who were particularly keen to see the work of the two organizations separated was Frederick Osborn, founding member of the American Eugenics Society, the Pioneer Fund, and Princeton University's Office of Population Research.[40] In 1952, Osborn co-founded the Population Council with the help of a major grant from John D. Rockefeller, III, grandson of the wealthy oil magnate. A supporter of multiple causes in his family's tradition, Rockefeller himself adopted a moderate, optimistic view of population growth and its control. On the subject of the "population crisis," he told a conference in Dallas:

> A nation's growth often depends on the number and qualities of its people. Our own American history is proof: two hundred years ago, we were a few hundred thousand people lately come to the edge of a vast, untapped continent. Not until our population grew in numbers, in education and in skill, could we fully develop the riches of our land. Even in today's world, areas and nations – like Australia, Alaska, and Canada – recognize the need for the vital force of population.
>
> Our contact goal must be the enrichment of human life, not its restriction. Above all, we must not confuse ends with means. We must not think of population as some mere abstract factor in an economic equation. It is not a soulless sum to be reduced or subtracted simply to facilitate the easier balancing of that equation. We must think in no such negative terms – but [rather] with the positive purpose of moving toward economic decisions that are to serve and to enrich human life and individual dignity. . . . In this matter of perspective, it is essential that we see the population problem in its full and true dimensions.[41]

The Population Council itself did not share his optimism. It saw the matter in starker and essentially neo-Malthusian terms. Osborn's successor as director was Frank W. Notestein. Born in 1902, Notestein entered population studies in the late 1920s. He took a position as research associate at the Milbank Memorial Fund after earning his Ph.D. at Cornell. His research agenda during the 1930s might be described as reform eugenicist: he was vexed by the differential fertility rates between classes, but he hesitated in drawing any direct policy implications from them.[42] When Frederick Osborn convinced Albert G. Milbank of the Milbank Memorial Fund to establish a formal training center for demographers

---

[39] *New York Times*, "Puerto Rico Acts on Birth Control," November 1, 1959, 11/98, AVS. On this, see Schoen, *Choice and Coercion*, chapter 4.

[40] Frederick Osborn to Clarence Gamble, October 2, 1957, 42, Gamble Papers.

[41] "A Citizen's Perspective on Population," speech given on May 16, 1960 in Dallas, TX by John Rockefeller, III, before the closing dinner session of the National Conference on the Population Crisis, 42, Gamble Papers.

[42] See Frank W. Notestein, "Class Differences in Fertility," *Annals of the American Academy of Political and Social Sciences*, 188, no. 1 (1936): 26–36. On the same topic, see E. J. Lidbetter,

at Princeton University (the Office of Population Research), Notestein became its first director.[43] Whereas Rockefeller, the layman, displayed scholarly caution in drawing conclusions on population issues, Notestein, the scholar, displayed a near-boundless enthusiasm:

> Mankind has the technical ability to reduce the toll of sickness and poverty throughout the world to an extent never dreamed of even a few decades ago. Yet we teeter on the brink of self-destruction by allowing population growth to outstrip economic advance. ... The problem is most intense in those parts of the world where average incomes are barely above the minimum of subsistence, illiteracy is prevalent, the traditional agrarian economy is largely intact, and scant use is made of modern technology in either industry or agriculture. The area, encompassing over a billion people, includes all of non-Soviet Asia except Japan, the South Sea Islands, all of Africa north of the Union, and virtually all of Latin America except Argentina.[44]

And that is exactly where birth control activists turned their attention. Notestein, like others in the Population Council, recognized the implications of the rights revolution. "Compulsory sterilization" was a "barbarity," and whatever the effectiveness of "brutal ways of checking population growth, ... we must find those that are acceptable in humanitarian terms, those that we would advocate if our own lives and those of our children were involved."[45] They had to pursue a strategy of what Frederick Osborn called "crypto-eugenics."[46]

The members of the Pathfinder Fund, the Population Council, and other population-control lobbies discussed later did react to a real and measurable social change: the increase in birth rates after the war. They also responded to real and measurable social problems, namely hunger, poverty, maternal death, and high infant mortality. At the same time, however, Gamble, Notestein, and Osborn interpreted these developments through a set of ideas framed by an overly simple neo-Malthusian perspective and limited their remedies to ones in the narrow field of reproduction. In the former, they both generated and responded to the new neo-Malthusian climate. Within the halls of the academy, many researchers were glad to help them.

Heredity and the Social Problem Group (London: E. Arnold, 1933) and Raymond B. Cattel, *The Fight for Our National Intelligence* (London: P. S. King & Son, 1937).

[43] Dennis Hodgson, "Notestein, Frank W.," in *The Encyclopedia of Population*, ed. Paul Demeny and Geoffrey McNicoll (New York: Macmillan Reference USA, 2003), 1: 696.

[44] Frank W. Notestein, "Poverty and Population," *The Atlantic Monthly*, 1959–60, 84. 42, Gosney Papers.

[45] Ibid., 85.

[46] Connelly, *Fatal Misconception*, 163.

THE MISANTHROPY OF THE INTELLECTUALS: POPULATION
CONTROLS' ACADEMIC ACTIVISTS

In 1954, a young Cal Tech geochemist, Harrison Brown, published a book
entitled *The Challenge of Man's Future*. Plummeting death rates, Brown argued,
would lead to a sharp increase in population, mass starvation, a sharp decline in
living standards, and an eventual return to agricultural economies with lower
standards of living.[47] Hunger could be combated through greater use of algae-
and yeast-based foods, but Brown regarded such a development as unlikely. In
the same year, the IUSSP (founded by Pearl in 1928 and reconstituted in 1947)
helped sponsor an international conference in Rome.[48] The President of the
International Union since 1949, Liebmann Hersch, was elected president of the
conference. Somewhat gingerly, Hersch told reporters that, whereas the last four
population conferences had been chiefly concerned with declining birth rates in
highly developed Western countries, the new concern was the rapidly increasing
population in Asia, above all in India.[49]

Sir Charles Galton Darwin (grandson of Charles Robert Darwin) put his
famous name at the service of the population growth alarmists in the 1950s. In
1952, Darwin published a book entitled *The Next Million Years*.[50] The book
was partly neo-Malthusian, partly eugenic. In the former vein, he made the
familiar argument that population would outstrip the planet's food supply. In
the latter, voluntary contraception created the equally familiar class differentiation in fertility:

> [I]t is indisputable that the more prosperous members of the community are not
> producing their share of the next generation, so that selection is now operating
> against the prosperous. As an example, if the list of candidates is examined, who are
> applying for any office of high or even mediocre importance, it will be found that
> something like nine-tenths of them have either no children, or one, or two. Of
> course, if everyone had exactly two children, and both these children married and
> had exactly two more, the population would be exactly steady, but as things are, it
> is a fair guess that, in each thirty years of a generation, this part of our population is
> reducing itself to something between a half and two-thirds. This signifies that
> within a century, there will at most be quarter as many people of this type as
> there are now. There will of course be some compensation by the rise from other
> levels, but, as I have pointed out, to found our hopes on them is to take a worse

---

[47] Bruce Bliven, "A Gloomy Prophecy," *New Republic* 131, no. 7 (August 16, 1954): 20.

[48] During the early 1930s, the International Union funded research commissions dealing with
"population and food supply," "differential fertility," and "vital statistics of primitive races."
"Populations Problems," July 4, 1931, RF 190–200, 3, 2, 9, RAC.

[49] Robert K. Plumb, "Science in Review: Rome Conference on Population Trends Raises Serious
Questions Concerning Food Supply," *New York Times*, September 9, 1954.

[50] Charles Galton Darwin, *The Next Million Years* (London: R. Hart-Davis, 1952).

instead of a better chance. The whole thing is a catastrophe which it is now almost too late to prevent.[51]

There was thus no contradiction between his frequent calls for birth control and his own five children.[52] The coercive inflection in his argument was by no means unique: for decades, the American birth control movement's rhetorical commitment to voluntarism reflected political necessity rather than conviction.

In other interventions, Darwin's tone was harsher. In 1959, he visited the United States for a centennial celebration of the publication of *Origin of the Species*. He gave an interview to *U.S. News and World Report* and told his interviewer that, for birth control to be any good,

> [I]t's got to be worldwide. If you just have it in this country, what's the result? Most of the world in fifty years will be black ... or more likely yellow. ... But you've got to have an educational system. You've got to teach a billion – at least a billion – grown-up people how to use it. ... You've got to have over a million teachers just to show people how to use this. You can't get that going in fifty years. So it's out.[53]

For Darwin, the closest thing to a success story was Japan: "Japan is one of the most conscious countries about it. They've succeeded in overcoming some parts of the problem by legalizing abortion, and for the last few years ... they had a million abortions a year." When the interviewer suggested that a nuclear war might affect population trends, Darwin's objection was numerical, not moral. "[W]hat does it mean – 100 million dead. One hundred million are replaced in three years. You've got to have a war like that every three years, you see. You must keep to arithmetic on this thing." Darwin had been Master of Christ's College, Cambridge, and director of the United Kingdom's National Physical Laboratory, the country's chief laboratory for applied physics, during the war. He was hardly a marginal figure.

Nor was John Maynard Smith, one of the United Kingdom's most influential evolutionary biologists. Writing about eugenics in 1965, Maynard Smith's tone was moderate and liberal.[54] He distinguished humane and inhumane policy options and dismissed old-style eugenicists' interest in differential fertility among social classes as unimportant.[55] Yet Maynard Smith was also more optimistic about what he called "negative selectionist eugenics" than its "positive selectionist" cousin. That is, although efforts to increase a population's intelligence through selective breeding (e.g., inseminating willing women with the sperm of the highly intelligent) would have little effect, there was a limited case for preventing some people from breeding:

---

[51] Darwin, *The Next Million Years*, 140–1.

[52] See for example, "Darwin Grandson says there are too many of us," *The New York Post*, March 11, 1956.

[53] Quoted in David Lawrence, "Will World Be Black or Tan in the Next Fifty Years?" *The Evening Independent*, December 4, 1959.

[54] John Maynard Smith "Eugenics and Utopia," *Daedalus* 94, no. 2 (1965): 487–505.

[55] Ibid., 488–9, 502–3.

Selectionist eugenics involves altering the relative number of offspring born to particular kinds of individuals or pairs. In most cases, these procedures are likely to be too ineffective to be worth bothering with. But it is also worth making an effort to prevent individuals who carry deleterious dominant mutations which manifest themselves late in life from having children and to prevent the carriers of the same recessive lethal or deleterious gene from marrying one another, although the latter measure would have dysgenic effects in the long run. In the positive field, selectionist eugenics is again likely to be relatively ineffective.[56]

Maynard Smith's article is most interesting for its conclusion:

[R]ecommendations of positive eugenic measures can at the present only distract attention from more urgent and important questions. The most urgent message which biologists have to convey to the public is that if something is not done to arrest the present increase in world population, then that increase will be arrested by war, disease and starvation. Eugenics can wait, birth control cannot.[57]

It would be a leap to suggest that these public intellectuals' concern with population policy was purely opportunistic and nothing more than a way of updating eugenics. More plausibly, Osborn, Pearl, Notestein, Sanger, and others saw the new problem of population growth through the old lens of sterilization and contraception. They framed the problem of population growth as the problem of human fertility rather than, for example, economic arrangements or political institutions. The argument developed and popularized by the Nobel laureate economist Amartya Sen in the 1970s, that famine resulted from political failure rather than an absence of food, would never have occurred to them.[58]

Mainstream scholars' fears overpopulation growth were founded firmly in eugenic concerns with IQ, class-based heredity, and national decline. In 1951, Robert C. Cook, a former professor at the George Washington University and editor of the *Journal of Heredity*, published a book titled *Human Fertility: The Modern Dilemma*.[59] As *New York Times'* reviewer Orville Prescott noted, the book concerned itself as much with too many of the wrong people as with too many people.[60] "Intelligent parents *tend* – on the average – to have intelligent children, and parents with low I. Q.'s *tend* to have stupid ones," Prescott quoted from the book. "And," he reprovingly continued, "in this country the people of superior intelligence, energy and accomplishment do not have enough children

---

[56] Ibid., 502.

[57] Ibid., 503.

[58] Amartya Kumar Sen, *Poverty and Famines: An Essay on Entitlement and Deprivation* (Oxford: Oxford University Press, 1981).

[59] Cook was also a good friend of Alexander Graham Bell, who had worked closely with Davenport and who had, with David Fairchild, asked Cook to edit the *Journal of Heredity*. See Joan Cook, "Robert C. Cook, 92, A Longtime Scholar Of Human Genetics," *New York Times*, January 9, 1991.

[60] Orville Prescott, "Books of the Times," *New York Times*, April 17, 1951, 27, 11/93, AVS.

to reproduce themselves."[61] The book also included an old eugenic argument about the perverse consequences of social policy:

> Economic and educational success works eugenic miracles in reverse in an industrial society. The price for success is a slow, steady, remorseless biological extinction. Those who move up the ladder are, in effect, fractionally sterilized in direct ratio to the degree of success they achieve ...
>
> Today, in the United States, the intelligent get degrees, and the diligent and competent get houses and bank accounts and stomach ulcers. But it is the poor and unschooled who beget.[62]

Cook would later join the Population Reference Bureau, another institution created by disillusioned eugenicists moving into population policy. Guy Irving Birch, an unabashed admirer of National Socialist eugenics, and Raymond Pearl founded the Bureau in 1929. Throughout the 1950s, more than a decade before Paul Ehrlich published his bestselling book, *The Population Bomb*, Cook continued to warn of the perils of population growth.[63] The term, in fact, had been coined by Hugh Moore, inventor of the paper Dixie cup, who used the same title for his influential pamphlet on the perils of rapid population increase. Moore, who would go on to lead two of the main anti–population-growth organizations in the 1960s, justified his support for population control squarely in posteugenic concerns about differential fertility. He told his supporters in 1966:

> You doubtless have read that the population of the United States is expected to exceed three hundred and fifty million in the next thirty years. The effect on life and living of this doubling of our numbers is beyond calculation. Over-crowded cities, polluted air and water, countless unwanted and suffering children, skyrocketing taxes for welfare! Half of the babies born in some cities are from indigent families on relief. Need we say more?[64]

Moore cited William Vogt, critic of the World Health Organization's misguided humanitarianism, as his inspiration.[65] Moore would later become an enthusiastic supporter of abortion and of the Supreme Court decision in *Roe v. Wade* that created a national right to it – not because it provided a "choice" for women but because it was a way of reducing population.[66]

By the early 1950s, then, almost all major eugenicist thinkers from the 1930s had moved into population policy. Funded by rich patrons, they entered the 1960s with large budgets, dedicated staff, and a highly educated and articulate

---

[61] Ibid.
[62] Robert C. Cook, *Human Fertility: The Modern Dilemma* (New York: W. Sloane, 1951), 238.
[63] Robert C. Cook, "Wonder Drugs Made Birth Rate a Global Peril," August 2, 1959, 11/96, AVS. In his book, Ehrlich gives credit for the title to Hugh Moore. Paul R. Ehrlich, *The Population Bomb* (New York: Ballantine, 1968), unpaginated page preceding title page.
[64] Quoted in Kluchin, *Fit to Be Tied*, 33.
[65] Dowbiggin, *Sterilization Movement*, 114.
[66] Ibid., 116.

TABLE 10.1 *Anti-Population Growth Foundations (as named in the 1960s)*

Organization	Date Founded	Founders & Key Members	Previous Name
International Union for the Scientific Study of Population	1928	Raymond Pearl	International Union for the Scientific Investigation of Population Problems
Population Reference Bureau	1929	Guy Irving Burch, Raymond Pearl	–
Association for Voluntary Sterilization (1965)	1937	Marian S. Olden	Sterilization League of New Jersey, Sterilization League for Human Betterment, Birthright, Inc., Human Betterment Association of America; later, *Human Betterment Association for Voluntary Sterilization, Association for Voluntary Surgical Contraception, EngenderHealth*
International Planned Parenthood Foundation	1952	Margaret Sanger	Planned Parenthood Federation of America
The Population Council	1952	J. D. Rockefeller Frederick Osborn	–
Pathfinder International	1957	Clarence Gamble	Pathfinder Fund
World Population Emergency Campaign (folded into Planned Parenthood Federation of America in 1961)	1960	Hugh Moore	–
Population Action International	1965	Hugh Moore	Population Crisis Committee
Zero Population Growth (renamed Population Connection in 2002)	1968	Paul Ehrlich, Richard Bowers Charles Lee Remington	–

leadership ready to warn the American public of the perils of population growth. Over the next decade and a half, new organizations would be added to the list. Table 10.1 provides an overview of the major organizations, their history, and their funders.

By the mid-1960s, seven major organizations were warning of the perils of population growth; all were founded by individuals with deep, if varying, commitments to eugenic principles; and all channeled arguments and resources into convincing legislators and the public of the need for birth control, including sterilization. As these quotations made clear, eugenics informed these organizations' purpose, remit, and recommendations long after 1945.

These organizations had two effects. First, they shifted discourse. Through their tireless lobbying in favor of birth control and above all sterilization, they made sterilization largely acceptable to mainstream American opinion by the 1970s as a sensible option for those wishing to have no more children or none at all.[67] Second, most dramatically, these organizations contributed greatly to the spread of often-coercive sterilization in the developing world. Family planning organizations' interest in using sterilization in population policy extended out from the East Coast of the United States into the South and, from there, on to the developing world. This paternalist project resulted in a litany of abuses.

One such abuse involved the intrauterine device, the IUD, inserted into women's uteri to prevent pregnancy. Clarence Gamble and Robert Dickinson were enthusiastic advocates of the IUD, but it was only in the late 1950s that the U.S. medical establishment endorsed it after the original wire was replaced with easier-to-use plastic.[68] Alan Guttmacher, who became president of Planned Parenthood-World Population in 1962, was another enthusiastic supporter and had supported tests on a new plastic IUD while at Mt. Sinai Hospital in New York. "As I see it," Guttmacher wrote to the president of pharmaceutical firm G. D. Searle in 1964, "IUDs have special application to the underdeveloped areas where two things are lacking: one, money, and the other sustained motivation. No contraceptive could be cheaper, and also, once the damn thing is in, the patient cannot change her mind. In fact, we can hope she will forget it's there and perhaps in several months wonder why she has not conceived."[69]

American foundations immediately opened their hearts and their wallets. The Population Council put up $2.5 million for testing.[70] India received special attention. The country had long been a source of horrified fascination for Western advocates of birth control. Confusing narrow streets with narrow planets, "[s]uch figures as Julian Huxley, Claude Lévi-Strauss, Dwight Eisenhower, and Paul Ehrlich – whose population bomb provided the most famous account – all reported a feeling of being overwhelmed by the corporeality of Indian crowds at close quarters, and a new commitment to the cause of

[67] See ibid., chapter 5.
[68] Dowbiggin, *Sterilization Movement*, 138.
[69] Alan Guttmacher to John Searle, Planned Parenthood Federation of American Papers ("PPFA Papers"), Sophia Smith Collection, Smith College, Northampton, MA. Quoted in Andrea Tone, *Devices and Desires: A History of Contraception in America* (New York: Hill and Wang, 2001), 271.
[70] Dowbiggin, *Sterilization Movement*, 139.

population control."[71] Expert consultants sent to India by the United Nations (UN), World Bank, and Ford Foundation recommended unanimously in favor of IUDs. This was hardly surprising: the Population Commission had worked with the World Bank and the Ford Foundation to select the experts.[72]

There was certainly support within the Indian government for action on population policy, but pressure in favor of the IUD came from the West. As the Population Council sent more than 1 million of the devices to India in the 1960s (before building a factory in the country itself), USAID, the World Bank, the UN, and the Ford Foundation, which collectively provided most of India's annual $1.5 billion in aid, pressured India to accept them.[73] The U.S. government piled on further pressure. Convinced by his advisors that a dollar spent on population growth was worth one hundred spent on industrial development, President Lyndon B. Johnson effectively linked food aid with bold action on population growth.[74] Putting it as only he would, LBJ told a member of his staff, "I'm not going to piss away foreign aid in nations where they refuse to deal with their own population problems."[75]

India "dealt" with its problem in the manner prescribed. The Indian government set targets for IUD insertions, government doctors received bonuses for every 150 IUD insertions, and private practices received five rupees per insertion. Delhi – under pressure again from the Ford Foundation, UN, IPPF, and World Bank – began paying both states (11 rupees) and individuals (3–7 rupees) for every IUD insertion beginning October 27, 1966.[76]

The results were tragic. Individuals accepting IUDs were choosing between contraception and starvation, especially as India teetered on the brink of famine in the later 1960s. Medical staff were ill trained and hugely rushed. And follow-up medical visits were rare, if not nonexistent. Reports of serious complications were relayed back to the Population Council, the single greatest backer of IUD programs: women were suffering perforated uteri at fifteen times expected rates, ectopic pregnancy (in which a pregnancy develops outside the uterus), and, in some areas, heavy bleeding in fully 50 percent of all cases.[77] The data are incomplete, but estimates suggest that tens of thousands of women experienced pain and bleeding, pelvic inflammatory disease, ectopic pregnancies, and septic abortions.[78]

The results were also largely predictable. As a doctor told a Population Council conference in 1962, IUDs are "horrible things, they produce infection, they are outmoded and not worth using."[79] But for all that, they were worth it,

[71] Connelly, *Fatal Misconception*, 89–90.
[72] Ibid., 216.
[73] Ibid.
[74] Ibid. 212–21.
[75] Quoted ibid., 221.
[76] Ibid., 222–7.
[77] Ibid., 223–4.
[78] Ibid., 220.
[79] Dowbiggin, *Sterilization Movement*, 138.

for when they went wrong, the consequence was sterility (a condition the Population Council hardly opposed on principle), birth defects, or death. If a patient developed an infection and lost her ovaries and fallopian tubes, the doctor continued, "How serious is that for the particular patient and for the population of the world in general? Not very. . . . Perhaps the patient is expendable in the general scheme of things, particularly if the infection she acquires is sterilizing but not lethal."[80] Nothing in the Population Council's subsequent behavior suggests that it disagreed.

The IUD experiment did not end in the 1960s; on the contrary, the population lobby redoubled its efforts in the early 1970s. The Pathfinder Fund (Gamble's organization) issued 59,000 Dalkon Shield IUDs to thirty-five family planning projects across the developing world between 1972 and 1974. The result was at least eighteen deaths, 200,000 miscarriages, and a large number of birth defects.[81]

The population lobby was equally enthusiastic in pursuing putatively voluntary but actually coerced sterilization. Both Planned Parenthood and the Association for Voluntary Sterilization identified Japan, China, and India early on as ripe for both mass birth control and mass sterilization. Planned Parenthood's website tells us today that "for nearly 100 years, Planned Parenthood has promoted a commonsense approach to women's health and well-being, based on respect for each individual's right to make informed, independent decisions about health, sex, and family planning."[82] The website neglects to mention that the IPPF was in fact formally founded in Bombay, India on November 29, 1952 and that its chief concern throughout the 1950s was not individual choice. Its focus was, rather, population control. In 1959, Planned Parenthood held its "International Conference on Planned Parenthood," again in India, this time in Delhi. C. P. Blacker, former secretary of the Eugenics Society, was its first director, and William Vogt met with Blacker and Margaret Sanger in the latter's suite at the Taj Mahal Hotel to work out the details.[83] The conference delegates provided strong support for the sterilization of couples with three children.[84] They were not disappointed. Within a month of the 1952 conference and with strong IPPF support, India launched a family planning program.[85] The numbers of sterilizations were high in the 1960s but never reached a million. Then, following the suspension of civil liberties in 1975, sterilizations sharply increased: 2.7 million in 1975–6 and 5.6 million in 1976–7.[86] IPPF, meanwhile, joined forces with a Hugh Moore-funded

[80] Ibid. 138–9.
[81] Ibid., 139.
[82] "Planned Parenthood: Who we are," http://www.plannedparenthood.org/about-us/who-we-are-4648.htm (accessed January 22, 2013).
[83] Connelly, *Fatal Misconception*, 168.
[84] "Wide Support for Sterilization: Family Planners Form Study Groups," n.d., 11/96, AVS.
[85] Dowbiggin, *Sterilization Movement*, 122.
[86] Trombley, *Right to Reproduce*, 222.

organization, the World Population Emergency Campaign, to form Planned Parenthood-World Population.[87] Not for the first time in the history of eugenics, sterilization, and birth control, there was much in a name.

India's sterilization program developed in three phases. In the first phase, from 1966 to 1970, the number of "voluntary" sterilizations ramped up following the decision to pay for sterilization along with IUDs. In 1967–8, 150,000 peopled opted for paid sterilization. States suffering from drought-induced hunger (Madhya Pradesh, Uttar Pradesh, and Orissa) showed a particularly high rate of "acceptance" of the procedure.[88] The Ford Foundation, Population Council, and World Bank promised that sterilization would prevent starvation; in fact, starvation led to sterilization. This period also showed the first signs of the patterns of abuse that would become the hallmark of later phases: eighty-year-old men and "uncomprehending mental patients" were sterilized, and an unknown number of people died from complications.[89]

In the second phase, from 1970 to 1974, mass vasectomy camps were set up across the country. While singers and musicians entertained family members, and as electric scoreboards kept track of the tally of sterilizations, one man after another went into tents for a vasectomy.[90] In one camp alone, almost 63,000 men were sterilized in a single month in 1971. USAID, prodded by the Association for Voluntary Sterilization, provided much of the funding.[91]

The third phase of the program began in 1975, on the eve of Indira Gandhi's announcement of a state of emergency. Subtle coercion became full-blown. Roaming bands of paid "motivators" harassed, intimidated, and at times physically forced men to agree to vasectomies. Some clients died from infections.[92] Afterward, motivators and doctors would often split the bulk of the payment, leaving the victim with only a paltry sum.[93] Indian states ratcheted up the pressure further. The Punjab state government withdrew medical aid subsidies after the third child. The Maharashtra state government passed a bill for compulsory sterilization of the father and compulsory abortion for the mother after the third child.[94] The Bihar government withdrew food ration cards for third and any subsequent children.[95] As in the IUD campaigns, Western institutions and Western money underwrote the programs: USAID, the Ford Foundation,

---

[87] Connelly, *Fatal Misconception*, 189.
[88] Ibid., 226.
[89] Ibid., 227.
[90] Dowbiggin, *Sterilization Movement*, 190.
[91] Ibid., 10, 190.
[92] Trombley, *Right to Reproduce*, 219–20.
[93] Dowbiggin, *Sterilization Movement*, 191.
[94] Penelope ReVelle and Charles ReVelle, *The Global Environment: Securing a Sustainable Future* (Boston: Jones and Barlett, 1992), 137.
[95] Trombley, *Right to Reproduce*, 221.

the IPPF, and the British government provided millions in funding,[96] all convinced of the urgency of the population growth crisis.

In a review of a critical history of the population control movement, *New York Times* reviewer Nicholas D. Kristof concludes, "[i]t's certainly fair of [historian Matthew] Connelly to dredge up the forced sterilizations, the casual disregard for injuries caused by IUDs, the racism and sexism and all the rest – but we also need to remember that all that is history."[97] "All that" is, in fact, not merely history. For all the rhetoric about reproductive choice and empowering women, the major population lobbies are still in the business of social engineering. In the 1990s, the Association for Voluntary Sterilization (AVS) – which changed its name in 1984 to the Association for Voluntary Surgical Contraception (AVSC) – was still active, once again with USAID funds. It provided training and equipment to the Peruvian branch of Planned Parenthood, an organization intimately involved in that country's aggressive sterilization campaign between 1996 and 2000.[98] That campaign had the worst elements of the Indian campaign: festivals, roundups, and anesthetic in, at best, half the cases. Some 200,000–300,000 poor, indigenous women were forcibly sterilized.[99] In the midst of this abuse, international population organizations applauded the reduction Peru's population growth rate from 3.2 percent to 1.17 percent.[100]

The population lobby's approach to fertility has been, at worst, coercive and, at best, simplistic. On the latter, organizations such as the AVS and IPPF oversimplify the problem they claim to wish to solve, and they do so in a manner that carries the argument in favor of sterilization. That is, they locate the complex effects of poverty, hunger, and instability in the now-familiar single cause of excessively high fertility rates. From the 1950s to today, the major foundations concerned with population policy have maintained that developing countries' wealth depends on low birth rates. They failed to consider the possibility that the relationship was the reverse: that wealth, underpinned by procedural justice, trustworthy, democratic institutions, property rights, education (above all for women), proper sanitation, and a developed infrastructure all might lead to low birth rates. This oversight is curious for two reasons. First, they and their

[96] Michael E. Latham, *The Right Kind of Revolution: Modernization, Development, and U.S. Foreign Policy from the Cold War to the Present* (Ithaca: Cornell University Press, 2011), 108; Trombley, *Right to Reproduce*, 215; Oscar Harkavy and Krishna Roy, "Emergence of the Indian National Family Planning Program," in *The Global Family Planning Revolution: Three Decades of Population Policies and Programs*, ed. Warren C. Robinson and John A. Ross (Washington, DC: The World Bank, 2007), 319–20.

[97] Nicolas D. Kristof, "Birth Control for Others," review of *Fatal Misconception* by Matthew Connelly, *New York Times*, March 23, 2008.

[98] Dowbiggin, *Sterilization Movement*, 219.

[99] Ibid.

[100] Brita Schmidt, "Forced Sterilization in Peru," *Committee on Women, Population, and the Environment*, July 12, 2006. Available at: http://www.cwpe.org/resources/popcontrol/forcedsterperu.

predecessors incessantly decried low birth rates among rich Americans and Europeans, which should have suggested that there was a basic relationship between higher levels of wealth and lower levels of birth. Second, birth rates in the West were on the verge of a dramatic fall at precisely the moment that the population lobby screamed loudest about them. The charts that Paul Ehrlich and other anti-population growth activists showed the academic conferences and talk show hosts accurately showed that the vast majority of births on the planet had occurred since 1800, and indeed since 1945. What those charts failed to show is that birth rates in the West were on a sharp, downward trajectory from the 1960s[101]; a chart tracing birth rates from, say, 1960 to 1975 would have shown a line moving sharply in the opposite direction: downward. And this is indeed precisely what has happened: across Europe, North America, and the more affluent parts of East Asia, birth rates have plummeted.

Since anti-population growth advocates were convinced that the lines of causality ran from low birth rates to high wealth levels rather than the other way around, they drew no conclusion from rapidly dropping birth rates. Putting this another way, as population control enthusiasts viewed the problems they wished to solve with but single framing, they have also pursued only a single solution: reducing the birth rate, often at any cost. Coerced sterilization has been the most obvious cost, and it has been heavily paid by society's poorest and most vulnerable members.

But population control advocates have not been shy about suggesting other options. Paul Ehrlich's *The Population Bomb* was and is probably the most famous book on the topic and is still in print. Ehrlich's predictions about population numbers proved broadly correct.[102] Erhlich's recommendations for how to reduce population pressure partly involve what are now, perhaps partly thanks to him, standard environmental programs: returning bottles, recycling, reducing phosphates in detergents, and cutting carbon dioxide emissions through increased public transportation, smaller houses, and so forth. But he also recommends some illiberal measures: denying people the right to choose the number of children they have and encouraging if not requiring abortion, contraception, and sterilization.[103] Ehrlich urged his readers to join Zero Population Growth, a lobby group he co-founded with Richard Bowles and Charles Lee Remington. Bowles pressed for coerced sterilization and abortion. "Voluntarism," he declared, "is a farce."[104] Ehrlich also recommended a neo-imperialist project in which the rich West sorts the developing world into

---

[101] On this, see Susan Greenhalgh, *Just One Child: Science and Policy in Deng's China* (Berkeley: University of California Press, 2008), 143–52.

[102] Kenneth R. Weiss, "Fertility Rates Fall, but Global Population Explosion Goes On," 5 parts, *Los Angeles Times*, July 22, 2012.

[103] Ehrlich, *Population Bomb*, chapters 2 and 4.

[104] Quoted in Kluchin, *Fit to Be Tied*, 34.

countries that can be saved and those that are lost.[105] The former are to be given aid for industrialization and other forms of development; the latter are to be abandoned. "Apparently brutal and heartless" but necessary for "cutting out the cancer" of population growth, selective sterilization under Ehrlich's recommendation would give way to selective starvation.[106]

---

[105] Ehrlich, *Population Bomb*, chapter 4.
[106] Ibid., 152.

PART C

II

The Sterilized

Voices from Alberta and Oregon

When we spoke to her in 2003, Laverne Throckhorn was a sweet, somewhat plaintive woman in her eighties living with her two brothers in a small bungalow in Portland.[1] She was born in St. John, Oregon, in 1920, and had three brothers: Harry, Sydney, and Alvin. Her parents also had two babies who died at or shortly after birth. All of the children suffered some degree of developmental disability. Laverne's family was poor, and her mother's serious illness made their difficult lives worse yet. For reasons that are not clear, social services – or what passed for them in the 1920s – learned of the Throckhorns' situation and attempted to take the children. The family moved several times, always staying just one step ahead of the state. In the small town of Hood River, however, the state caught up with them. Alvin, who had a serious speech impediment, was walking through the streets one day unaccompanied. Someone reported him to social services. The authorities picked Alvin up, made him show them where the family was living, and took the four children away.

Social services delivered them to the State Institution for the Feeble Minded, later renamed Fairview Hospital and Training Center in 1933.[2] The institution had been opened in 1907, outside the state capital of Salem, to house the state's "mentally defective." There, as elsewhere, this label lumped together two categories of people: the cognitively impaired (or mentally handicapped) and the mentally ill (e.g., those suffering from severe depression). This institution looked like most institutions that existed before the Second World War. It was attractive, grand even, and set on well-landscaped grounds; in the 1920s, it was approached via a winding road known locally as "Lunatics' Lane." In its day, Fairview resembled a feudal estate: the director lived in a grand house on the site, and all permanent staff was housed in detached buildings surrounding the large,

---

[1] This paragraph is based on an interview with Laverne Throckhorn, Portland, OR, July 17, 2003.
[2] Mental Health and Developmental Disability Services Division Overview, 2. Oregon State Archives (OSA).

columned main structure. Also housed in the peripheral buildings were the estate's workers – but in place of serfs, it was patients who were set to work at planting vegetables, milking cows, gathering eggs, and raising other animals.[3] Such farms were common. As early as 1910, most such institutions had farm colonies adjacent to or near the main facility, and many institutions bought or rented additional farms.[4] This self-sufficiency meant that colonies could save a considerable amount of money on both food and labor.[5] Against the bitter protests of most families of the remaining patients, the state closed the institution in 2000. The once-bustling complex now sits in empty, eerie silence.

When Fairview was closed, some parents had to confront the fact that they had been lied to and that their children, with a few exceptions, could survive perfectly well in assisted community care and, thus, at home with their families. The guilt created by this realization may have motivated some parents to oppose the closing.[6]

Most parents did follow their doctor's or social worker's advice, but Laverne's family was either stronger or more stubborn. Against advice, her parents visited the children and sought their release. Their only success was with Laverne. Before turning thirteen, the doctors sterilized her. She was told that she was not to have babies, that they would inherit her disabilities, and that she had to sign a document. Laverne did not know what she was signing, but it was a consent form. She was sterilized on the basis of her condition and as a prerequisite to her release, although this release did not occur until much later. Because they were all part of the same family, Sydney, Alvin, and Harry – whom Laverne was not allowed to see while at Fairview – were also sterilized.[7] Although the destruction of the institution's records makes it impossible to verify, it is unlikely that any of the cases went before the Oregon Eugenics Board, which would mean that the sterilizations had been performed illegally. Laverne's sterilization nonetheless secured, eventually, her release from the institution, and she was able to join her parents at home. Harry, Sydney, and Alvin were not so lucky: they stayed at Fairview for more than fifty years. This was not unusual. Even though sterilization was usually a precondition for release, it did not constitute an entitlement to the same; discharge from these institutions was a matter of superintendent discretion.

During the wave of deinstitutionalizations in the 1970s, Harry and Sydney were released. They found Laverne, helped their ailing father, and eventually moved in together. Alvin had little contact with his family, and, when he died in the early 1980s, his and their penury meant that his body lay at the morgue for

---

[3] Interview with Dennis Heath, Salem, OR, July 19, 2003.
[4] Trent, *Inventing the Feeble Mind*, 107.
[5] Ibid., 105–7.
[6] Interview with Bill Lynch (Council on Developmental Disabilities), Portland, OR, July 16, 2003.
[7] Interview with William West (Adult and Family Case Coordinator, Association for Retarded Citizens [ARC]), Portland, OR, July 16–18, 2003.

six weeks without being buried. Only the intervention of activists and public outrage brought this final indignity to an end, and it led Oregon to reinstitute state-sponsored burials for the indigent.[8] When Alvin died, the three other siblings were living together on one meager income in a house without heating – proof, were any needed, that deinstitutionalization hardly solved all of the problems of people with developmental disabilities.

PARENTAL INDIFFERENCE: THE CASES OF VELMA HAYES AND LEILANI MUIR

Velma Hayes

In early 2002, a Pulitzer Prize-winning journalist, Julie Sullivan, was working for *The Oregonian*, an Oregon newspaper that, almost ninety years earlier, had championed the cause of eugenic sterilization and showered praise on the state's first sterilization law.[9] A colleague told her about a Virginia campaign in favor of an official state apology for coerced sterilizations. Sullivan began researching the issue, published pieces on Oregon's eugenic experience, and joined a coalition pressing the state's departing governor, John Kitzhaber, to issue a formal apology.[10] The state eventually apologized for its past actions; the newspaper did not. One woman read of the sterilizations and the apology, and she called Sullivan.[11] Her name was Velma Hayes.

Velma Hayes was born on May 15, 1934, one of twin sisters. She entered her second year with a vengeance: crying, biting, and refusing all food but sweets. She was an irritating two-year-old, but in this she was hardly alone. Her mother, who had wanted a boy, decided for whatever reason to get rid of her. Telling her relatives that Velma needed her legs straightened,[12] she committed her two-year-old daughter to Fairview. From that moment on, the twin sisters – Velma and Thelma – lived parallel lives. Their mother raised Thelma in a stable if at times cool family. Velma saw none of her relatives. Thelma was put through high school. Velma was denied any academic education, and, when she finally left the institution, she could not read or write. Thelma played in parks, playgrounds,

---

[8] Ibid.

[9] On November 2, 1913, the paper argued: "The intent of the [sterilization] legislation is to prevent low criminals and bestial men and women from propagating their species.... Severe laws to prevent the deterioration of breeds in horses and cattle are passed and enforced without protest. But laws for the protection of the human race are denounced as an invasion and inversion of human liberty!" Quoted in Largent, "'Greatest Curse,'" 198.

[10] Julie Sullivan, "Eugenics Records Shredded: Documentation of the Last 20 Years of Forced Sterilizations in Oregon Has Mysteriously Disappeared from State Archives," *The Oregonian* (Portland, OR), July 30, 2002.

[11] Interview with Velma Hayes, Portland, OR, July 21, 2003.

[12] Julie Sullivan, "Misjudged and Sterilized, Velma Hayes Remains Haunted by All She Lost," *The Oregonian*, August 11, 2002.

and gardens; Velma knew only Fairview's regime of work, discipline, and routinized punishment. Thelma married and had three children; Velma's fallopian tubes were cut.[13]

Velma heard of the operation while she was at Fairview. Girls would disappear and come back a few days later with a scar across their abdomens. In 1948, it was her turn. On November 1, the Board of Eugenics had its quarterly meeting and invited nine members of the Board of Health and the superintendents of the Oregon State Hospital, Eastern Oregon State Hospital, Fairview Home, and the State Penitentiary to participate.[14] Until 1967, the targets of sterilization were not permitted to appear before the Board. The Board's minutes were shredded in the 1980s, so there is no record of how long it took to deliberate.[15] The Board decided to sterilize her. Velma pleaded hopelessly with attendants as her mother signed the consent form. The date was arranged for two years later. A fire had destroyed Fairview's infirmary, so Velma was dispatched to the Oregon State Hospital for the procedure.[16] As one more element of bitter irony, Velma was working in a very particular capacity for then-Superintendent Irvin B. Hill in his large and elegant home when he recommended her sterilization: she was caring for his children.[17] On January 12, 1950, as a snowstorm blew in, doctors cut and tied Velma's fallopian tubes in a seventy-two-minute operation.[18] The snow meant that she could not leave, and the doctors left her to recover among the seriously mentally ill who screamed and moaned around her.

In all likelihood, Velma should not have been at Fairview at all. The state placed her in three foster homes, and, in all three, the husbands sexually harassed her. Her final placement was a happy one, but her mother then finally decided to bring her home. Velma asked to go to school, but her mother forced her to work. She escaped about a year later, when she married at the age of nineteen. Her first marriage was disastrous, but she went on to have two happy ones, both ended by her husbands' deaths. Despite a complete absence of schooling, she earned a living for more than fifty years. During her second marriage, she had two operations to reverse the sterilization. They appeared to work: she became pregnant, but she later miscarried. The doctors told her that the damage she had suffered from the sterilization operation made childbirth impossible. She lives a life that is outwardly cheerful but inwardly scarred by all that her mother and the state denied her: an education, a professional career, children, and

---

[13] Velma Hayes, interview.
[14] Sullivan, "Misjudged and Sterilized."
[15] Interview with John Murphy (President, Portland Habilitation Center), Portland, OR, July 21, 2003; Julie Sullivan, "State of Oregon Shredded Forced Sterilization Records," *Oregonian*, July 30, 2002.
[16] Velma Hayes, interview.
[17] Sullivan, "Misjudged and Sterilized."
[18] Ibid.

grandchildren. She is a monument, one of many, that recalls sterilization policy at its most arbitrary and abusive.

Leilani Muir

Leilani Muir was born in 1944 in Calgary, Alberta, to a mother who had not wanted any girls. An alcoholic, she beat Leilani frequently. Like Velma's mother in Oregon, the local mental institution allowed Mrs. Muir to be rid of her daughter: in July 1955, Leilani's mother committed her, at age ten, to the Provincial Training School. At the school, Leilani worked quietly and avoided contact with other patients, although her attempt at keeping a low profile did not spare her sexual harassment at the hands of attendants.[19] When she was fourteen, in 1959, she was asked to appear before the Alberta Eugenics Board for classification and recommendation for – or, presumably, against – sterilization. In its record, the Board noted that Leilani was a "pleasant-looking child," who spoke "easily and volubly."[20] The space where the IQ score was to be filled in was left blank. Based on this information, the Board members had to decide whether Leilani was "backward," "dull," "borderline," or a "moron"; they underlined the last. After the Board meeting, Leilani was held back at eleven in the morning while the others went to school. The attendants told her that she was going to have her appendix removed. Conditioned by a mother who told her never to ask questions, she did not object. Leilani and four other girls were taken to the institution's infirmary and sterilized. The reasons: "the danger of the transmission to progeny of mental deficiency" and their incapacity for "intelligent parenthood."[21]

Still believing that her scar resulted from an operation to remove her appendix, Leilani was discharged from the Provincial Training School in March 1965, at the age of twenty. When she was twenty-four, her mother, who still tyrannized her daughter, told Leilani that no man would ever love her. The phrase turned over and over in Leilani's head, and she decided to visit her doctor. The physician examined Leilani and told her that all but a quarter inch of her fallopian tubes were gone.

In the early 1970s, Leilani began a lobbying effort that would eventually blow Canada's sterilization history wide open, an effort that took more than twenty years. In the meantime, Leilani sought to rebuild her life. She married twice, the second time happily. Fearing rejection, she told her husbands about neither the Provincial Training School nor the sterilization. Discreetly, she had two

---

[19] Interview with Leilani Muir, Edmonton, AB, October 18, 2003.
[20] Interview with Jon Faulds (attorney for plaintiffs in class-action suit), Edmonton, AB, October 17, 2003.
[21] *The Sterilization of Leilani Muir*, directed by Glynis Whiting (Montreal: National Film Board of Canada, 1996).

operations to reverse the tubal ligation. Neither succeeded: one doctor told her in 1974 that her internal organs looked like they had been through a slaughter-house. She then turned to adoption, which looked like it might succeed until the agency uncovered her history at Provincial Training School (now called Michener Services) and rejected her application. Her marriage began to break down, and her husband turned to alcohol. Leilani's life would eventually improve, but, before that, it got worse: by 1989, she had lost her job, her husband, and her dignity. In December of that year, she emptied the contents of three prescriptions onto a table next to the phone. In a decision that would eventually transform her life and that of hundreds of sterilization victims, she chose to pick up the phone. She told her story to lawyers who launched a suit that both brought substantial financial compensation to Leilani and other victims and, equally importantly, drew the attention of journalists, activists, and scholars to Alberta's experiment in coerced eugenic sterilization.

ADOPTION: KEN NELSON

One of the practices of the mental health field in the first half of the twentieth century, and the broader society that this field mirrored, was to take external physical abnormalities as an indicator of inner intellectual and moral inadequacies. Severe – or not so severe – physical handicaps, blurred speech patterns, and epileptic fits were all bases of decisions to commit people to mental health institutions. Subsequently, they were usually the same bases for the decision to sterilize. Ken Nelson, who suffers from cerebral palsy, is a typical case in this respect.

Ken was born in Edmonton, Alberta, in 1943. He did not know his mother, who put him up for adoption almost immediately.[22] Ken was nervous in school, and his adoptive parents, the Nelsons, gave up and sent him to a home for problem children. One day, the director told him he was going to Red Deer, a city about 150 kilometers south of Edmonton. The next day, he entered the Provincial Training School. At it, he observed the catalogue of staff-on-patient and patient-on-patient abuse, both random and routinized, that was common in mental health institutions. The boys, segregated from the girls, spoke to each other, and one of the things they spoke about was "the operation." They talked of "being fixed" or "snipped." A few saw it as a badge of pride, a mark of their entry into a group of insiders, or as a license for uninhibited sexual experimentation.[23] Ken must have heard these conversations, but he did not understand their content. One day, in 1957, he was told that he would have an operation to have his appendix removed – the standard story at the Provincial Training School. Another boy ominously told him, "I know where you'll be for the next three days." Ken asked, "How do you know?" The answer: "I just know."

[22] This paragraph is based on an interview with Ken Nelson, Edmonton, AB, October 14, 2003.
[23] Interview with Gordon Bullivant (Executive Director, Foothills Academy), Calgary, AB, October 16, 2003.

Ken was led to the School's infirmary with several other boys. They were given an injection and put to sleep.[24] He awoke after the operation and looked at himself. His scar horrified him. He asked the attendants: "Why did this happen to me? How come you let this happen?" They told him: "That's none of your business. We're in charge here. We run the place. ... And you're not human."[25] Although sterilization was often sold as a ticket to release, it was so not for Ken. He remained at the School until he was thirty, when the new policy of deinstitutionalization led to his release.

THE HOPE FOR SOCIAL MOBILITY: RUTH MORRIS

Some parents, perhaps those of a more naïve or literalist bent, saw in these institutions the opportunity for an education that they could not otherwise afford to give their children.[26] Such was the experience of Ruth Morris. Ruth was born in eastern Oregon on April 30, 1942. As a young girl, she would travel with her mother to Salem, where she would meet officials at what looked like a hospital. Ruth learned later that her mother was attempting to secure her a spot at Fairview, although she only succeeded in securing a spot on the institution's waiting list. In the late 1950s, there were as many as 500 children on the waiting list at any given time, and wait periods of up to two years were not uncommon.[27] One day, when Ruth was thirteen, her mother went to bed with a violent cough and died soon thereafter. Ruth tended house for a few months for her father, until they received word that a place had opened up for her at Fairview. Her father, believing it was a sort of state boarding school, took her.

During the intake process at Fairview, Ruth saw her father sign some papers, which she was also asked to sign. She did so without reading them. They were her commitment papers. Within a few hours of her father leaving, Ruth felt that she had been left in prison, and she made a futile attempt to secure her release. She was stuck. For the next two years, she kept her head down, did her chores, and avoided contact with other patients. That she was quiet no doubt helped her: many witnesses emphasized the importance of saying little, of not responding when spoken to by staff, and – above all – of not complaining. Those who were quiet had less trouble. The strategy worked for Ruth.

She suffered the collective punishments – usually mild ones in the form of denied movies or dances – silently, and she suffered no individual ones. Every morning, she jumped out of bed before she heard the familiar sound of clanging keys as the ward nurse approached. Anyone brave, foolish, or simply unfortunate enough to be still in bed when the nurse arrived would find themselves

---

[24] The next paragraph is based on the interview with Ken Nelson.

[25] Ken Nelson, interview included in *The Sterilization of Leilani Muir*.

[26] The next three paragraphs are based on an interview with Ruth Morris, Eugene, OR, July 22, 2003.

[27] "In Our Care" (video, 1959), http://vimeo.com/365508 (accessed January 22, 2013).

yanked out of bed. The closest thing to a skill Ruth learned at Fairview was physical therapy, but it provided her no real basis for practicing the profession on the outside. Her experience was nonetheless much better than most, perhaps because her cottage – Magruder – lacked the brutality that seemed to characterize another cottage, Kozer. But, at its best, Fairview was still pretty grim, and she received one day the inevitable call. In 1969, her father accompanied her to a meeting of the Board of Social Protection. The chairman told her father that Ruth would be operated on. Her father accepted it with the same passivity that he had accepted all else that happened to his daughter, and he signed on the indicated line. Not understanding the document or knowing what the operation was for, Ruth signed it also. The attendants took her to the operating room and gave her an injection; she fell asleep and awoke to the words "get up." She could not move. Her abdomen ached.

About a month after her operation, Fairview transferred Ruth out to a half-way home and then into the community. She began working in unskilled jobs and has been regularly employed ever since. It was not until several years after leaving Fairview that her father told her that she was unable to have children. Although she had not planned to have children, Ruth nonetheless felt that half her life had been stolen. She never told her father how she felt, and he never apologized.

THE DISCIPLINARY SLIPPERY SLOPE: RITA FAYANT

In some cases, mental health institutions became a sort of net that caught people who had been spat out of other institutions. When there was no room at other institutions, or when the workers in them were unwilling or unable to cope with difficult children, the kids found themselves passed on to mental health institutions. This was the experience of Rita Fayant.

Rita was born in rural Alberta. Her mother gave her up for adoption at the age of three, and she went to live with an elderly First Nations couple. Her life was one of work and grinding poverty, and she hated it. After running away several times, her foster parents placed her in the care of the nuns at the Good Shepherd Home, a Roman Catholic Mission in Edmonton. She ran away a few more times, performed poorly on standardized tests, and her reports got increasingly worse. The nuns felt they could not deal with Rita, and they, too, passed her on. One day, when she was about fifteen, they told her she was going to the Provincial Training School in Red Deer for, as the name might have implied, training. There was something extraordinary about Catholic nuns sending their ward to an institution that continued to practice eugenic sterilization. To be charitable, the nuns might not have known what happened there. But Rita and her peers did: they knew of other girls who had gone there and were sterilized. Rita arrived terrified but clung to the hope that she might escape the operation.

She did not. Rita's stay at the institution was short. After several months of tedious work (cleaning, feeding the mentally ill), she and several other girls were summoned and told that they would have their appendixes taken out. Rita guessed what was about to happen, and she began crying and kicking. There was a van waiting outside with men in white coats to restrain the girls. They were placed in the van and taken for the operation. The next morning, Rita awoke with a large and painful gash on her lower abdomen. She knew she would never have children. She was told she had tuberculosis, although she felt no symptoms, and was sent to another institution outside of Edmonton. She stayed there for another six months, was subjected to psychological testing, and then released. She went on to work and was able later to adopt a child.

The short histories of these individuals' lives cast light on some of the pathways to mental institutions. The poverty of some victims' families and their need for support in raising cognitively impaired children brought them into the sights of a state that saw no middle ground between laissez-faire indifference and taking children from their parents (Laverne Throckhorn). Others were placed there by parents unable to cope with their child's neurological, psychological, or physical conditions (Ken Nelson). Others were placed there by parents who had never wanted their children, or who wished to be rid of those they had (Velma Hayes, Leilani Muir). Still others saw in the institutions the only hope for their children's social and economic advancement (Ruth Morris). Most were given an IQ test in some form at one time or another. Many tests were incompetently administered by individuals with inadequate training.[28] Those tests that were properly administered did not take into account the degree to which the tests presumed social and cultural experience that patients could, by definition, not have, and they did not consider the possibility that IQ scores increase with maturity and education.

Throughout the twentieth century, almost all eugenic sterilizations occurred within state institutions for the feebleminded, of which Fairview was one. There were several paths to these institutions. Social services could deliver patients, as they did Laverne and her three brothers. Parents or other relatives who were unable or unwilling to cope with cognitively impaired or simply difficult children could commit them. Other parents, with touching naiveté, believed that such state-led "training centers" could offer an education that they themselves could not afford to provide. Still other parents never saw the children who ended up in these institutions. When a physically deformed or "mentally defective" child was born in prewar America, particularly in severe cases, doctors would often whisk the child away before the mother could see it. She and the father would later be told the truth and advised that it would be best for them and the child if he or she were taken care of by others. The parents were assured that the child's best interests were at stake, that any contact with the child would only make the

---

[28] Gordon Bullivant, who worked at the Provincial Training Center in Red Deer, Alberta, was handed a pamphlet on the Wechsler intelligence test and told to give it to the patients. The test produced a score that was crucial in the decision to sterilize; he very much doubts to this day that he properly administered the test. Gordon Bullivant, interview.

situation worse, and that they should put the whole matter out of their minds. In some such cases, the child was already dead: in the 1910s and 1920s, doctors in Chicago and elsewhere took it upon themselves to end the lives, in the name of mercy, of "defective" children.[29] But, in most cases, the child was sent to a mental health institution, where it lived out its days.

These victims' narratives put personalities and human lives to statistical data and trends in the second half of the twentieth century. Among cases in which large numbers of sterilizations occurred, only California stands out as a state with an essentially prewar program: 17,399 people were sterilized up to 1945, and 20,108 up to 1980. For America's other states, the postwar decades entailed active sterilization programs and routines.[30]

Postwar sterilizations acquired a firmer eugenic foundation in that the main source of eugenic concern – the supposedly feebleminded – became by far the most important target of sterilization within institutions. Before the Second World War, sterilization centered on people with mental illnesses and developmental disabilities in broadly equal measure. But, by 1948, two cognitively impaired patients were sterilized for every one mentally ill patient.[31] Seven months before the Nazis adopted their sterilization law in July 1933, 16,066 had been sterilized in the United States on eugenic grounds.[32] By 1945, twelve years later, 45,127 people had been sterilized in the United States under eugenic sterilization laws.[33] Of them, 18,830 were men, 26,297 women; 21,331 were insane and 22,153 were classified as feebleminded.[34] In October 1947, Superintendent Fred O. Butler, then at California's Sonoma facility, expressed disappointment that Delaware had surpassed California in the number of annual eugenic sterilizations conducted, and he offered assurance that "as far as we are concerned, there has been no intentional let-up in our sterilization, as we still carry through our policy of insisting on sterilization for those where the operation is indicated."[35] By 1960, another 16,000 people had been sterilized, bringing the total to 61,700.[36] By the time of the final sterilization – possibly in Oregon in the early 1980s – some 63,000 Americans classified as cognitively

---

[29] See, e.g., Martin S. Pernick, *The Black Stork: Eugenics and the Death of "Defective" Babies in American Medicine and Motion Pictures Since 1915* (Oxford: Oxford University Press, 1996), 3–6.

[30] Largent, *Breeding Contempt*, 79–80.

[31] Trent, *Inventing the Feeble Mind*, 222.

[32] "Operations for eugenic sterilizations performed in state institutions under state laws up to January 1, 1933," January 1, 1933, HBF, 51/49, AVS.

[33] "Sterilizations officially reported from States having a Sterilization law up to January 1, 1946," January 1, 1946, SW015.1/72, AVS.

[34] Ibid. The latter figure leaves more than 4,000 unaccounted for. These were sterilizations for other reasons, such as criminality.

[35] F. O. Butler, Sonoma State Hospital, to Robert L. Dickinson, National Committee, October 24, 1947, 13/109, AVS.

[36] 24,340 men, 37,200 women; 27,346 mentally ill, 31,391 "mentally deficient." "Sterilizations reported in the United States up to January 1, 1960," January 1, 1960, 10/90, AVS.

impaired or mentally ill had been sterilized. When Canada is added, where the vast majority of sterilizations occurred in Alberta, the official figure for North America rises to 66,000 (2,822 people in Alberta and about 300 in British Columbia). The number of sterilizations carried out unofficially, without legal authority, or voluntarily in institutions is not known.

These sterilizations were undertaken in institutions to which patients were confined. Sterilizations before and after the war were inseparable from these institutions. Coerced sterilization was, by definition, an arbitrary policy in that, with a few exceptions such as North Carolina, it only affected the institutionalized. It was no coincidence that there was a high probability, if not a certainty, that an individual would be sterilized once confined to certain state institutions. Institutions were virtual vacuums in which the notions of individual rights and informed consent were only cursorily acknowledged, craftily circumvented, or simply ignored.

As the eugenicists never tired of pointing out, however, approximately 90 percent of the population they wished to sterilize lived within the community. If the state did not identify them, or if their parents had either the guile to hide them or the wealth to care for them privately, they could escape sterilization (although some affluent parents did approve sterilizations in private clinics). Physicians working in mental institutions, nominally isolated from the rest of society both geographically and intellectually, continued the mundanely cruel routine of coerced sterilization. It was only in passing through the institution's door that sterilization became, again with few exceptions, a probability, if not a certainty. It is largely the case that once these individuals were institutionalized, their fate was sealed. Until the 1960s, and then only in certain parts of the world, sterilization was intimately bound up with the institution in which it occurred.

All of this means, of course, that American policy makers, mental health superintendents, and other supporters of coerced sterilization were unaffected by the news of Nazi German excesses, justified on eugenic grounds and reported in North American newspapers from 1945. The decisive international response to these crimes similarly seems not to have trickled into the world of North American eugenics.[37] The international community's postwar criticism and dismissal of "race" and alleged interracial differences, including inherited traits, was swift, symbolized by the influential United Nations Educational, Scientific, and Cultural Organization (UNESCO) Declaration on "The Race Question." There was no scientific basis, the Declaration's authors explained, for the "conclusion that inherited genetic differences are a major factor in producing the differences between the cultures and cultural achievements of different peoples or groups." Furthermore, no scientific proof existed that the "groups of mankind" differ in their "innate mental characteristics, whether in respect of intelligence or temperament."[38]

---

[37] Edwin Black writes of how American eugenicists "rushed" to amend their names "from eugenics to genetics" after 1945, but, in practice, there is little evidence of a comparable alacrity to change. *War Against the Weak*, xvii.

[38] UNESCO, *The Race Concept: Results of an Enquiry* (Paris: UNESCO, 1952), 100, 102.

12

Postwar Sterilization

Institutions and Abuse

Mental health institutions – geographically separated from the rest of society,[1] hierarchically ordered, and governed by a clear system of rewards and punishments – were established in an aura of optimism about their capacity for improving mental health.[2] Indeed, the basic tenets of the moral treatment philosophy from which they took inspiration were antithetical to cruelty and abuse.[3] Their grand architecture, attractive grounds, and professional ethos seemed to make them the very opposite of the torture chambers of the past. We have no reason to doubt that their superintendents were, overall, scrupulous individuals committed to their work. Yet, within the space of a few decades, these institutions became sites of intimidation, fear, abuse, and even torture.[4]

Abuse within institutions for the mentally ill and/or cognitively challenged is nothing new. Some of the earliest reports from inside institutions attest to it. In 1906, a drifter named John W. McCarthy on the American West Coast had run out of money.[5] He related his dire straits to an acquaintance at his boarding house who was a former attendant at the Southern California State Hospital at Patton. The man told him that the hospital was always looking for workers, and McCarthy secured on this advice a position. Not long after, he visited a journalist and friend, Arthur L. Dunn, in Los Angeles, and reported that the institution

[1] Reformers urged that mental health institutions be established outside the cities because the congestion of urban areas was viewed as a threat to mental health. Noll, *Feeble-Minded in Our Midst*, 25.

[2] Ira Hardy, who tirelessly lobbied for North Carolina's first institutions for the "feebleminded," was motivated by the death of his mentally handicapped daughter at the age of ten. He feared that children like his daughter would be neglected were the state not to provide for them. See ibid., 21.

[3] Trent, *Inventing the Feeble Mind*, 117.

[4] Edouard Séguin was himself accused of terrible abuses at Bicêtre and may, indeed, have been fired for them. Ibid., 41–3.

[5] This paragraph is based on "Testimony in Investigation into Management of Southern California State Hospital: Cross-Examination of Dunn," October 14–18, 1907, F 3617: 381, CSA.

was a hellhole of abuse. Dunn, sensing an angle, spoke to his editor at the *Los Angeles Record*, who told him to investigate the story. Dunn traveled to Patton and applied for a job at the hospital. Dunn had no references and no experience. When asked why he wanted the position, he said that he had "learned of the place from a cigar man at Colton."[6] He got the job.

During a brief, two-week tenure, Dunn witnessed a catalogue of abuse ranging from ridicule and humiliation to what can only be described as torture: knocking the legs out from patients, jumping on them, kicking them as they lay on the ground, and denying medication. He witnessed paresis patients (those with movement loss or partial paralysis) being dumped out of their beds, as well as patients who were strapped and choked. Throughout his two weeks, Dunn saw patients "wasting with disease," their limbs bare and covered in flies; a recently committed patient who was a mass of bruises and discolorations; and the fear of large, healthy patients as they slinked past attendants like "whipped curs."[7]

After these two grueling weeks, Dunn left without notice. He wrote about what he saw in the *Record* and asked that charges be filed against the institution. The state Lunacy Commission investigated the matter and called for the firing of all the attendants identified by Dunn, with two exceptions.[8] At the same time, however, the Commission concluded that the abuse inflicted on patients was "occasional" and unrepresentative; that beatings of patients were a necessary part of an attempt to subdue violent patients (although such beating could be reduced if two attendants rather than one did the subduing); and that, therefore, the charges made by Dunn in his articles about routinized cruel treatment were "misleading and untrue."[9] The Commission was fulsome in its praise of the superintendent and his officers' skill, efficiency, kindness, and results.[10] The Commission rejected, in other words, the possibility that cruelty and abuse in any way inhered in the institution itself. It is a possibility to which we will return.[11]

---

[6] "Testimony in Investigation into Management of Southern California State Hospital: Statement submitted by A. W. L. Dunn," October 14–18, 1907, F 3617: 381, CSA.

[7] Ibid.

[8] "Testimony in Investigation into Management of Southern California State Hospital: Final Report and Findings of the Lunacy Commission," October 14–18, 1907, F 3617: 381, CSA.

[9] Ibid.

[10] Ibid.

[11] Shortly after the Patton investigation, a discharged patient from Agnews State Hospital made similar accusations of beatings, kicking, choking, broken bones, the liberal use of straightjackets, and even the murder of a patient at the hands of attendants. Some of the accusations were backed up by other patients. One of the main witnesses, a Dr. Fred Webster, did his case few favors by writing a series of letters to the institution's chief, Dr. Leonard Stocking, in which he compared the doctor's mother to ladies employed in the world's oldest profession, suggested the doctor was himself a sexual deviant of the lowest sort, and threatened to kill him. Whether the patient was murdered by staff, or, as the hospital argued, by another patient, was not resolved, but the document nonetheless provides evidence of an (indeterminate) amount of physical abuse at

NEW YORK: 1971

Sixty-five years after Dunn's investigation, a New York doctor, Michael Wilkins, phoned a young journalist named Geraldo Rivera. Wilkins told Rivera that he had been fired from the Willowbrook State School on Staten Island for urging parents to lobby for better conditions for their children housed there. At the time, there were 5,300 inmates at the institution, which shared no similarity with a school other than the name. Wilkins sneaked Geraldo into the facility. There, he filmed boys and young men lying naked and unattended, feeding themselves soup with their fingers; tied in straitjackets, strapped to beds; whimpering alone on floors; smearing their own excrement and wasting away, three to a bed.[12] A woman lay naked on the floor in solitary confinement. Some children were emaciated, with distended bellies. Others huddled in shabby clothes on radiators. There were children with ugly, open wounds on their shoulders, heads, and backs; men and women with puffy, bruised faces; an ankle badly broken and unset. The film leaves the viewer with little doubt that Willowbrook was a place of torture and neglect. The pictures were broadcast nationally in the 1972 documentary, *Willowbrook: The Last Disgrace*. They created a national scandal and, following intensive lobbying efforts, the institution was finally closed in 1987.

It is impossible to know how common Patton was in 1906 or Willowbrook in 1972. They were perhaps the extremes, although, as late as 1989, the Sonoma Development Center in California (where Fred Butler oversaw the sterilization of thousands) was the center of a by-then familiar scandal. A young boy, Billy Coffelt, the victim of a rare chromosomal order and given to violent fits, was found bleeding from a head wound on the floor of a communal toilet and hospitalized for a week.[13] After four families filed complaints with the U.S. Office for Civil Rights in San Francisco, the institution promised staffing and management changes to "ensure that their children are safe"; the next day, a twelve-year-old boy was sexually molested.[14] Even more recently, the BBC series *Panorama*, in May 2011, smuggled a secret camera into Winterbourne View, a home for people with mental disabilities in Bristol, England. The ensuing furor resulted in six care workers being jailed for mistreating or neglecting their

---

Agnews. "In the manner of the investigation of Agnews State Hospital," June 12, 1919, F 3617: 37 9 AD, CSA.

[12] Pictures from Willowbrook are available in "Willowbrook: A Photo Essay" by William Bronston, M.D., http://www.mncdd.org/extra/wbrook/willowbrook.html (accessed January 23, 2013).

[13] Ellen Robinson-Haynes, "Retarded Children Neglected, Abused at Sonoma Center," *Sacramento Bee*, October 9, 1989.

[14] Ibid. Court proceedings, instigated by Billy's parents, led to a reduction in the state's institutionalized population. Nancy Weaver, "Settlement Opens Doors for Teen," *Sacramento Bee*, January 27, 1994.

charges and another five given suspended sentences.[15] The camera recorded attendants slapping and jumping on patients, pinning a woman under a chair and ignoring her cries, and encouraging one woman to throw herself from a window. As in Willowbrook and Patton, a brave whistleblower broke the story to the media.

Accusations and proof of cases like Patton, Willowbrook, or Sonoma are common enough,[16] but there were, of course, other institutions that attempted to achieve the ideals originally envisaged by Philippe Pinel. Even under the best of circumstances, however, institutions often became islands of abuse. It occurred on all levels: in admissions and selection for sterilization, in routinized punishment and treatment, and in medical experiments on patients – even after the Second World War.

ADMISSIONS AND STERILIZATION SELECTION

An important minority of individuals – children found on the street or removed from families by social services – were involuntarily committed by state officials. Other, voluntary admissions occurred under heavy pressure from doctors. Physicians would whisk deformed children off immediately after birth, never to be seen by the parents, and they often encouraged parents to commit children whose cognitive or mental difficulties appeared later in life. In both cases, the institutions would ensure that as little contact as possible existed between families and patients. Robert Perske, a chaplain at the Kansas Neurological Institute in the 1950s, recalled the standard message given to parents: "All you need to do is forget this situation and go on making a life for yourselves and the rest of the family. Johnny is in our hands now."[17] In other cases, such as those of Velma Hayes or Leilani Muir, whom we met in Chapter 11, parents saw commitment to institutions as a way of ridding themselves of unwanted children. Many others were driven to the decision by poverty. The absence of extrainstitutional care also meant that even middle-class parents would lack the funds or training to cope with physically disabled, cognitively impaired, or mentally ill children. Despite this, only a minority, well below 50 percent, of the mentally ill population in the United States was institutionalized. The reasons for commitment were thus random and arbitrary: whether children were born to the rich or

---

[15] Steven Swinford and Andy Bloxham, "BBC Panorama abuse care home 'should be shut down,'" *Daily Telegraph*, June 1, 2011, http://www.telegraph.co.uk/health/healthnews/8549547/BBC-Panorama-abuse-care-home-should-be-shut-down.html. For details of the convictions see, "Winterbourne View: Care Workers Jailed for Abuse," October 26, 2012, http://www.bbc.co.uk/news/uk-england-bristol-20092894.

[16] Noll, *Feeble-Minded in Our Midst*, 58. On the ubiquity of such practices today, see Michele Hanson, "I cried through Panorama's program on care home abuse," *Guardian*, June 3, 2011.

[17] Robert Perske, *Hope for the Families: New Directions for Parents of Persons with Retardation of Other Disabilities* (Nashville, TN: Abingdon Press, 1981), 42.

the poor, to the caring or the indifferent, it was essentially a matter of luck that determined if they would end up in an institution. And that, depending on the equally arbitrary matter of the state in which they lived, in turn determined whether they became candidates for sterilization.

Once in the institution itself, the selection process, particularly as the decades wore on, became routinized, rushed, and imprecise. Eugenics boards were composed of individuals who sought the best for patients and society, but they were unpaid and had many other demands on their time. In some cases, there was no interview with the patient, which led to quick, careless decisions based on admissions paperwork and faulty IQ tests. In cases in which interviews did take place, the panel had usually made up its mind beforehand, and the questions asked were formulaic and not designed to ascertain important information. In Alberta in the 1960s, the decision to cut boys' vas deferens and girls' fallopian tubes took five minutes.[18] In a context of repetition and high volumes, any decision – confinement, sterilization, even physical abuse – can become mechanized. A similar development can also be observed in jurisdictions where sterilizations were rare. In British Columbia, from 1935, when the law came into force, until 1965, around four or five people were sterilized on eugenic grounds per year, with 1940 (eleven people) and 1952 (fifteen people) constituting peak years; a total of sixty-four people were sterilized during the period.[19] Yet, the chairman of the province's eugenics board gave more thought to the content of his briefcase than the decision to rob people of their ability to produce children. In 1944, one Isobel Harvey, a superintendent of child welfare and a member of the eugenics board, wrote the following account of its chairman:

[N]either Dr. Minorgan [another member of the Board] nor I would like to have to go on a witness stand and give evidence that these hearings are held in the proper manner. In the first place, [Chair] Judge Manson's mind is never on Eugenics, and I am telling you the literal truth when I say that I do not think he listened to me for more than half a minute consecutively this morning and, as I have to present the cases and give the medical evidence, I felt that he really did not know when the patient came in to be interviewed exactly what was the matter with him. He telephoned, – he went several times to the filing cabinets and hunted for files, – he made entries in his book, – he searched his law bag for things he wanted, – all the time while we were trying to discuss these cases. There were four cases seen, and I do not suppose there was one minute's discussion of them aside from what Dr. Minorgan and I did privately while we waited for the judge.[20]

---

[18] Jon Faulds, interview.
[19] Tables 2 and 3, n.d., GR 542 11/4, BCA.
[20] Isobel Harvey to E. W. Griffith, Asst. Deputy Provincial Secretary, March 18, 1944, GR 542 11/4, BCA.

CRUELTY

Once in the institutions, patients found themselves – again, even under the most ideal conditions – subject to treatments and punishments that constituted routine cruelty and abuse. One of the most common abuses involved physical restraint. Attendants would target patients who had been violent, unruly, or who refused orders and would place them in straightjackets or tie them to beds.[21] There they would be left, unable to move, for hours and, at times, even days. Otherwise, they would suffer "lock-down." In Alberta, this took the form of the greatly feared low-grade ward, in which patients were locked for hours, days, or weeks in a small room with only a rubber bed and a small, sometimes covered window. The confinement in Alberta, however, was not always solitary: patients placed in the low-grade ward were often locked up with those suffering from serious mental illness or motor control problems.[22] The rooms were pitch black or contained only a small, closeable window. Patients would scream, twitch, rave, and drool. The result for the other patients was both clear and intentional: it terrified them. After her operation, Velma Hayes was left to recover on such a ward. At other times and places, patients were locked into a solitary room. At the Willowbrook State School in New York, patients were left, sometimes naked, to sleep on the floor.

At other institutions, attendants would punish patients for failing to get up, for defecating in their trousers, for wetting their beds, or even for walking too slowly. At Fairview, the diet in the 1950s and 1960s was made up of heavy, starchy foods. The result in many patients was constipation. The attendants would give them laxatives, with the inevitable result being diarrhea. If patients defecated in their clothes, the attendants would place them in freezing-cold baths, sometimes leaving them there for several hours. Ted McNeil, a patient at Fairview, suffered this punishment.[23] If patients wet their beds, the sheets would be soaked in freezing or searing hot water, and the attendants would force the patients to sleep in them.[24] In another punishment reminiscent of punitive work camps, attendants at Fairview and at the Provincial Training School would strap patients who misbehaved to a heavy wooden block and force them to push it around the floor for hours.[25] Others had bars of soap forced into their

---

[21] Whereas restraint was a sensitive and controversial issue, the consensus among psychiatrists before the Second World War was that its abolition, if desirable in theory, was impossible in practice. Grob, *Mental Illness and American Society*, 18.

[22] Gordon Bullivant, interview.

[23] Written report by Ted McNeil of conditions at Fairview Institution, Salem, OR [copy given to authors during interview, May 16, 2003].

[24] Ibid. The latter was the fate of Donald Morehouse at Fairview.

[25] Confidential interview, Portland, OR, July 18, 2003.

mouths.[26] If these cruelties had any justification at all, they were designed to correct behavioral problems that resisted regular punishment.[27]

The abuse did not stop with these public, legal, and accepted punishments. There was a much broader pattern of illegal yet widespread abuse that was often physical and, at times, sexual. When Ken Nelson refused to clean excrement from the floor at the Alberta Provincial Training School, an attendant slammed his hand in a door; he lost the end of one finger.[28] When Rita Haggerty refused to clean up human waste – most likely from the wall and the floor – the attendants placed her in solitary confinement.[29] At the same institution, John Smith had a broom handle broken over his back and, when moving too slowly on a group walk, found his face pushed into the shower.[30] John also witnessed an act of manslaughter. An attendant, whose first name was Gregory, wrapped a belt around the neck of a small, mute patient named Joey and dragged the child around the room. Ken Nelson began to scream; this anxiety stopped the attendant, but after removing his belt, he hit Joey squarely in the chest. The next day, among the other patients, the boy collapsed and died.[31] Yet another patient at Fairview, Terry Schwartz, was hung in a bag for hours, unable to see out, as punishment for some indiscretion; passing attendants would occasionally hit the bag with a stick.[32] The majority of former patients we spoke with told us of beatings, although a minority – often the least difficult ones – said they had been treated with kindness. Patients were hit with belts, brooms, and fists. If patients at Oregon's Fairview hospital played with their own excrement, the attendants forced them to eat it.[33] Ted McNeil provided a very moving description of this sort of abuse in a transcript written years after his release from Fairview Hospital:

> I am telling the whole thing, the truth about what happened in Fairview Institution Hospital in Salem. The first [abuses] start[ed] on Kozer cottage. The employee worker put some ... pooh in [my] plate and make me eat pooh and I was small, about 6 years old. ... Mr. Harry White was beat[ing] everyone up and so mean. ... Mr. McGary put me and [another patient] in cold bath for [messing] the pants, for punishment. ... I, Teddy, and [the other patient were] both [in the] cold bath. [We] both can't help it, [and were punished] with a cold bath. ... Some attendants ...

---

[26] Ted McNeil, written report.

[27] Interview with Bruce Uditsky (CEO, Alberta Association for Community Living), Edmonton, AB, October 18, 2003.

[28] Ken Nelson, interview.

[29] *The Sterilization of Leilani Muir.*

[30] Confidential interview. The name of this interviewee has been changed for the purpose of anonymity.

[31] Ibid.

[32] Dennis Heath, interview. Another patient, Laverne Throckhorn, told us in her interview that she was hit in the face. Terry Schwartz himself did not relate this story to us.

[33] William West, interview.

been using the bars of soap [in] each patient's mouths and using [the] strap and belt ... on some patients.[34]

Ingesting excrement can be fatal, so it is unclear if this incident occurred in the manner recalled by Mr. McNeil. The attendants may have claimed the substance was excrement to bully McNeil. The rest of the story, in any case, is consistent with what others have claimed about Fairview and similar institutions.

Patients, former attendants, and fieldwork supervisors often spoke, usually on condition of anonymity, of sexual abuse at Fairview and Alberta's Provincial Training School. At the latter, Leilani Muir suffered unwanted sexual attention.[35] Members of staff impregnated two patients at Fairview.[36] The sexual abuse of female patients by male attendants was sometimes mentioned discreetly as a justification for coerced sterilization.

Many superintendents, such as the efficient, cold, and incorruptible Dr. Le Vann of the Provincial Training School in Alberta, probably knew little about what occurred in the darker corners of their institutions. It is impossible to know how widespread this form of abuse was. Many patients claimed to have experienced or witnessed it, some said they saw others abused but experienced none themselves, and still others saw nothing but heard rumors. The degree to which the institution responded to such abuse varied from institution to institution and superintendent to superintendent. That it occurred, however, cannot be doubted.

SHOCK TREATMENTS AND MEDICAL EXPERIMENTS

One treatment, intended as such, was understood by patients themselves as a form of abuse: shock treatments. As the name implies, the treatments involved subjecting patients to massive cardiovascular and neurological shocks that led to shaking, twitching, convulsions, and sometimes vomiting. Frequently, particularly in the early years of the treatment's application, these patients would suffer fractures from smashing into bedposts or tables or from falling onto the floor.

There were three types of treatment: insulin shock, Metrazol (pentylenetetrazol), and electric shock. A Viennese doctor, Manfred J. Sakel, publicized insulin shock in 1933 (not a good year for residents of mental health institutions).[37] He published an influential article on the topic, and the practice was almost immediately adopted across the United States and Europe. The method involved injecting the patient with large doses of insulin, which would lead to convulsions. The doctors would end the shock treatment by knocking the patient

---

[34] Ted McNeil, written report.

[35] Leilani Muir, interview.

[36] Two confidential interviews. The names of these interviewees have been withheld by mutual agreement.

[37] Western State Hospital to Alphia Hlavac, April 1, 1938, Box 89: "Western State Hospital Treatment, File, Shock Treatment 1937–1941," Washington State Archives (WSA), Olympia, WA.

out with a large injection of glucose. The treatment was repeated thirty or forty times; if it had not achieved the desired effect by then, the doctors despaired of it.[38] Also in 1933, a Hungarian physician named Ladislas von Meduna began experiments with chemically induced convulsions. Meduna communicated his results to the mental health establishment in 1937.[39] Both techniques led to typically exaggerated predictions of their curative effects, and psychiatrists debated the merits of the two techniques: insulin was milder but required five to nine hours of hospitalization and follow-up care; Metrazol was faster and cheaper, but led to sharper convulsions, which could result in broken bones.[40]

The experience of Metrazol treatment was described in a letter to the Eugenics Record Office in 1938.[41] The patient signed only her initials: TGS. She hoped that the Office might be able to answer her questions and do something about her treatment. There is no record of a reply. In a typed document of thirty-eight pages, TGS described her treatment at the Elgin State Hospital in Illinois (previously known as the Northern Illinois Hospital for the Insane), to which her sister had committed her. One might doubt the validity of the source, except that so much of it repeats what other patients have said about Fairview and the Provincial Training School, and it is consistent with what we know about Willowbrook. Much of the letter describes conditions and experiences that occurred repeatedly at mental health institutions across the West: upon arriving at the asylum, TGS was dragged to her quarters, and her clothes were taken from her. Her life became a dull routine of rising early, tedious work, bad food, and tyrannical attendants. All of this, however, paled in comparison to the Metrazol treatments.

The treatments occurred on Thursday mornings, and, the night before, there was much preparation: the patients were bathed, their clothes were taken away from them and marked, and they were given a dose of salts.[42] The morning itself did not begin with the usual breakfast, lest the patients vomit it over their doctors. The patients spared the treatment prepared about twenty-six beds for those who were about to undergo it. They covered each bed with a clean sheet, folded a blanket into the foot of the bed, and placed a small pillow and a towel at its head. The patients ranged in age from about seventeen to around sixty-two. The patients laid down, and the doctor inserted a needle into a vein near the elbow.

> After the Metrazol has been injected you lie there a few seconds . . . fully conscious, waiting for the medicine to strike the brain. A nurse or attendant stands on each

---

[38] Ibid.

[39] Renato M. E. Sabbatini, "The History of Shock Therapy in Psychiatry," n.d., http://www.cerebromente.org.br/no4/historia/shock_i.htm (consulted January 23, 2013).

[40] Ibid.

[41] Manuscript sent by TGS to the Eugenics Record Office, December 18, 1938, Coll. 77, ERO HLF #4, APSA.

[42] This paragraph is based on the same manuscript, ibid.

side of the bed holding your hand and arm down … to hold your head still when the convulsions should come. … As the Metrazol strikes the brain, the patient begins to yell and holler, some crying, moaning, and groaning; others screaming at the very top of their lungs. Others act wild, and rave, and have to be tied down in bed. One patient … omitted a sound which was exactly like the barking of a dog … Cries of "Chester!" "Chester!" – probably someone's husband. One girl, left quiet, suddenly jumps out of bed and runs around, having to be put in restraint. Patients cry, "I am not going to live through another. This is my last!" "What can I do to protect my body?" "O God, please take me home." "Take me home, Lord Jesus, take me home."[43]

Metrazol and insulin treatments came gradually to be replaced by electroconvulsive shock therapy, or ECT. One of the more famous patients treated with ECT in its early stages was the poet Sylvia Plath, who was treated at McLean Hospital – a "genteel" psychiatric institution for the rich.[44] In the meantime, however, patients continued to suffer the pain and danger of Metrazol and, to a lesser degree, insulin shock therapies. Although many doctors believe ECT is itself a useful last resort, its antecedents take their place among the instruments of abuse inflicted on the mentally ill and mentally handicapped.[45] The treatment was used repeatedly in the course of a single day to restrain unruly patients, and attendants threatened its use to discipline and punish patients.

As brutal as these methods were, their avowed intention was to help the patients. That could hardly be said of other techniques. Throughout the United States, doctors used resident populations of feebleminded and mentally ill patients for medical experimentation. At the Fernald School for the Feebleminded in Massachusetts, Dr. Clemens E. Benda, a psychiatrist and neuropathologist in charge of the School's on-campus medical laboratory, secretly injected boys' oatmeal with radiation in order to study its effects on the body for Cold War research.[46] At Massachusetts General Hospital, researchers cut open the head of brain tumor victim Jacob Lifton.[47] Ostensibly intended to cure the patient, the researchers' actual interest was in understanding the tumor itself. As they cut into it, the tumor and healthy tissue spilled down onto the floor.[48]

The horrors of medical experimentation were not unique to institutions for the feebleminded, and the topic connects the particular study of institutionalization and sterilization to wider postwar developments that were also connected to the Cold War. From 1960 to the 1970s, researchers at the University of Cincinnati's College of Medicine worked partly under contract for the U.S. Department of Defense. They exposed ninety patients, mostly African

---

[43] Ibid. Emphasis in original.
[44] Michael D'Antonio, *The State Boys Rebellion* (New York: Simon & Schuster, 2004), 51.
[45] The medical/psychiatric community is divided about whether shock therapy should be used at all.
[46] Ibid., especially chapters 11 and 12.
[47] Ibid., 263.
[48] Ibid.

Americans, to high levels of radiation. The cases followed a basic pattern: individuals suffering from cancer would come to the hospital for treatment.[49] Telling them that the procedure was designed to cure them, doctors would bathe their bodies in high doses of radiation. The patients' white blood cell count would plummet and, with a few exceptions, they died. In most cases, they suffered great pain. One of the most gruesome cases involved an African-American boy, David Jungnickel, who had been diagnosed with bone cancer at the age of fifteen. Two years later,

> [H]e fell into the hands of the doctors at General Hospital and was irradiated over his whole body. At the age of nineteen David died. The total body exposure of 1963 was a very painful event for him, according to the doctors' own notations, and then the frequent high doses of radiation to the tumors around his body left him with burns and sores. Indeed, David's last four years of life seem to have been a nightmare of pain, excruciating treatments, and day-to-day neglect in hospitals, both before and after his whole body exposure. ... He begged for pain medicine and most often [his family] found him in tears because he had not received it.[50]

The radiation tests were themselves nested in a longer history of American military tests on human subjects.[51] With varying degrees of consent, soldiers were subjected to malaria and mustard gas tests; some tests were fatal.

The Cincinnati studies based scientific study on hastening death. Others involved slowly tracing it. In perhaps the most infamous example of medical abuse, doctors at the Tuskegee Institute and the U.S. Public Health Service in Macon County, Alabama studied the effects of syphilis on hundreds of poor, illiterate African-American men over the course of forty years.[52] Sold implicitly or explicitly as therapeutic, the tests were diagnostic: they involved tracing the destructive effect of syphilis on the men's bodies. Then, using burial funds as a carrot to secure familial consent, the officials performed autopsies on the dead.

Still other medical experiments had a commercial as well as a medical motivation. Dr. Albert M. Kligman of the University of Pennsylvania's School of Medicine conducted a decades-long research project at Holmesburg Prison in Pennsylvania.[53] With prison superintendents' permission, and securing patient consent through payment, researchers tested perfumes, deodorants, sprays, and jellies on the patients' skin; inserted gauzes into open wounds; and exposed patients to flu strains. They gave patients diet milkshakes to drink; exposed them

---

[49] See Martha Stephens, *The Treatment: The Story of Those Who Died in the Cincinnati Radiation Tests* (Durham: Duke University Press, 2002), chapter 8.

[50] Ibid., 163–4.

[51] See Jonathan D. Moreno, *Undue Risks: Secret State Experiments on Humans* (New York: Routledge, 2001), chapter 2.

[52] James H. Jones, *Bad Blood: The Tuskegee Syphilis Experiment* (New York: The Free Press, 1981).

[53] Allen M. Hornblum, *Acres of Skin: Human Experiments at Holmesburg Prison; A True Story of Abuse and Exploitation in the Name of Medical Science* (New York: Routledge, 1998).

to extreme temperatures; and, after the latter, cut open their armpit glands for examination.[54] They also performed liver and lung biopsies.

These experiments connect homes for the feebleminded to the wider history of medical experimentation. At Willowbrook, mentally retarded children were injected with a hepatitis virus. Scientists separated them into test and control groups and gave the former a gamma globulin vaccination. The results were reported in the *New England Journal of Medicine*. The unvaccinated or inadequately vaccinated children suffered from enlargement of the liver, vomiting, and anorexia, but no one died. The parents had to consent to the experimentation, but, from 1964, the institution would not admit children without their parents' consent. At the Provincial Training School in Alberta, physicians tested, without consent, steroids and antipsychotic drugs on up to 100 children. One of the drugs, norbolethone, is now illegal because it can damage the liver.[55] At the same institution, doctors operated on several boys with Down's syndrome. They sterilized them and removed one testicle for medical examination. Boys with Down's syndrome, as has been widely known since the 1940s, are sterile at birth. When Margaret Thompson – founder of the Genetics Society of Canada, a leading Canadian geneticist, a recipient of the Order of Canada, and a member of the Eugenics Board of Alberta – was asked in court to justify the operations, she replied: "What was lost?"[56] In the 1960s, she sat on the Alberta Eugenics Board. In this context of such attitudes, it is perhaps unsurprising that the superintendent and his employees saw little wrong in sterilizing their subjects.

Reviewing the history of medical experimentation, it is difficult to identify common patterns. Victims were both black and white, with developmental disabilities and with average (and quite possibly above-average) intelligence, soldiers and civilians, and those living in institutions and those living in society. In the history of sterilization, the isolation of patients in institutions in which they had few rights and no power made medical experimentation easier, but institutionalization itself is not in any simple way a prerequisite to abuse by medical or scientific researchers.

Where institutions appear to play a much more consistent role is in the encouragement, or at least facilitation, of abuse.

STERILIZATION AS ABUSE AND CRUELTY

In institutions throughout North America, many of the mentally ill and feebleminded were tied down, locked away, beaten, subjected to shock therapy, and exposed to extreme heat and cold. And they were, of course, sterilized. Although not conceived of as such, the operation was itself the culmination in a process of

---

[54] Ibid., 120.
[55] Marina Jimenez, "Alberta Sterilization Victims Also Used as Guinea Pigs," *National Post* (Toronto), October 28, 1998.
[56] Jon Faulds, interview.

abuse, and it was itself an abuse. Patients were threatened with it when they misbehaved. Officials made release from the institution depend on it. One such case in Alberta was that of Rita Haggerty. Rita was placed in the institution because she had suffered polio and had a speech impediment.

The Eugenics Board told her that if she wanted to be released, she would have to submit to sterilization. She told them: "you leave me the way I was created. You don't touch my body."[57] Their response: "then you don't get out." She was confined there for almost fifty years; she died shortly after her release.

In almost all cases, the attendants, doctors, and/or superintendents did not tell the patients what would happen to them. The patients were told that they were going for an appendectomy or some other routine operation, or they told them nothing at all. Many only learned of their sterilization through other patients, from a cruelly offhand comment made by an attendant or family member, or following a doctor's exam after their release.

Until the 1960s, and perhaps even later, treatments such as restraint and shock therapy existed alongside unfettered physical and sexual abuse, were employed at times to create intimidation and fear, and, equally importantly, were viewed by many patients as a punishment rather than a cure. When we spoke with some of these patients, they expressed as much if not more outrage about this random and routinized abuse as they did about their coerced sterilization.

The patients' equation of abuse and sterilization provides a clue to understanding how such operations could continue after the war's end and after the revelation of related Nazi crimes. In the context of an institution organized around rigid hierarchies of power, in which abuse was woven into the fabric of everyday life, sterilization, like other forms of cruelty, became routine and unremarkable. It was entrenched as an idea influencing a routinized set of procedures and surgeries. The attendant who told Ken Nelson "you're not human" made a statement that was at once outrageous and accurate. It was outrageous because it violated our principles about how people are to be treated; it was accurate because these principles did not apply to institutional life. As patients entered the door, they were robbed of their humanity, isolated from their families, cut off from society, and governed by the norms of an autonomous community.[58] Patients ceased to be rights-bearing agents as they entered the institution.

The institution's role in dehumanizing subjects was related to the attendants themselves. There can be little doubt that the attendants' work conditions – in

---

[57] Jimenez, "Alberta Sterilization Victims"; interview with Judy Lytton, Edmonton, AB, October 19, 2003; *The Sterilization of Leilani Muir* (National Film Board of Canada, 1996); *Thou Shall Not Procreate* (Edmonton: Alberta Association for Community Living, 1993). The last two are documentaries.

[58] We are grateful for Dennis Heath for suggesting this idea to us. As ever, the establishment of institutions for the feebleminded in rural settings had a positive aim: doing so would avoid the city's predisposition to crime and immorality. Trent, *Inventing the Feeble Mind*, 100.

overcrowded, underfunded institutions – led to frustration, short tempers, and violence.[59] More important, patterns of institutional behavior replicate themselves. Young or otherwise new attendants entered work at institutions in which the patients were treated as subjects without rights, in which other attendants were contemptuous of and violent toward their patients, and in which abuse was part of the pattern of daily life. They inevitably internalized these norms. Most attendants were poorly educated and ill equipped to deal with a complex and often difficult group of patients.[60] There are often difficult challenges in understanding and sensitively dealing with the various behavioral and speech problems faced by the mentally handicapped. The level of training and basic understanding needed to effectively confront these problems were wholly absent in the 1950s and 1960s and, in cases such as Winterbourne View, are apparently still lacking. In addition, a lack of patience in meeting these sometimes frustrating challenges could itself lead to abuse. And although most violence originated in attendants and was directed toward patients, the reverse could also be true. Patient attacks were, in the words of one observer, "not uncommon," and patient assaults led to the deaths of several superintendents, assistant physicians, and attendants.[61]

Naturally, these generalizations cannot account for all cases: there were attendants who were kind, correct, and admirable in the behavior. Steven Noll reports the response of a female attendant from North Carolina's Caswell Training School, who was asked how she could bear nine years of poor pay, long hours, and substandard housing. "When people ask me how I've stood nine years in my Junior Ward," she replied, "I always reply by saying I would not swap my job for any other job in the place . . . [T]he appeal of helpless children brings forth the most sympathetic treatment one can give."[62] But even the best motivated – some of whom are now active in the deinstitutionalization movements and community care – participated in a system of oppression. As Noll also notes, superintendents at the Florida Farm Colony dismissed dozens of employees in the 1920s and 1930s for striking or "otherwise inappropriately handling" patients.[63] Similarly, not all institutions were of a piece: funding varied between states (with the South less generous than the North), and conditions in some institutions were far more abusive than others.[64] But institutions consistently struggled to contain abuse. As Gerald Grob delicately puts it, "[t]he human frailties of patients, attendants, and physicians created a precarious balance that was easily upset; the threat of disruption and disintegration were ever-present."[65]

---

[59] On this, see Grob, *Mental Illness and American Society*, chapter 1.
[60] Noll, *Feeble-Minded in Our Midst*, 52.
[61] Grob, *Mental Illness and American Society*, 22.
[62] Noll, *Feeble-Minded in Our Midst*, 144.
[63] Ibid., 142.
[64] Grob, *Mental Illness and American Society*, 26–7.
[65] Ibid., 22.

Such a line of argument might lead to the thought that the boundary between perpetrators and victims in an institution was a highly blurred one. If the victims did not become perpetrators, then one might think that all those who entered the institutions – the superintendent, the doctors, the nurses, and the mass of attendants – were, on some level, victims of the cruel logic and practice of the institutions. Such a line of argument is unpersuasive and may be morally suspect. Actors, not structures, created the institutions, defined their aims, and oversaw their operation.

The hierarchical nature of the insane asylums and homes for the feebleminded meant that the most consequential actors were primarily their superintendents. Their role is part of the key to understanding how coerced sterilization could continue in the period after the war. The very operation that made eugenic sterilization possible emerged out of these institution: Hoyt Pilcher's castrations and Harry Sharp's vasectomies laid the foundation for eugenic sterilization. The superintendents were crucial in influencing both the intellectual climate and the lobbying campaigns that led to eugenic sterilization laws; they were intimately involved in the Indiana, Kansas, and Wisconsin campaigns. A 1930 questionnaire sent to 317 superintendents indicated, based on 227 replies, that 80 percent of them favored sterilization.[66] The superintendents decided whether a particular institution would apply a state or province's eugenic sterilization law, and the superintendents themselves performed many of the sterilizations.[67] It was their position, their power, and their decisions that led to the application of eugenic sterilization before 1945; it was they who decided whether such sterilizations would continue thereafter. And it is with them that responsibility for abuse ultimately rests.

---

[66] Trent, *Inventing the Feeble Mind*, 201.

[67] In other cases, they were performed in hospitals. Sterilizations could be a lucrative business for the doctors performing them: over a three-year period from 1929 to 1931, one doctor at Edmonton's University hospital, Dr. L. C. Conn, received $4,300 in remuneration. The figure is equivalent to about $59,000 in 2012 figures. "Re: Chapter 37–1928 Statutes of Alberta," GR 1970, 33/1311, Doc. 1, PAAB.

13

Welfare, African Americans, and Coerced Sterilization

Legal segregation, racism, and the relative economic backwardness of Southern states, which resulted in smaller government revenues for social spending, largely spared African Americans the experience of compulsory eugenic sterilization in the pre-Second World War years. Segregation denied them access to most of the institutions where sterilization occurred and to the welfare programs that would have brought them into contact with the state.[1] In the postwar period, this situation changed. During the 1950s and especially the 1960s, as the number of eugenic sterilizations of the mentally ill and handicapped decreased (but did not end), sterilization emerged from institutions and, in a new federal policy, targeted African Americans.

Four factors account for this shift. First, the legal basis for such sterilizations had been there for decades: since the early 1930s, for example, North Carolina authorized the sterilization of people outside institutions, which made it an exception in the United States. Second, Clarence Gamble's bankroll helped Birthright, Inc. set up a network of birth control clinics throughout the South beginning in the 1930s. Third, the Civil Rights Act of 1964 established full rights of citizenship for African Americans. This entitlement included access to federal welfare programs, and, partly as a result of activist mobilization, African-American take-up rates for welfare rose.[2] The result, in the 1950s and 1960s, was a perfect storm: large numbers of African Americans on welfare in some of the most conservative states, laws that made their sterilization legal, and social workers trained in a cultural context in which wealthy donors and foundations

---

[1] On the former, see Steven Noll, "Southern Strategies for Handling the Black Feeble-Minded. from Social Control to Profound Indifference," *Journal of Policy History* 3, no. 2 (1991): 130–51. For a discussion of exclusion from welfare programs, see Robert C. Lieberman, *Shifting the Color Line: Race and the American Welfare State* (Cambridge: Harvard University Press, 1998).

[2] Kluchin, *Fit to Be Tied*, 74. This take-up also resulted from the efforts of civil and welfare rights reformers.

had propagated eugenic arguments for decades.[3] Finally, the federal government allowed family planning funds, distributed from 1964, to be used for sterilization. Because of who implemented sterilization policy and how, the result was the coerced sterilization of African Americans across the United States.

EUGENICS AND WELFARE

Among eugenic thinkers, the argument that welfare could have "dysgenic" effects was hardly novel by the 1960s. Three decades earlier, Paul Popenoe had warned that welfare payments encouraged the reproduction of a eugenically inferior class, and he issued a call for contraception and voluntary sterilization.[4] During the 1930s, as unemployment climbed ever higher (in 1932, 28 percent of American households lacked a single breadwinner),[5] and as the tally of those dependent on welfare and charity went up, the indigent appeared to be both the cause and the effect of the country's problems. When state officials visited rural parts of states such as North Carolina, they saw shabby residences, often a dilapidated trailer with a lean-to built into it, into which large families were crowded.[6] To complete the picture, the children were dirty and the parents uneducated and often inarticulate. That the federal government sought to address poverty through the 1935 Social Security Act only exacerbated the problem for these critics: Marian S. Olden wrote in 1943 that the Act made the need to "control the quality of life which is to be made [through welfare] secure" all the more pressing.[7]

For sterilization advocates ill accustomed to framing public policy in terms of individual rights (such a framing was, in any event, very much a postwar phenomenon), recipients of public assistance were virtually fixed targets for eugenic intervention. Indeed, the intellectual origins of the eugenics movement seemed to be resurrected, as advocates argued for intervention in order to prevent the West from being overrun by teeming hordes of welfare dependents.

The main eugenic lobby groups were quick to pick up the argument. In 1943, the Sterilization League of New Jersey became Birthright, Inc. In 1949, Birthright took over the papers of Gosney's Human Betterment Foundation, and the organization transformed itself into the Human Betterment Association of America (HBAA). During these six years, it was the only national

---

[3] The donors included the Scaife Family Foundation, the Sunnen Foundation, the Chichester duPont Foundation, and, of course, wealthy individuals such as Hugh Moore and Clarence Gamble. Dowbiggin, *Sterilization Movement*, 136.

[4] Paul Popenoe and Ellen Morton Williams, "The Fecundity of Families Dependent on Public Charity," *American Journal of Sociology* 40, no. 2 (1934): 214–20.

[5] Johanna Schoen, *Choice and Coercion*, 91

[6] Ibid., 90.

[7] "Platform of Birthright, Inc.," 1943, 9/9, AVS.

organization sponsoring sterilization.[8] After 1950, the organization moved away from promoting coerced eugenic sterilization and toward voluntary sterilization. By 1952, HBAA claimed that, although it worked with eugenic organizations, it was itself not one.[9] In 1962, the organization again modified its name to emphasize the turn away from its compulsory past while reemphasizing its commitment to sterilization: it became the Human Betterment Association for Voluntary Sterilization. In 1965, under the leadership of Hugh Moore, it became the Association for Voluntary Sterilization (AVS). Throughout the 1950s and 1960s, the organizations lobbied for the legalization of sterilization, paid for people who wished to be sterilized but who lacked the financial means, and provided funds for local clinics offering sterilization.[10]

Some couples appreciated the opportunity and wrote in to tell of better marriages, more sexual gratification, and improved relationships with children. But the AVS's motivation was different: it wished to encourage the sterilization of the poor.[11] From the early 1960s, the organization had pursued this goal by linking its pro-sterilization agenda with welfare.

H. Curtis Wood Jr., president of the HBAA, outlined the link between welfare and sterilization at a meeting of the Kentucky Gynecological Society in late September 1963. The speech began by quoting the misanthropic suggestion that the world had cancer and that "people [were] the cancer cells, growing in an uncontrolled and destructive manner."[12] This problem, Wood continued, was exacerbated by welfare policies that left 7.25 million dependent. But this figure was not the greatest threat; it was, rather, the figure's tendency to increase. In a passage reminiscent of the family history studies of the Jukes and Kallikaks popularized in the 1910s and 1920s, Wood speculates on the reproductive future of two fictional families: the educated, hard-working, and reproductively responsible Smiths with three children; and the uneducated, mostly unemployed, reproductively irresponsible, and welfare-dependent Joneses with ten children:

> Let us ... make a big assumption and say that the Smiths were a three child family and the Jones a 10-child family, each continuing in this pattern for two generations. *Under such circumstances Mr. Smith would have 27 great grandchildren who would have to work hard and pay taxes to support the 1,000 great grandchildren*

---

[8] Fred O. Butler to Karl Bowman, January 11, 1951, 4/34, AVS; William P. Richardson and Clarence J. Gamble, "The Sterilization of the Insane and Mentally Deficient in North Carolina," *North Carolina Medical Journal* 8, no. 1 (1947): 19–21.

[9] Lila Rosenblum, Secretary, Human Betterment Association of America, to Mr. Clarte, August 19, 1952, 4.36, AVS.

[10] Kluchin, *Fit to Be Tied*, 40–2.

[11] Ibid., 42–3.

[12] H. C. Wood, "A Prescription for the Alleviation of Welfare Abuses and Illegitimacy," *Journal of the Kentucky State Medical Association* 61, no. 4 (1963): 319–23. See also "Comments by H. Curtis Wood as to Sterilization Rules and Regulations in Hospitals," June 27, 1960, 4.38, AVS. Wood was quoting Alan Gregg, who was director of the Rockefeller Foundation's Division of Medical Science from 1929.

of the unfortunate Mr. Jones. There are many inadequacies in such an over-simplification of the problem ... but I feel it does illustrate how rapidly those on welfare may outnumber those who feed them. *There are also many who feel that intelligence is strongly heredity and that there is an alarming fall in our national intelligence level for these same reasons.*[13]

It all sounds very familiar.[14] In the early 1960s, one of the main birth control organizations was recycling prewar eugenic arguments by making them primarily about welfare and, secondarily, but still importantly, about intelligence. As this was going on, the Pioneer Fund, founded by Wickliffe Draper in 1937, supported the research of Stanford's William Shockley. Shockley was a professor of electrical engineering of such distinction that he received the Nobel Prize in Physics in 1956 for his invention of the transistor. His physics research was paralleled by arguments about race, intelligence, and eugenics. Shockley adopted an all-too-predictable and recurring twentieth-century thesis, arguing in 1970 that African Americans had, controlling for environment, a low IQ and that welfare "unaided by eugenic foresight may contribute to the decline in human quality."[15] The same year, Shockley founded (with Pioneer Fund support) the Foundation for Research and Education on Eugenics and Dysgenics – known by the felicitous acronym, FREED.[16]

In most states, welfare recipients remained beyond the reach of sterilization advocates until the 1960s because eugenic sterilization occurred within institutions, and involuntary sterilization outside them was illegal. The important exception to this general trend was North Carolina.

EUGENIC STERILIZATION IN NORTH CAROLINA

In February 1948, William P. Richardson, a member of the State's Board of Health, and Clarence Gamble co-wrote an article about North Carolina's eugenics program. Gamble had funded birth control programs in North Carolina since 1937.[17] The article is important both for the information it contains on sterilization in North Carolina and for its authorship. That a sterilization activist such as Gamble co-authored a piece with a doctor from

---

[13] Wood, "Prescription." Emphasis added.
[14] In a 1940 piece, Wood outlines the argument that demographic developments were leading to a decline in the national IQ. See Wood, "The Physician's Responsibility in the Decline of National Intelligence," 1940, RF 632–44, 3, 59, 635, RAC. The article rejects the suggestion that the increasing birth rate would prevent the old eugenic menace of "race suicide," arguing that the differential birthrate (a prominent fear in the 1910s) renders the development dysgenic.
[15] Kluchin, *Fit to Be Tied*, 81.
[16] Ibid.
[17] Schoen, *Choice and Coercion*, 33. Gamble worked with Elsie Wulkop, a social worker, to establish North Carolina's first birth control programs and helped do the same in Michigan, Indiana, Iowa, Nebraska, Kansas, Missouri, Florida, and South Carolina. Clarence J. Gamble to H. Curtis Wood, April 8, 1947, 2/14, AVS.

the State Board of Health highlights again the importance of lobbyist-state links in the history of coerced eugenic sterilization.

The two men expressed pride and disappointment: pride in North Carolina's place as one of a number of "forward looking" states that sterilized psychotic, feebleminded, and epileptic citizens,[18] and disappointment at the fact that the state performed a mere eighty-three sterilizations in 1946. It was ranked at best in the middle of the twenty-seven states with eugenic sterilization laws. Worse, 2 percent of the state's population was feebleminded; there were 70,000 such individuals living in the state and 1,400 more arriving every year. This figure was seventeen times the number of sterilizations in 1946.[19] North Carolina, they concluded, needed to expand its sterilization program energetically.

It did. Against the broader national trend of gradually decreasing but nonetheless continuing sterilizations, North Carolina intensified its own sterilization program in the first two decades after the Second World War. By 1980, North Carolina was responsible for the third-largest number of sterilizations in the United States – California was first with 20,108; Virginia second with 7,325; and North Carolina third with 5,993.[20] North Carolina adopted its sterilization law in 1929 and replaced it with a new statute in that fateful year of 1933.[21] Introduced by a former member of the Burke County Board of Public Welfare, the 1929 law allowed the sterilization of

> those who are feeble-minded and the mentally diseased, who would be likely to transmit their defects to their children or who are entirely incapable of rearing children. Under the provisions of the North Carolina statute any mentally diseased, feeble-minded, or epileptic inmate or patient of the State or county institutions, or any mentally diseased, feeble-minded or epileptic resident of a county, not an inmate of a public institution may be sterilized ... if it is believed to be for the best interest of the mental, moral, or physical improvement of the patient inmate or non-institutional individual; or when it is believed to be for the public good, or when such patient, inmate or non-institutional individual would be likely, unless sterilized, to procreate a child or children who would have a tendency to serious physical, mental, or nervous disease or deficiency.[22]

[18] Richardson and Gamble, "The Sterilization of the Mentally Handicapped in North Carolina," *North Carolina Medical Journal* 9, no. 2 (1948): 75–8.

[19] Ibid.

[20] Largent, *Breeding Contempt*, Tables 3.2 and 3.4, 77, 79–80. Johanna Schoen suggests that North Carolina was actually second after California in the number of sterilizations carried out. She cites a figure of "over eight thousand" for North Carolina, with Virginia in third place with "over seven thousand." *Choice and Coercion*, 82.

[21] A law passed in 1919 was not enforced because of doubts in the state about its legality.

[22] R. Eugene Brown, *Eugenical Sterilization in North Carolina: A Brief Survey of the Growth of Eugenical Sterilization and a Report on the Work of the Eugenics Board of North Carolina through June 30, 1935*, 8–9, AVS Pamphlets Misc. Box 24, AVS.

For intellectual justification, the law drew on Francis Galton and Harry Laughlin.[23]

As another example of the institutions-policymakers nexus, a member of the Board of Directors of Caswell Training School personally introduced the 1933 law to the state legislature.[24] The law remained operative until 1973. In April 2003, the state legislature voted to overturn North Carolina's eugenic sterilization law, which was by then seventy-four years old.[25]

The 1933 law established the North Carolina Eugenics Board. The Board's members were senior public health officials, including superintendents in the state's institutions. They reviewed all recommendations for sterilization to be carried out, approving 8,000 petitions. Of these, 7,600 were implemented, 85 percent of whom were women and 15 percent, men.[26] Most of these occurred after 1945. The Board had five members: the Board of Charities and Public Welfare Commissioner, the secretary of the State Board of Health, the chief medical officer of a state institution for the feebleminded or insane, the chief medical officer of the State Hospital at Raleigh, and North Carolina's attorney general.[27] At least three of the five members had to support any recommendation for sterilization for the operation to be approved, although this approval was often granted by deputies sent to the meetings by the senior officials.[28] Within institutions, the superintendent made the petition.[29] If the recommendation was for someone not confined in a state institution, an authorized state agent, such as the county superintendent of welfare, would prosecute the case.

Petitions included a medical history of the person proposed for sterilization and a signature from a medical officer in support of the recommendation. The Board's decision was then sent to the person affected or to a guardian or next of kin. An appeal process allowed for review in the state courts up to the state supreme court.

North Carolina's distinctive "contribution" to the application of eugenic sterilization policy was its direct link with welfare. Here again, the pro-sterilization lobbyists were influential. As a superintendent at the Department of Public Welfare put it,

[23] Ibid., 7.
[24] The member was Rep. W. A. Thompson, who also created the Eugenics Board of North Carolina. See ibid., 8, and Noll, *Feeble-Minded in Our Midst*, 67.
[25] Kaelber, "North Carolina," http://www.uvm.edu/~lkaelber/eugenics/NC/NC.html.
[26] Katherine Castles, "Quiet Eugenics: Sterilization in North Carolina's Institutions for the Mentally Retarded 1945–1965," *Journal of Southern History* 68 (2002): 849–78; and Johanna Schoen "Reassessing Eugenic Sterilization: The Case of North Carolina," in Paul Lombardo ed. *A Century of Eugenics in America: From the Indiana Experiment to the Human Genome Era* (Bloomington: Indiana University Press, 2011).
[27] Brown, *Eugenical Sterilization in North Carolina*, 8.
[28] Katherine Castles, "Quiet Eugenics," 852.
[29] Brown, *Eugenical Sterilization in North Carolina*, 13.

Here in North Carolina I am quite convinced that we are greatly indebted to Dr. Gamble, not only for the financial assistance he has provided, but for the very flexible policies which he has consistently advocated for in the past ten years and more. It was definitely upon his stimulation that North Carolina became the first state in the nation to incorporate contraceptive services within its State Department of Public Health.[30]

The superintendent particularly appreciated Gamble for his sensitivity to local concerns: the Department had worked with Margaret Sanger and the American Birth Control League but found them insufficiently deferential to local leadership.[31] A few years later, in 1951, Ellen Winston, North Carolina's Commissioner of Public Welfare, urged the state to follow up on families that received assistance from the Aid to Dependent Children program. This tactic targeted families in which one member had already been sterilized in order to see if other family members might benefit from the surgery.[32]

By bringing sterilization out of the institutions and into society, North Carolina provided the legal basis for a new wave of coerced sterilization. In the early years after the law's passage, the impact on African Americans was minimal. North Carolina, like other Southern states, limited African Americans' access to Aid to Dependent Children and other New Deal programs.[33] Federal and judicial pressure forced a change after the war, and African-American welfare rates rose accordingly. As they did, so did coerced sterilization: before 1950, thirty-one years into the program, 2,401 people had been sterilized under North Carolina's law.[34] Within a decade, the figure had more than doubled: 5,051 had been sterilized by 1960, and, five years later, a total of 5,993 individuals had been sterilized.[35] Whereas African Americans made up 23 percent of North Carolina sterilizations in the mid-1930s, this figure rose to about 65 percent by the mid-1960s.[36]

The decision-making procedure in North Carolina was similar to that of other jurisdictions, such as Oregon and Alberta. Sterilizations were carried out

[30] George H. Lawrence, Superintendent, Department of Public Welfare, North Carolina, to H. Curtis Wood, May 8, 1947, 2/14, AVS.

[31] Ibid.

[32] Schoen, *Choice and Coercion*, 109.

[33] Kluchin, *Fit to Be Tied*, 75.

[34] Largent, *Breeding Contempt*, 79.

[35] Ibid.

[36] Schoen, *Choice and Coercion*, 108. In the early 1950s, African Americans remained the minority of those sterilized. In the two years from July 1950, 704 people were sterilized in the state, 531 whites, 171 "Negroes," and two Indians. *Biennial Report of the Eugenics Board of North Carolina, July 1 1950 to June 30, 1952*. From then, the ratios began to shift, in the first instance because the number of whites declined: in the 1952–4 period, the figures were 357 whites, 198 "Negroes," and one Indian. *Biennial Report of the Eugenics Board of North Carolina, July 1, 1954 to June 30, 1956*. By 1958–60, African Americans were in the majority: 209 whites were sterilized vs. 315 "Negroes" and eleven Indians. *Biennial Report of the Eugenics Board of North Carolina, July 1 1958 to June 30, 1960*. All reports found in SW015.1/73, AVS.

after approval by the Eugenics Board. Doctors and social workers filed petitions for sterilization and were supposed to include the results of an IQ test, although this was not always the case. The Eugenics Board decided each case by a simple majority vote.[37]

In practice, the arguments made before the Eugenics Board in North Carolina rarely focused on hereditary conditions. The justifications relied instead on, first, whether the sterilization candidates or their parents could be economically dependent and, second (and secondarily), whether female candidates were likely to be sexually promiscuous.[38] Witnesses who wanted to defend themselves or their children from the operation needed to frame their arguments in these terms: that they were economically productive, or could be, and that they were not promiscuous, or would no longer be. As Johanna Schoen crisply puts it:

> No witness wasted many words on the question of whether the sterilization candidate in fact had a hereditary condition. ... Despite the fact that Eugenics Board members and the families of sterilization candidates came from vastly different economic and educational backgrounds, they shared the same language and understanding of the nature of eugenic sterilization cases. All knew that money and sex, rather than eugenics, lay at the heart of the matter.[39]

ENTER THE NATIONAL STATE

After a sharp increase in sterilizations in the two decades after the Second World War, North Carolina's program seemed to end abruptly; no coerced sterilizations occurred under the state's eugenic sterilization law after 1965. The reason was simple: federal law provided new avenues through which doctors and health and social workers could secure the sterilization of their patients.

Before the mid-1960s, the federal government had played no direct role in coerced sterilization. President Dwight Eisenhower, responding to a statement of the National Conference of Roman Catholic Bishops that opposed public funding for both domestic and overseas birth control, stated that funding should come from private agencies instead of U.S. government appropriations.[40] When the HBAA opposed the bishops and Eisenhower, it did so essentially on eugenic grounds. Ruth Proskauer Smith, executive director of HBAA from 1955 to 1964 and a supporter of eugenic sterilization, abortion, and euthanasia, wrote that tax money was "being poured into a bottomless pit here in New York" in order to "support ... quantity production of these many unloved, unwanted, and

---

[37] This paragraph is based on a confidential telephone conversation with a former member of the Board who wished to remain anonymous.

[38] See the study of a sample of the Eugenic Board's 8,000 case files undertaken by Johanna Schoen, *Choice and Coercion*, chapter 2.

[39] Ibid., 97–8.

[40] James Reston, "Foreign-Aid Problems: Birth Control Issue and Coordination with Allies Trouble the Administration," *New York Times*, December 10, 1959. For reactions, see AVS 11/96 and 11/98.

neglected children" instead of being used to pay for "facilities for education, recreation, and housing for a smaller group with higher potential."[41]

The situation changed in 1964 when President Lyndon B. Johnson made tackling American poverty a defining element in his domestic agenda. The expansion of federal spending and of state agencies under his administration created new opportunities for a federal role in family planning. In the absence of a family planning agency as such, however, the responsibilities were fragmented among several bureaus. In 1964, the Economic Opportunity Act created the Office of Economic Opportunity (OEO), with R. Sargent Shriver as its first director. Following intensive lobbying from the Population Council, the Population Crisis Committee, the Planned Parenthood Federation of America, and the Ford Foundation, it authorized the granting of federal funds to public and private agencies for "social programs and assistance."[42] As it did, Ruth Proskauer Smith wrote directly to President Johnson, urging him to use sterilization as a weapon in his war on poverty:[43] "Federal aid to depressed areas such as the Appalachian region, without birth control information[,] is futile. . . . One major cause of poverty is uncontrolled human fertility."[44] The HBAA was to be disappointed, at least initially. Although the wealthiest, best-connected, and most articulate Americans threw their weight behind population control and were backed up by an often-uncritical media, Johnson remained sensitive to powerful Roman Catholic opposition and acted cautiously.[45] The funds included grants for community-based agencies providing family planning advice and contraceptives.[46] The agencies were keen on sterilization, but Shriver was a practicing Roman Catholic (and husband of the equally devout Eunice Kennedy Shriver). He prohibited the public funds he administered from being used for sterilization outside of marriage.[47]

The funds were very limited in any case. Congress, with an eye to that same Roman Catholic opposition, was hesitant to provide much funding for birth control. Often the funds available for it went unspent: North Carolina doctors, who were prescribing contraceptives for a fee, opposed federal government clinics out of a fear that they would interfere with the business they were transacting.[48] In

---

[41] Dowbiggin, "'Rational Coalition,'" 242.

[42] Thomas B. Littlewood, *The Politics of Population Control* (Notre Dame, IN: University of Notre Dame Press 1977), 44. On lobbying efforts, see Critchlow, *Intended Consequences*, chapter 2.

[43] On Smith, see Dowbiggin, *Sterilization Movement*, 85–8.

[44] Quoted in Kluchin, *Fit to Be Tied*, 47.

[45] In the early 1960s, alarmist popular publications published articles with titles such as "Population Increase: A Grave Threat to Every American Family," "Intelligent Woman's Guide to the Population Explosion," and "Why Americans Must Limit Their Families," flooded the American market. See Critchlow, *Intended Consequences*, 54–5.

[46] See Gary D. London, "Family Planning Programs of the Office of Economic Opportunity: Scope, Operation, and Impact," *Demography* 5, no. 2 (1968): 924–30.

[47] Item 7, Exhibit IV, CA memo 37-A relating to the Economic Opportunity Act of 1964; cited in Kluchin, *Fit to Be Tied*, 96.

[48] Littlewood, *Politics of Population Control*, 45.

the mid-1960s, the situation changed. A coalition of the above-mentioned population activists, liberals who genuinely believed in providing choice, and conservatives who wanted to reduce welfare dependency led Congress to open the financial taps. Among the activists, the Population Council was the wealthiest – with a budget of $3 million per year – but AVS and Planned Parenthood also lobbied assiduously.[49]

At the midpoint of the decade, everything began to go their way. In 1965, the U.S. Supreme Court affirmed a right to privacy found in the Fourteenth Amendment in the seminal *Griswold v. Connecticut*. The decision overturned state restrictions on the sale of contraceptives by finding unconstitutional an 1879 Connecticut law that made it a crime if someone used a drug or instrument to prevent conception during sexual intercourse.[50] Two previous challenges to the state law had failed in 1943 and 1961. It was a classic case of judicial activism about what Justice Stewart called an "uncommonly silly law." The constitutional basis of the decision was secondary to the political aim of legalizing contraception. In *Griswold*, the Planned Parenthood League of Connecticut's executive director and medical director were appealing a conviction for violating the 1879 law by distributing information about family planning to married persons. They won their appeal by a 7–2 margin and a judgment. In the words of Justice William O. Douglas's majority decision, "specific guarantees ... have penumbras, formed by emanations from those guarantees that help give them life and substance."[51] A concurring judgment by Justice Arthur Goldberg and Chief Justice Earl Warren celebrated a right to privacy. Two dissenting justices, Hugo Black and Potter Stewart, did not like the Connecticut law but failed to see the statute as a violation of any constitutional rights or guarantees. The first draft of the majority opinion, penned by Douglas, located the right to use contraceptives in the First Amendment freedom of assembly.[52] It was only when Hugo Black ridiculed the idea – "the right of a husband and wife to assemble in bed is a new right of assembly to me" – that Douglas seized on his law clerk's suggestion to base a right to contraception on a right to privacy implied by the Third, Fourth, and Fifth Amendments.[53]

Soon after the *Griswold* decision, John Rague, the executive director of AVS from 1964 until 1972, wrote to President Johnson in protest against Shriver's ban, arguing that it was unconstitutional.[54] In the same year, 1965, USAID allocated $2.7 million to combat overpopulation abroad.[55]

The Department of Health, Education, and Welfare (HEW) also began funding family planning through a series of measures. First, drawing on the Maternal and Child Care and Mental Retardation Planning Amendments of 1963, HEW

---

[49] On the Council's budget, refer to Critchlow, *Intended Consequences*, 65.
[50] *Griswold v. Connecticut*, 381 U.S. 479 (1965).
[51] Ibid. at 484.
[52] Critchlow, *Intended Consequences*, 59.
[53] Ibid.
[54] Dowbiggin, *Sterilization Movement*, 149.
[55] Kluchin, *Fit to Be Tied*, 69.

provided funds for family planning at the state level through the Children's Bureau. By 1965, limited grants of $1.75 million went to twenty-four states.[56] Second, in 1967, Congress added amendments to the Social Security Act that governed Aid to Families with Dependent Children (AFDC), the most expensive federal welfare program (dating from 1935) and the one widely associated with African Americans, although, in fact, it was consumed proportionately more by poor whites. The amendment:

- Required that of all HEW funds available for maternal and infant care at least 6 percent be earmarked for family planning;
- Directed all the states to offer family planning services to present, past, and potential AFDC recipients in an effort to reduce illegitimate births and resulting welfare expenses;
- Authorized states to purchase family planning services from nongovernmental providers;
- Established a limit on the proportion of children under eighteen who could qualify for AFDC in any state; and
- Provided matching federal grants to the states for family planning (the federal portion was set at a high 75 percent, then raised in 1972 to a still higher 90 percent).[57]

Third, Congress increased funding for family planning further through the Family Planning and Population Research Act of 1970, an amendment adding Title X to the late New Deal-era Public Health Services (PHS) Act of 1944, to provide $382 million for family planning services, research, and training.[58] HEW administered the PHS Act. By comparison, HEW had spent $26 million in 1969, itself a sharp increase from four years earlier.[59] In addition, under intensive lobbying from AVS, the states began allowing Medicaid to reimburse up to 90 percent of the costs of a sterilization operation. By 1971, forty-eight states were doing so.[60]

Through this complex of new legislation and expanded funds under existing programs, money flowed through HEW and the OEO to the states. Federal funds could not be used to support private agencies prior to 1967. The AFDC amendment of that year allowed federal funds, for the first time in U.S. history, to be granted to and spent by nongovernmental actors.[61] As it was nested in

[56] Critchlow, *Intended Consequences*, 73.
[57] Littlewood, *Politics of Population Control*, 51; Kluchin, *Fit to Be Tied*, 95. HEW and the Department of Labor jointly administered AFDC.
[58] Kluchin, *Fit to Be Tied*, 95.
[59] Critchlow, *Intended Consequences*, 81.
[60] Alexandra M. Stern, "Sterilized in the Name of Public Health: Race, Immigration, and Reproductive Control in Modern California," *American Journal of Public Health* 95, no. 7 (2005): 1128–38; Kluchin, *Fit to Be Tied*, 68–9.
[61] Critchlow, *Intended Consequences*, 79.

larger workfare proposals, the family change went unnoticed politically, so little attention went to birth control.[62]

But the result was nothing short of revolutionary. Over decades, Birthright and Planned Parenthood had built up an extensive network of family planning clinics across the United States. By 1966, Planned Parenthood clinics provided contraceptive services to 320,000 women, of whom 35 percent were in receipt of some public welfare support.[63] The organization thus worked as both an intermediary and as an activist. It maintained many of the clinics through which federal funds passed, and it worked to expand the number of poor Americans eligible. It successfully urged Congress to include as eligible for funds 3.4 million women who were above the poverty line and another 1.7 million sexually active teenagers.[64] The number of people receiving publicly funded birth control services doubled to 10 million.[65] All the while, the Roman Catholic Church condemned sterilization both as a tool of birth control and as a policy instrument designed to reduce welfare costs.[66] As early as 1962, the secretary of the National Conference of Catholic Charities, Monsignor Raymond Gallagher, came out against the sterilization of welfare recipients.[67] Beyond theological objections, Gallagher made two arguments against the practice: first, it commodified black women's labor, seeing them chiefly as family breadwinners; and, second, it ignored the effect of family breakdown and absentee fathers on young black men.

As the federal government ramped up the funding, it quietly dropped its prohibition on funding sterilization. Congress lifted the restriction on funding sterilization for HEW and PHS programs in 1970, and OEO followed a year later.[68] Shriver, whose views on birth control had in any case moderated, had left the OEO and was eventually replaced by a young Donald Rumsfeld, who in turn was replaced in 1971 by Frank C. Carlucci. With Carlucci's support, the OEO moved into the sterilization business.[69] On May 18, 1971, the agency issued Instruction 6130–1, which removed the prohibition on federal funding for sterilizations (but left the prohibition on abortions).[70] From that moment forward, federal funds could pay for sterilization procedures. The consequences of this decision became clear for two African-American girls from Alabama.

---

[62] Dowbiggin, *Sterilization Movement*, 148.

[63] Critchlow, *Intended Consequences*, 78.

[64] Littlewood, *Politics of Population Control*, 64.

[65] Ibid.

[66] See the clippings in AVS 12/1010.

[67] See, for example, "Mother of 8, Priest Disagree on Sterilization," *Jet Magazine*, October 4, 1962, 50.

[68] Kluchin, *Fit to Be Tied*, 95–6.

[69] Littlewood, *Politics of Population Control*, 112. There is some debate about whether Carlucci intended the funds to be used for only male or both male and female sterilization. Ibid.

[70] Office of Economic Opportunity, Executive Office of the President, "OEO Instruction 6130–1, May 18, 1971," [copy provided to authors by Dr. Warren Hern]. The decision may have taken effect on May 19, 1971.

In June 1973, officials from the Montgomery Community Action Committee went to visit Minnie Relf, an illiterate mother of three children and a Medicaid recipient, who lived in a shabby apartment on $150 per month in welfare support.[71] The girl's father, Lonnie Relf, a former field hand incapable of work due to an auto accident, was away from home.[72] The family planning nurses, who were also African American, told Mrs. Relf that her daughters needed some shots and that the girls should accompany them. Their mother drew an "X" on a form she could not read, although the nurses later stated that the form's content had been explained to her.[73] The two youngest girls, Mary Alice, age twelve, and Minnie, age fourteen, went quietly; their sixteen-year-old sister locked herself in her room and refused to leave.[74] The two younger girls were taken to the Professional Center Hospital, kept overnight, and, the next day, their fallopian tubes were cut.[75] Their mother had signed the consent form, and OEO funds paid for the operations.

The Relf girls, too young to articulate their objections or to furnish informed consent, would have remained forgotten victims had others not intervened. A Roman Catholic charity worker regularly visited the Relf household. When she learned of the sterilizations, she reported them to the Southern Poverty Law Center (SPLC), then a small civil rights firm.[76] SPLC launched a lawsuit, *Relf v. Weinberger*,[77] which exposed the story of welfare-based coerced sterilization. In his 1974 decision in the case, federal Judge Gerhard Gesell concluded that between 100,000 and 150,000 women, most of them poor and/or African-American adults, were sterilized without giving their informed consent. Observing that the distinction between eugenics and contraception remained "murky," Gesell concluded:

> [T]here is uncontroverted evidence in the record that minors and other incompetents have been sterilized with federal funds and that an indefinite number of poor people have been coerced into accepting a sterilization operation under the threat that various federally supported welfare benefits would be withdrawn unless they submitted to irreversible sterilization.[78]

He prohibited both the use of federal funds for involuntary sterilizations and the practice of coercing women on welfare to volunteer for sterilization.[79] Federally funded family planning sterilizations, the ruling continued, were permitted only

---

[71] A. M. Stern, "In the Name of Public Health."

[72] "Sterilized: Why?" *Time*, July 23, 1973.

[73] Kluchin, *Fit to Be Tied*, 100.

[74] Franks, *Margaret Sanger's Eugenic Legacy*, 185.

[75] "Sterilized: Why?" *Time*, July 23, 1973.

[76] Gregory Michael Dorr, "Eugenics in Alabama."

[77] *Relf v. Weinberger*, 372 F. Supp. 1196 (D.D.C. 1974). Vacated and remanded back to the District Court for dismissal September 13, 1977. 565 F.2d 722 (D.C. Cir. 1977).

[78] *Relf v. Weinberger*, 372 F. Supp. 1196 at 1199.

[79] Southern Poverty Law Center, "Relf v. Weinberger," http://www.splcenter.org/get-informed/case-docket/relf-v-weinberger (accessed January 23, 2013).

where there is informed and uncoerced consent on the part of competent individuals. The judge included a commentary on the federal government's entrée into social engineering:

> Surely the Federal Government must move cautiously in this area, under well-defined policies determined by Congress after full consideration of constitutional and far-reaching social implications. The dividing line between family planning and eugenics is murky. And yet the Secretary, through the regulations at issue, seeks to sanction one of the most drastic methods of population control – the involuntary irreversible sterilization of men and women – without any legislative guidance. Whatever might be the merits of limiting irresponsible reproduction, which each year places increasing numbers of unwanted or mentally defective children into tax-supported institutions, it is for Congress and not individual social workers and physicians to determine the manner in which federal funds should be used to support such a program. We should not drift into a policy which has unfathomed implications and which permanently deprives unwilling or immature citizens of their ability to procreate without adequate legal safeguards and a legislative deter-mination of the appropriate standards in light of the general welfare and of individual rights.[80]

It was a landmark decision. The related publicity dealt a serious blow to the pro-sterilization lobby, which had been on the crest of its greatest success before the story broke. The pro-sterilization coalition "of individuals and organizations in favor of birth control, population control, legal abortion, and women's rights," writes Ian Dowbiggin, "abruptly dissolved when radical feminists and minority activists began protesting the high rates of sterilization among low-income, minority, and immigrant women."[81] AVS and Planned Parenthood could only seethe in anger when the New York chapter of the National Organization for Women (NOW) broke ranks and endorsed a series of limits on sterilization, including a thirty-day waiting period and a ban on federal funds for sterilizing anyone under the age of twenty-one. "Oh God, those stupid women in NOW," their members moaned when NOW's support was announced.[82]

There is some question about whether the sterilization of the Relfs and many other poor people resulted from poor oversight or from other sources. Dr. Warren Hern, then chief of the program evaluation and development branch at the OEO, was charged with drawing up the guidelines. Dr. George Contis, director of the OEO's family planning program, issued instructions to regional directors and comprehensive health center directors that sterilizations were not to be undertaken until Hern had outlined the necessary procedural safeguards.[83] Hern, who would eventually resign over the issue, maintains to this day that the

---

[80] *Relf v. Weinberger*, 372 F. Supp. 1196 at 1205.
[81] Dowbiggin, *Sterilization Movement*, 183.
[82] Ibid., 183–4.
[83] Littlewood, *Politics of Population Control*, 112.

overzealous implementation of the sterilization program resulted from the OEO suppression of official guidelines.[84] According to Hern:

> [M]aterials from Planned Parenthood, the Association for Voluntary Sterilization, the University of Michigan Medical Centre and a comprehensive report from the Population Council were incorporated into a draft set of guidelines. OEO's Family Planning Division then issued the draft guidelines to OEO regional health specialists for their review and comment. . . . A new Special Condition was to be attached to all OEO health program grants which specified procedures to protect the legal rights of OEO clients seeking voluntary sterilization and to assure informed consent.[85]

The final draft of the guidelines, accompanied by consent forms, was completed by late 1971,[86] received its OEO General Counsel clearance on December 27,[87] and was officially approved by OEO Deputy Director Wesley Hjornevik on January 10, 1972.[88] Approximately 25,000 copies of the guidelines were printed for distribution.

Two weeks later, their distribution was halted. On February 2, a new deputy director of the Office of Health Affairs, E. Leon Cooper, ordered the guidelines withdrawn from circulation.[89] On February 4, the 200 copies held by Hern were collected, counted out, and put in a safe.[90] The order came from the White House from future Secretary of the Treasury Paul O'Neill.[91] Hern spent the next two months lobbying both the White House and the OEO to release the guidelines. He received a reprimand when he tried to contact the office of White House Counsel John Dean.[92] On May 2, Cooper ordered Hern to refrain from discussing the sterilization guidelines and other "sensitive program areas" with other

---

[84] Interview with Dr. Warren Hern, Boulder, CO, June 2003. See also Bill Kovach, "Guidelines Found on Sterilization," *New York Times*, July 7, 1973 and "H.E.W. Chief Issues Guidelines to Protect Rights of Minors," *New York Times*, July 20, 1973. We are grateful to Dr. Hern for sending these and many other documents to us.

[85] Hern, "Sterilization and the Poor," unpublished paper provided by Dr. Hern to the authors.

[86] Office of Economic Opportunity, "OEO Instruction 6130 -.2[] Voluntary Sterilization Services," n.d. [copy provided by Dr. Hern].

[87] Office of Economic Opportunity, "Executive Office of the President, Memo from Deputy Director Wesley L. Hjornevik, January 10, 1972. For distribution to OEO Assistant and Associate Directors, Regional Directors, Community Action Agencies, Comprehensive Health Centres, and State Economic Opportunity Offices" [provided by Dr. Hern].

[88] Warren M. Hern, "Statement Concerning the Suppression of OEO Instruction 6130–2 (Guidelines for Voluntary Sterilization Services)." *Presented before the Senate Health Subcommittee*, Senator Ted Kennedy, Chairman, July 10, 1973.

[89] "OEO Official Withdraws Sterilization Guidelines," *Denver Post*, May 31, 1972. Cooper cited costs and the operation's long-term psychological effects as the reasons. See George Vecsey, "Federal Sterilization Program in Doubt," *New York Times*, May 28, 1972. Rebecca Kluchin dates the withdrawal to January 31, 1972. *Fit to Be Tied*, 95.

[90] Hern, "Statement Concerning the Suppression of OEO Instruction 6130–2."

[91] E-mail correspondence with Dr. Hern, October 18, 2012 and January 9, 2013.

[92] Littlewood, *Politics of Population Control*, 123.

government agencies or individuals outside the OEO, such as the family planning clinics to whom the guidelines were to be sent.[93] The program was "sensitive" because 1972 was a presidential election year, and the White House staff was torn between two competing impulses: (a) to move Roman Catholics disillusioned with the "amnesty, abortion, and acid" Democratic candidate, George McGovern, into the Republican column; and (b) to build a coalition in Congress by linking pro-abortion liberals with conservatives interested in using contraceptives and abortion to control, discipline, and reduce the number of the poor.[94] On June 2, Hern resigned. In July 1973, OEO transferred its responsibilities for sterilization to the HEW.[95]

Hern has insisted to this day that the distribution of the guidelines would have prevented the sterilizations.[96] It is possible. His argument is strengthened by the fact that at least one OEO official, unaware that the guidelines had not been distributed, went on record stating that the sterilization of the Relf girls was a violation of the guidelines' consent requirements.[97] That said, there are also good reasons to think that the guidelines would have proved insufficient. One's eyebrows should be raised by the AVS and Planned Parenthood's involvement in drafting them. AVS's interest in sterilization was not grounded in a primary concern for poverty, still less in a regard for the privacy or rights of America's poor. It was a lobby group for increased sterilization, one that had only reluctantly – if at all – abandoned eugenic principles. For decades, AVS and its predecessors had put the solution (sterilization) before the problems (mental disability, overpopulation, poverty, crime, welfare abuse, and so on). Similarly, the birth control zealotry of Guttmacher and Planned Parenthood was scarcely concerned with the fertility or the civil liberties of the poor or so-called mentally incompetent. In the mid-1960s, Planned Parenthood had three policy goals.[98] The first was ensuring the availability of contraceptive devices and drugs. The third aim was creating access to abortion. Instituting the government-financed sterilization of the poor became the second objective.

Planned Parenthood did not draw on rights-based arguments until 1967.[99] Until at least 1968, when the organization's board of directors articulated an explicit call to repeal all abortion bans, Planned Parenthood "was more of a

[93] Office of Economic Opportunity, "Memo from E. Leon Cooper to Warren Hern on 'Unauthorized External Communications'" [provided by Dr. Hern].

[94] Littlewood, Politics of Population Control, 62.

[95] Kluchin, Fit to Be Tied, 99. On the suppression of the OEO guidelines generally, see Mark Bloom, "Sterilization Guidelines: 22 Months on the Shelf," Medical World News, November 9, 1973.

[96] Kovach, "Guidelines Found on Sterilization"; Dr. Warren Hern, interview.

[97] Kovach, "Guidelines Found on Sterilization."

[98] Littlewood, Politics of Population Control, 108.

[99] David J. Garrow, Liberty and Sexuality: The Right to Privacy and the Making of Roe v. Wade (Berkeley: University of California Press, 1998); Mary Ziegler, "The Framing of a Right to Choose: Roe v. Wade and the Changing Debate on Abortion Law," Law and History Review 27, no. 2 (Summer 2009): 303.

population control and public health organization than a group dedicated to abortion reform."[100] The latterly discovered rhetoric of individual choice masked the fact that AVS and Planned Parenthood were interested in meaningful choices only for the middle and upper classes; the poor were to have only one option. Not once did AVS, Planned Parenthood, or any other pro-sterilization lobby group outwardly entertain the idea that there might be policy instruments other than sterilization – such as comprehensive, publicly financed health care, sanitation, or investment in infrastructure and education – that could help end poverty in West Virginia and elsewhere in the United States. Finally, from the earliest days, the implementation of sterilization policy was discretionary. A formal requirement that advice be noncoercive and that consent be voluntary meant little if state officials such as doctors, nurses, welfare officers, and family planners bombarded uneducated people with "expert" arguments in favor of sterilization.

The distinction between voluntary and involuntary sterilization was difficult to maintain at the best of times. The distinction became nearly nonexistent when those responsible for its maintenance were unabashed sterilization advocates.

Whatever might have happened under the umbrella of guidelines, what happened in their absence was clear. Dr. Louis Hellmann of HEW estimated that 100,000 sterilizations paid for with federal funds during 1972 and 1973 were coerced, which represented one-half the 200,000 paid for overall.[101] Coercion chiefly occurred within the hospitals and in private practices.

The experience of two African Americans in South Carolina, Shirley Brown, then twenty-three, and Mrs. Virgil Walker, then thirty, was typical. Both were patients of Dr. Clovis Pierce, the only obstetrician in Aiken County. According to court testimony, Dr. Pierce would tell his patients on the maternity ward, "Listen here, young lady, this is my tax money paying for this baby, and I am tired of paying for illegitimate children. If you don't want this operation, find another doctor."[102] Doctors such as Pierce, as one scholar put it, "literally inscribed their politics upon their patients' bodies."[103] Pierce told Brown not to come back until she was willing to be sterilized, and he threatened Walker with the severance of her welfare payments.[104] The day after she gave birth, Walker signed the form and was sterilized. By the end of 1973, fully one-third of the welfare mothers in Aiken County had been sterilized.[105] As Judge Gesell noted in his above-cited opinion, "[p]atients receiving Medicaid assistance at

[100] Ziegler, "Framing of a Right," 306.
[101] Littlewood, *Politics of Population Control*, 109 and 113. The most conservative estimates are a few thousand, the most generous a few hundred thousand. Kluchin, *Fit to Be Tied*, 74.
[102] Quoted in Littlewood, *Politics of Population Control*, 109.
[103] Kluchin, *Fit to Be Tied*, 112. A North Carolina physician put it more bluntly: "A doctor who had just got his income tax back and realized it all went to welfare and unemployment was likely to push for sterilization harder." Ibid.
[104] Littlewood, *Politics of Population Control*, 109.
[105] Ibid.

childbirth are evidently the most frequent targets of this pressure, as the experiences of the plaintiffs Brown and Walker illustrate."

Black women were not the only victims: physicians in New York, California, Colorado, South Dakota, and Michigan pressured Native Americans, Mexicans, and Puerto Ricans into consenting to sterilization, often exploiting poor English language skills in the case of the last two groups to do so.[106] Between 1970 and 1976, the Indian Hospital Service and its affiliates sterilized a dramatically high 25 to 42 percent of all Native American women of childbearing age.[107]

Many if not most of these cases combined voluntarism in theory and coercion in practice. Aiken County is one example, but there are others. Dr. Bernhard Rosenfeld, the primary author of Ralph Nader's Health Research Group's (HRG) *Study on Surgical Sterilization* (1973), recalled his chief resident's instructions when a woman scheduled for a tubal ligation refused it: "Go [into the postpartum delivery room] and see if you can talk her into it."[108] When she refused to be talked into it, there was a third option: doctors would add a "certificate's state of emergency" next to the phrase "no consent signed." "This," it read, "is to certify that the delay necessary to obtain complete consent for treatment would endanger this patient's life or chance of recovery. We believe emergency operation is necessary."[109] Under no circumstances could sterilization, as distinct from a cesarean section, be considered an "emergency," and under no circumstances would sterilization save a patient's life.[110]

The role of Planned Parenthood and AVS in these sterilizations is unclear. Extensive funding flowed through the family planning clinics; by 1969, Planned Parenthood operated two out of every five projects delegated by community action agencies.[111] Given the organization's enthusiasm for population control and its late-1960s turn toward sterilization, it is doubtful that the advice provided by Planned Parenthood affiliates was neutral. But, in the absence of further documentation, little more can be said. It is clear, however, that Planned Parenthood, AVS, and the population lobby more generally were cheerleaders for the creation of the policy instruments leading to the coercive sterilizations. AVS spent the 1960s and 1970s browbeating legislators about the dangers of welfare abuse and population growth, and Hugh Moore's Population Crisis

---

[106] See Kluchin, *Fit to Be Tied*, 102–11 and Schoen, *Choice and Coercion*.
[107] Kluchin, *Fit to Be Tied*, 108. Another scholar notes that some studies estimated the number of Native American women of childbearing age to be sterilized to be more than 50 percent. Jane Lawrence "The Indian Health Service and the Sterilization of Native American Women," *American Indian Quarterly* 24, no. 3 (2000): 410.
[108] Kluchin, *Fit to Be Tied*, 104.
[109] Ibid., 105.
[110] Ibid., 104, citing documents from the USC LA Country Medical Center from 1970 to 1974.
[111] Critchlow, *Intended Consequences*, 87.

Committee (PCC) paid for full-page newspaper ads urging passage of the Family Planning Services and Population Research Act of 1970.[112] Moore himself claimed that PCC's direct lobbying of Speaker of the House John McCormack (D-Massachusetts) was decisive in getting the bill passed there (it had been introduced into the Senate).[113]

During the 1960s, multiple processes converged to create a new wave of coerced sterilization. First, in North Carolina, the reform of New Deal welfare legislation to include African Americans brought, given the state's reliance on extrainstitutional sterilization, large numbers of people under the purview of the state's eugenic sterilization law.[114] Second, federal funding of contraception and sterilization vastly increased the resources of existing family planning clinics, hospitals, and private clinics. Third, legislators began looking to punitive sterilization, among other measures, as a way of criminalizing fecundity and illegitimacy in reaction to evidence of paternal desertion, illegitimacy, and disproportionate reliance on welfare among African Americans.[115] Many of these proposals got nowhere, but they reflected and encouraged a cultural context in which black fertility and single motherhood were deeply suspect and in which harsh medical responses, sanctioned by law or professional discretion, seemed normal and natural. As Julius Paul put it, "[t]he surgeon's knife (sterilization) still seems to have the same magical quality in the minds of some people for 'saving' America from its shame, squalor, and various miseries of human or social instigation (especially poverty) as it did over sixty years ago."[116] Finally, old eugenic organizations such as AVS and Planned Parenthood had spent the first three decades of the postwar period warning the world of the social, economic, and environmental threats occasioned by population growth. By the time Paul Ehrlich published *The Population Bomb*, mainstream magazines, late-night TV discussion shows, and senior politicians were forecasting war and mass starvation in the absence of drastic checks on population growth.

These deep-seated fears about population growth intersected with post-New Deal and post-Great Society fears about the social and fiscal consequences of high African-American birthrates and ensuing welfare dependency. In 1965, Harvard professor and Assistant Secretary of Labor Daniel Patrick Moynihan published a report that identified the breakdown of the African-American family

[112] Ibid., 93.
[113] Ibid.
[114] Johanna Schoen's work has attempted to show that some black women sought voluntary sterilization. The numbers, however, do not support such a strong claim: 446 out of 8,000. See *Choice and Coercion*, 112–3.
[115] For a contemporary overview, see Julius Paul, "The Return of Punitive Sterilization Proposals: Current Attacks on Illegitimacy and the AFDC Program," *Law & Society Review* 3, no. 1 (1968): 77–106.
[116] Ibid., 101.

as the key cause of the black community's economic and social challenges.[117]
The report's apparent judgmentalism of African-American family structures
overshadowed its policy recommendations, and the latter rapidly receded from
the pool of plausible state actions. In understanding the challenge, Moynihan
emphasized that the sudden granting of equal rights of citizenship could not be
taken as the equivalent to providing equal material income or future opportunity. But the sociological evidence he received and compiled while writing the
report continuously shifted his focus to African-American family structure and
what he fatefully labeled the "tangle of pathology": "the Negro community has
been forced into a matriarchical structure which, because it is so out of line with
the rest of American society, seriously retards the progress of the group as a
whole, and imposes a crushing burden on the Negro male and, in consequence,
on a great many Negro women as well."[118]

Moynihan's conclusions focused primarily on the family crisis: "in a word, a
national effort toward the problems of Negro Americans must be directed
towards the question of family structure."[119]

These factors converged to create the conditions for extensive sterilization
abuse: often, poorly educated patients found themselves at the mercy of doctors
conditioned to believe that excessive African-American fertility was a fiscal and
environmental time bomb. The presence of racism, expressed both in casual
ways and in institutional structures, ineluctably made coercion much more
likely. The bulk of sterilizations occurred in the South, but African Americans
in the Northern United States were affected as well. The above-cited report by
the Health Research Group, part of Ralph Nader's consumer protection empire,
concluded that coerced sterilization, generally in the form of thrusting a consent
form in a patient's face minutes before a caesarian was to be carried out,
occurred at hospitals in Los Angeles, Baltimore, Boston, and Chicago, as well
as in New Orleans, Nashville, and Louisville.[120] Were this not enough, poor
women, often African Americans, became little more than guinea pigs in
research hospitals. Interns gained useful practice by operating on welfare
patients, with "in many instances . . . little evidence of informed consent by the
patient."[121] As a physician at Atlanta's public Grady Memorial Hospital put it,
"If there's a doctor and a uterus around, the two will get together. Let's face it, if
you're a gynecology resident, how are you going to practice if you ain't yanked

---

[117] Littlewood, *Politics of Population Control*, 49. The report was published as *The Negro Family: The Case for National Action*, Office of Policy Planning and Research, U.S. Department of Labor, March 1965, http://www.dol.gov/oasam/programs/history/webid-meynihan.htm.

[118] *The Negro Family*, "The Tangle of Pathology," chapter 4.

[119] Ibid, "The Case for National Action," chapter 5.

[120] Bernard Rosenfeld, Sidney M. Wolfe, and Robert E. McGarrah, Jr., *A Health Research Group Study on Surgical Sterilization: Present Abuses and Proposed Regulations* (Washington, DC: Health Research Group, 1973).

[121] Ibid., 2.

some utes?"[122] It was common for physicians to persuade poor patients to accept hysterectomies instead of tubal ligation in order to practice the more complicated of the two surgeries.[123] "Let's face it," a staff doctor told HRG investigators, "we've all talked women into hysterectomies who didn't need them, during residency training."[124]

Residents who complained found their concerns brushed aside. When one Boston City Hospital resident questioned whether the size of a particular fibroid tumor really necessitated a hysterectomy, the chief resident replied, "We don't know. That guy that sent her in thought there might be. Besides, she's 42 and doesn't need a uterus."[125] Some physicians' disregard for patient welfare contrasted with an intense concern for their own. One hospital physician recalled a colleague covering his tracks after he lied to a patient about the need for a hysterectomy: "We're going to have to make sure that the pathology report does not get back to the woman and make up a reason why she needed to have it taken out."[126] With African Americans used for the medical equivalent of target practice, the claim made by black radicals in the 1960s that publicly funded birth control was an excrcise in "Black genocide" becomes easier to comprehend.[127]

The hundreds of thousands of sterilizations carried out in the late 1960s and early 1970s may not have been eugenic in the strict sense that they were developed on a theory of heredity, but it was nonetheless a massive program of intrusive social engineering driven largely by that old eugenic fear of differential fertility rates. As conservative columnist George Will commented at the time, "frugality is not the motive of those who want to sterilize little girls. The people who want this are not misers; they are idealists. They want to sterilize inferior people; they want to improve the population. Only such a great project demands such a grave government involvement."[128]

Those who wanted to sterilize young women or teenage girls did not disagree. In 1973, a practicing obstetrician who had performed such sterilizations published an article in *Contemporary Ob/Gyn*.[129] "Too many people," he wrote, "crowded too close together cause many of our social and economic problems. These, in turn, are aggravated by involuntary and un-responsible parenthood. As physicians we have obligations to our individual patients, but we also have

---

[122] Quoted in Littlewood, *Politics of Population Control*, 126. The prevalence of this operation led them to be known colloquially as the "Mississippi appendectomies." Kluchin, *Fit to Be Tied*, 6.

[123] Kluchin, *Fit to Be Tied*, 107.

[124] Ibid. quoting Rosenfeld, Wolfe, and McGarrah, *Health Research Group Study*, 3, 4, 11.

[125] Kluchin, *Fit to Be Tied*, 107.

[126] Ibid., 108.

[127] There is some evidence of black women rejecting this reading and demanding birth control facilities. See ibid., 66.

[128] George Will, "Sterilization and 'Population Improvement,'" *Washington Post*, July 23, 1973.

[129] H. Curtis Wood, "The Changing Trends in Voluntary Sterilization," *Contemporary Ob/Gyn* 1, no. 4 (1973): 31–40.

obligations to the society of which we are a part. The welfare mess, as it is called, cries out for solutions, one of which is fertility control."

The article was signed by "Dr. Wood." It was Dr. H. Curtis Wood, past director of the AVS and of its predecessor, the HBAA, which was in turn the successor to Birthright, Inc. and the Sterilization League of New Jersey. The article directly linked the fears that had successively driven the organization in its various institutional forms for decades: differential fertility, population growth, and then welfare. Eugenics and old-style eugenicists were at the heart of birth control and welfare policy in the early 1970s.

14

Those Who Sterilized

Reflecting on the stories of those affected by sterilization, it is hard to understand through our modern, rights-based perspective how these sterilizations could have occurred. At the same time, it is very easy to wax indignant about the operations and to conclude that they were ordered and performed by uncaring officials. We began by assuming the situation was more complicated: that such individuals were a product of a culture that emphasized social responsibility as much if not more than individual entitlement; that they performed jobs they regarded as socially important and morally defensible; and that they found themselves in situations that most reasonable people would regard as difficult. Even if one wishes to reject these assumptions, the fact remains that individuals were part of this history, and only a one-sided account would deny them their voice.

The first point to make is that the social problems, above all poverty, were undeniably real. And these problems did indeed convince many professionals that birth control was the solution.

WEST VIRGINIA

In early 1971, Warren Hern, the official at the Office of Economic Opportunity (OEO) who later wrote the federal government's guidelines on sterilization, published an article in the *New Republic* recounting his experiences in West Virginia. Hern had witnessed and been affected by the death of women from illegal abortions in Latin America in the mid-1960s.[1] He described what he had seen in autumn 1970:

---

[1] "Biography, http://www.drhern.com/biography.pdf [official biography] (consulted January 23, 2013).

An ugly industrial slum winds through the valleys from Huntington to Charleston and is overlain with a yellowish grey pall which impartially dissolves lung tissue and automobile finish. The rivers are a stinking cauldron of industrial poisons.

Into this mutilated landscape the newly born miner's child is thrust. The mother, battered by repeated child-bearing, struggles against sheer physical exhaustion. Years in the mines have left the father with a chronic cough, frequent chest pain, shortness of breath, and other symptoms of severe lung disease and early heart failure. If he is lucky enough to have a job at all ... he may earn as little as $2.75 a car loaded with 1 to 1.5 tons of coal; and he loads seven cars a day in a 12-to-14 hour day.[2]

Hern went on to recount his interview with a woman from one of the "poorest families near the unemployment-stricken village of Fratersville":

She invited us to sit on a tattered sofa in a room otherwise barren except for a potbelly iron stove and a broken wooden chair. She pulled the chair near the glowing stove and sat down. As we talked, she slumped forward and watched her children listlessly, her arms crossed. ... She was 38 years old. She and her husband had nine living children. She was six months pregnant. "I tuck them pills for a while, but my stumick got to botherin' me like it did afor I tuck 'em so I quit for three months. Couldn't afford 'em, no how." [A local man] told her that she could get free medical exams and birth control at the OEO clinic, and she said "Well I'll be down after the baby's got borned, but ain't there some way asides a pill?"

Similar requests from families with all the children they want are commonly heard by ... OEO outreach workers in Tennessee. Jeanette Smith, the nurse directing the OEO-funded Anderson County family planning project through the local Planned Parenthood organization, estimates that her office receives four to five sterilization requests a week from poor families. ... [More broadly], eight percent of [family planning] projects [serving 300,000 people] reported that they wanted to provide sterilization as part of their regular services.[3]

LUMBERTON, NORTH CAROLINA

As part of our research in North Carolina, we tried to meet with members of the Eugenics Boards and/or the doctors performing the sterilizations. This ambition met a stone wall of silence. This wall was built by a report in the *Winston-Salem Journal* drawing comparisons between North Carolina and Nazi Germany (a more promising analogy would be between North Carolina and Scandinavia, where the development of sterilization matched that of the welfare state).[4] After countless messages and unreturned calls, we found one person who was close to the events described earlier in Chapter 13 and who was also willing to speak:

---

[2] Warren Hern, "Biological Tyranny," *New Republic*, February 27, 1971, 15.

[3] Ibid., 16.

[4] Kevin Begos, "Against Their Will," part one, *Winston-Salem Journal* website, December 8, 2002. The newspaper's project website is available at http://againsttheirwill.journalnow.com (accessed January 23, 2013). We owe the point on Scandinavia to Johanna Schoen.

Dr. Ernest H. Brown, Jr., a retired gynecologist and obstetrician, whom we interviewed at his Lumberton home.[5]

After graduating from medical school at the University of North Carolina (UNC)-Chapel Hill, Brown completed specialist training at the Medical University of South Carolina in Charleston, South Carolina and served two years in the U.S. Air Force in North Dakota as chief of Obstetrics and Gynecology. In 1963, Brown set up a private practice in Lumberton. He practiced obstetrics and gynecology until 1991, gynecology until 1998, and performed regular sterilizations on patients referred to him by the welfare department and Eugenics Board from 1963 onward. From the start of his residency in 1958 throughout his career, Brown was well acquainted with the evolution of both voluntary and involuntary sterilization practice in North Carolina.

In 1958, he told us, for a woman to be voluntarily sterilized she had to have five living children, to have been pregnant seven times, or to have undergone a third caesarean section. These regulations were not stipulated by the state in the case of voluntary sterilizations but, rather, were set by individual hospitals. This was the rule in the hospital where he began his career as a qualified doctor. In 1963, when he began his practice in Lumberton, the local hospital stipulated the same conditions. In the 1960s, he recalled, there was concern about sterilizing people. North Carolina still had an agrarian economy, and large families were an important part of the labor-intensive agricultural workforce. As social mores changed, however, the regulations were relaxed, allowing sterilization after the birth of four and, later, three children. The operation was eventually made available to any legally responsible adult on request.

Patients determined to be "mentally retarded," Brown noted, were referred by the welfare department to the Eugenics Board for a decision on their sterilization. The doctor then received what Dr. Brown described as "essentially a court order" to perform the sterilization, which was paid for by the state. Sterilization cases were commonly initiated by social workers or by a patient's family. Moreover, Brown emphasized that he was not qualified to assess the level of retardation of his patients. He had no contact with the Eugenics Board whatsoever, and social workers referred patients to him only once a decision had been made. He was, as he put it, "purely just a technician, that's all." Candidates for sterilization were often children in their teens. He recalled seeing children "that were profoundly retarded, unable to take care of their own

---

[5] Dr. A. M. Stanton, who was responsible for sterilizing Nial Cox Ramirez, refused to talk; this case has received extensive media and scholarly attention. Despite several reminders, John Railey of the *Winston-Salem Journal* also failed to provide the contact details of Dr. Robert Albanese of Martinsville in Virginia, who recommended the sterilization of Bertha Dale Midgett Hymes. Finally, Dr. A. V. Blount, who is reported as remembering assisting at the sterilizations of several patients, but whose involvement was limited overall, was called six or seven times, but we were never put through to him. He also failed to call back when we left a number.

sanitary needs and then others that were sterilized that probably in retrospect with good testing were not retarded; they were socially inept maybe."[6]

He also remembered dealing privately with the severely retarded children of wealthy parents. When sterilization was felt to be required in these cases, parents were likewise advised to go through the state's sterilization procedures. Dr. Brown did not give any indication that there was ever a major problem in getting the sterilization of children from rich families authorized. This suggestion means that the Eugenics Board was not simply concerned with sterilizing young, poor, African-American girls. Rather, the overriding concern of those committed to the sterilization program was more broadly eugenic and not merely racist in nature. That the overwhelming majority of cases were poor African-American girls probably reflects that social workers played a major role in identifying cases for sterilization. The figures probably also reflect some of the racist assumptions of North Carolina's welfare system and demographic structure.

Although he was not responsible for deciding who should be sterilized, Dr. Brown did play a role in recommending the type of sterilization operation a patient should undergo. He explained that, for most of those considered unable to take care of their own sanitary needs, he carried out total hysterectomies, which eliminated the need to deal with menstruation. Dr. Brown recalled only three or four such cases in total, so even these were rare. The majority instead underwent more straightforward tubal ligation. Most of those Dr. Brown was called on to sterilize were cases of teenage pregnancy. He did not, however, recollect treating patients immediately after they had given birth and said he would not have sterilized them at that point but, rather, would have waited.

Dr. Brown was unable to remember the exact number of coerced sterilizations he performed, but he estimated that he carried out four or five such operations a year. He went on to say that, in all, there were four or five doctors in Lumberton who performed these operations, of whom he was the only one still living at the time of his interview. These doctors were responsible for territory that included the whole of Robeson County, which had a steady population of around 90,000 during the period, as well as parts of neighboring Blaydon and Columbus Counties. He was uncertain when the program formally came to an end but said that he did not recall seeing mentally retarded people for sterilization in last fifteen years of practice and believed that the program was simply phased out. He also stated that he had had nothing to do with people in mental hospitals.

Brown went on to make the point, if implicitly, that the discussion of sterilization cannot be divorced from the cases of severe mental deficiency, in which individuals can barely care for themselves, much less others:

> I'm aware currently of a girl who is so retarded she is essentially bed-ridden, [who] was raped – I think they actually have charged an orderly for raping her – [and] who is pregnant. Now, under that situation, we would sterilize her again. And more

---

[6] Interview with Dr. Ernest Brown, Lumberton, NC, September 5, 2003.

particularly, some who are even able to be up and about; these women – it's hard to say this and I hope [you] understand what I am saying – they are so retarded they aren't humans.[7]

Dr. Brown asserted that race had no bearing on the program in Robeson County, although he said he had read of allegations of racism in other counties. In Robeson, he noted, there are three main racial communities. The majority has changed in the last forty years from Caucasian to Native American, whereas African Americans have remained at about 25 percent of the overall population throughout this time.[8] Dr. Brown reiterated: "Race had nothing to do with it." He was aware of the controversy surrounding cases of sterilizations of people not truly mentally retarded "among some minorities." This allegation he suspected to be true but thought the press made too much of it. He suggested that the standards of diagnosis of retardation in 1965 were more primitive than they are now and asserts that tests have surely improved. He himself did not administer such tests, although he did always carry out a preoperative assessment of the physical health of a patient, which did not encompass her mental or intellectual health. Nonetheless, Dr. Brown said he did not recall performing any sterilizations that he did not feel were justified at that time. The patients he operated on were unable even to go to a supermarket and buy a bottle of Coke and ensure they received the appropriate amount of change.

On the subject of heredity, Dr. Brown said, "I think that at that time we thought that there was the possibility that mental retardation per se was a genetic problem. Now we know it isn't." He then provided us with some basic information on some of the conditions that were felt at the time to justify sterilization in order to avoid them being passed on to future generations: Huntington's chorea is known to be genetic, but he knew of no evidence to suggest that schizophrenia is genetic. Brown also said that he was sure that, at the time, many teenagers were sterilized who were mentally ill rather than mentally retarded. This, he suggested, would explain why some victims now seem to be normal, particularly if they are now aided by medication. In his opinion, the advance of medical science made a difference; people who can function with help now could not then.

A BEDROOM COMMUNITY OUTSIDE PORTLAND, OREGON

Journalists in Oregon, notably Julie Sullivan, have been more measured in their coverage than their counterparts in North Carolina – with a refreshing absence of the imprecise National Socialist comparisons in most analyses – and the people involved there were more willing to talk.[9]

---

[7] Ibid.
[8] U.S. Census Bureau, decennial census data from 1960–2010, http://www.census.gov.
[9] One past superintendent of Fairview refused to speak with us.

One was Dr. Everett Winslow Lovrien, who was asked by Oregon's governor to join the state's Board of Social Protection (created as the Oregon Eugenics Board in 1923) in October 1970.[10] The Board had six members with a mix of backgrounds: a couple of physicians, a social worker, and an administrator. There was also oversight from the Attorney General's Office to ensure that the procedures it applied were within the law. According to Lovrien, Oregon's governor wanted him because he, Lovrien, was a Roman Catholic eugenicist, which would have the double advantage of being able to cope with Catholic opposition and to provide some balance on the committee. The appointment achieved the latter, certainly, in that Lovrien became known as the "no vote" on the Board. His opposition was, however, neither ritual nor total; eugenic sterilization had its place.[11]

The Board of Social Protection met at Fairview Hospital, which housed approximately 3,000 patients. A local physician would present patients from the Hospital to the Board of Social Protection, whose members would hear their cases before they were discharged. Under Oregon law, the Board had broad authority to hear cases: any person could petition to have another person involuntarily sterilized under one of two clauses. Under Clause A, the person to be sterilized had to be likely to produce defective offspring because of inherited characteristics. Under Clause B, the person, even if "normal," was considered unable to take care of a child. The motivation could, therefore, be both eugenic and noneugenic at the same time. The superintendent in each institution petitioned the Board with a recommendation in favor of sterilization.[12]

Dr. Lovrien described the Board's procedures:

> [W]e would hear cases from Fairview before discharge. The head physician was present, and the dossiers were there. We would read them before we saw the patient, but it was always important to see the patient. ... Most [patients] said little of anything ... I can't remember them saying "I want a baby, I have a right to have a baby." If anything, they objected to the operation. We tried to be sure that they understood. There were instances in which the same person came before us several times because they seemed not to have understood. We had to make it clear to them.

"Everyone [i.e., all members of the Board]," Lovrien continued, "was thorough and well meaning. Our emphasis was to do what's correct for society and the individual." The "individual" referred both to a potential child and to the adult in question. Of the potential child: "we wanted the child to be successful. Usually a retarded person marries or has sex with another retarded person; if their

---

[10] Interview with Dr. Everett W. Lovrien, Fairview, OR, July 22, 2003.

[11] His opposite was a woman, the name of whom he had forgotten, who always voted in favor; she was known as "the Sterilizer."

[12] Irvin B. Hill, "Sterilizations in Oregon," *American Journal of Mental Heredity* (January 1950): 400, AVS Pamphlets Misc. Box 34, AVS.

condition is passed on to the child, it faces serious problems and we have done it no favors." Of the adult: "I have followed families where both are retarded but useful citizens"; without children, mentally handicapped people can more easily find their way. For society, the issue was not the gene pool but, rather, cost: a severely handicapped "child would become a burden. It would require care, would be expensive. We were protecting society from the costs of the child."

Throughout, Dr. Lovrien argued, the Board was trying to do what was best for society and for the retarded person. Invariably, "it was a difficult decision." This was not to say that the members felt any shame. On the contrary, "we felt quite good about the outcomes. The Board tried to do what was best for the individual before us. We might have made mistakes along the way, but you have to remember that, before the Board of Social Protection was created, there was sterilization at Fairview." It was simply unrecorded, unchecked, and no one was responsible. "There was no accountability before the Board; the Board was partially created to secure accountability."

Lovrien then commented on his own thinking: "If a person was thought to be defective and have a defective offspring, and [there was in fact evidence that] it would not happen, I would vote against sterilization." Whether the condition was hereditary or not was the key: "I don't have any problem with [eugenic sterilization] in principle." He continued:

> I often voted against sterilization, but sometimes for it. For example, if they presented a male Down's patient for sterilization, I would vote against, as they are sterile. Sometimes my vote was overridden and they were sterilized anyway.... [When we did decide in favor of sterilization], we felt that it was important for the client to understand why this is happening, that it was not just a matter of getting out of Fairview; that is, it would be easier for them if they didn't have to worry about a baby. If I was convinced there was no risk of pregnancy, I would also vote against sterilization.

Lovrien ended the discussion expressing his skepticism of what he regards as excessive individualism underpinning bioethics today. "There have been cases," he noted, "where two Down's patients got married. What is the purpose of this marriage? There isn't one. They don't love each other as we do; they don't know how to love. The whole thing is driven by individualism, a misconceived sense of freedom." In the case of Down's syndrome girls, he felt there remained an argument in favor of sterilization: "Female Down's patients should be sterilized. Even if the baby does not have Down's syndrome, it will be retarded; it will not develop normally in the womb."

The Board of Social Protection's work ended with a whimper rather than a bang. Problems began to emerge in the 1960s as the Board's work bumped up against an emergent rights culture. People who had been sterilized complained and wanted the operation reversed. Physicians feared lawsuits and looked to the Board for help. There was, however, no great defining lawsuit that settled the matter. Rather, the referrals dried up, and the Board had less and less to do. The last sterilization, according to Lovrien, was approved in 1982 or 1983, and the Board was dissolved. Its last project, which was never completed, involved drafting regulations on marriage.

When we spoke with him in 2003, Lovrien looked back with no regrets. The current condemnation was, in his view, simply the result of a new *Zeitgeist*. There had been no demand for apologies until activists thought of the idea. In Lovrien's view, Governor Kitzhaber, then the gubernatorial incumbent, "is not really sorry. He is simply trying to avoid condemnation or lawsuits."

RED DEER, ALBERTA

The Provincial Training School (now named Michener Services) was to Calgary and Edmonton what Fairview was to Portland and Salem: the province's main institution for the mentally handicapped. Although less architecturally attractive, its layout followed the same model: residents lived in different houses on the site, and the superintendent, Dr. L. J. Le Vann, lived on the grounds until 1974. At its peak, there were 1,000 residents living in the complex.

One young medical student, Dr. Robert Lampard, had worked at the Training School in the 1960s. When we caught up with him in 2003, he had returned as the medical director. The resident population was 400, the net result of deaths, discharges, and a limited admissions policy over the course of thirty years.[13]

Lampard was a strong defender of the province's selective sterilization policy, based on prevailing knowledge and values and the absence of birth control pills while the law was in effect. In his view, the moral revulsion that swept the academic, legal, journalistic, and pro-"community living" circles resulted from the people reading a rights culture of individual entitlement back eighty years. It also showed a disregard for the social good. He was particularly distressed over the University of Alberta's decision to end a lecture series, several awards, and a conference room named after John Malcolm MacEachran, in his day a distinguished philosopher and the founder of the philosophy and applied psychology departments of the University. He was the chairman of the Alberta Sexual Sterilization Board for thirty-seven years. "They destroyed a great man's reputation," Lampard remarked.

According to Lampard, the Board established two primary criteria for evaluating cases for sterilization before adjudicating its first case in 1929: (1) could the resident procreate? and (2) could he or she parent?[14] If the answer to the former was yes and latter was no (i.e., the resident's IQ was less than 70), the Board approved the application. To the degree that this was true, it reflected the extent to which Alberta's program had cast aside its eugenic origins and

[13] Interview with Dr. Robert Lampard, October 17, 2003, Red Deer, AB.
[14] Robert Lampard, "The Alberta Sexual Sterilization Act," in *Alberta's Medical History: "Young and Lusty, and Full of Life"* (Red Deer, AB: privately printed, 2008), 571–91.

focused on the social capabilities of the parents and (if indirectly) the rights of potential children to have competent parents. These criteria remained in effect until 1972, when the Act was repealed.[15]

The end of the sterilization program did not, Lampard continued, resolve the problems that generated it in the first place. He cited anecdotal evidence of an aboriginal woman who had four children, all with fetal alcohol syndrome, who refused contraception, and of a woman from a rural Alberta town whose penchant for frequent pregnancy was not matched with a competence to parent: she had seven children, all of whom became wards of the state. If these two women had been presented to the Board and met the criteria for sterilization, they likely would have been approved. By contrast, Lampard notes that the Training School had three pregnancies during his tenure of twenty-six years. Two were aborted. The third woman became pregnant during a group home trial. After delivery, she was not able to hold her baby safely, so the child was taken by social services.

For the residents themselves, Lampard continued, the new philosophy of deinstitutionalization did not end abuse:

> Within the institution, any cases of abuse were handled by a strict review policy. The institution's insistence on accountability by recording each instance and any response to it, and by referring the few cases to the restrictive practices review committee, which reduced their frequency and duration. "Time-outs" were replaced with transfers to the resident's room. Antipsychotic drug use was reduced by 66 percent per person through a systematic review of maladaptive behaviors, drug reduction trials, and increased staff tolerance. The quality assurance program that was developed became the basis for a provincial one. By contrast, abuse practices in the private group homes occurred out of the sights of the government's supervisory authorities until the QA program became part of the licensing requirements.

The implication is that it was the province's commitment to correcting procedural deficiencies – by documenting sterilizations and the reasons given for them, as well as by recording accusations of physical or sexual abuse – that contributed to its undoing. Eugenics left a paper trail.

---

[15] Ibid.

15

Conclusion

A Century of Coerced Sterilization

The history of coerced sterilization in North America remains a topic of enduring interest for at least two reasons. First, the operation was an illiberal assault on citizens' autonomy and control of their bodies. Second, the history of coerced sterilization is at once both narrow and broad. It is narrow in that it is in part the history of a small subset of the overall population: people with developmental disabilities.[1] It is broad in that their personal histories intersected with and were intertwined with some of the largest and most defining trends of the twentieth century: public health, demographic decline, institutionalization, privately supported policy research and lobbying, and social engineering based on the application of "scientific" findings to public policy.

Many protagonists of coerced sterilization viewed it through a public-health perspective: an attack on threats to a nation's or race's health from within (through better breeding and sterilization) was the counterpart to an attack on these threats from without (through sanitation and medication). The elimination of disease and premature death could only be accomplished with the elimination of heritable feeblemindedness. Sanitation and sterilization, as it were, went hand in hand. Similarly, concerns about racial health and degeneration fed into anti-immigration sentiment and fears about dysgenic and ecological doom. Those who held these fears made no sharp distinctions between individual, racial, and environmental vigor and decay. It is not a coincidence that racist eugenicists founded some of the earliest environmental organizations on the West Coast.[2]

Historians on both sides of the Atlantic have documented the close affinity between supporters of eugenic sterilization and support for progressive causes. Eugenic sterilization had widespread, almost total appeal to progressive social reformers because coerced sterilization, along with sanitation, education, better

[1] For a thoughtful and sensitive discussion of the former, see Joseph P. Shapiro, *No Pity: People with Disabilities Forging a New Civil Rights Movement* (New York: Times Books, 1993).
[2] See Stern, *Eugenic Nation*, 118–49.

working conditions, and so on promised to harness science and interventionist public policy to the cause of creating healthier, safer, and better-educated citizens. Eugenic sterilization was a social engineering project par excellence, and such projects attract more attention and support within democracies from the progressive left than they do from the right. Many such projects result in outcomes that we admire: public healthcare, compulsory education, and vaccination, to name but a few. Others – slum clearance and public housing projects, residential schools for aboriginal populations, coerced sterilization of people with disabilities – have been public policy disasters, leaving a trail of broken communities and broken lives in their wake.

Eugenics was, however, something more pernicious than public policy gone terribly wrong. In the cases of slum clearance, public housing, and residential schools, supporters misdiagnosed the problem and articulated a solution that only created new problems. In the case of eugenic sterilization – and, for that matter, of global population control – something else occurred: its supporters claimed mastery of the very laws of heredity and therefore knowledge of the precise course of human history. As such, schemes for mass sterilization to block both feeblemindedness and population growth amounted to "public policy based obsessively on mathematically calculated planning devices [which] only justify themselves to the extent that they can claim perfect or near-perfect knowledge of future outcomes (not to mention present information)."[3] And herein lay the sterilization and population control advocates' fatal flaw: "[s]ince neither present nor future information ... is ever vouchsafed us in perfect form, planning is inherently delusory, and the more all-embracing the plan, the more delusory its claims."[4]

Until the postwar years, the experiment in coerced sterilization was inseparable from the great experiment in the mass institutionalization of the feebleminded. As Chapter 4 showed, the institutionalization of people with developmental disabilities and mental illnesses was an ambitious and, at least at the start, well-funded effort to address two of society's deepest and most intractable problems: madness and feeblemindedness. This solution was, of course, no solution at all; instead, it encouraged in a path-dependent way the next assault on the problems. The concentration of the insane and feebleminded in institutions isolated them from society, and the strict hierarchies that governed confining institutions were preconditions to the sterilization of large numbers of such people. Sterilization could and did occur in other institutions – notably hospitals and prisons – but the home for the feebleminded was the institutional basis of it all. Without the mass institutionalization of people with mental disabilities, coerced sterilization could not have occurred to the extent it did for the simple but definitive reason that they would have been outside the reach of the state.

[3] Tony Judt, *Thinking the Twentieth Century* (New York: Penguin Press, 2012), 92.
[4] Ibid.

This observation, in turn, grounds the answer to the first question posed in this book: why did so many states adopt policies allowing coerced sterilization? In the account offered here, we have distinguished between what might be called background factors and triggering conditions. Eugenic ideas, which provided a theory of heredity, as well as eugenicist concerns about differential fertility, female licentiousness, and economic cost, were background conditions. They provided reasons justifying coercive sterilization. But sterilization could only occur where legislators were prepared to sanction it and where doctors were prepared to order and/or perform it. In the latter, and to a degree the former, the central figure was the superintendents at homes for the feebleminded. In Indiana, Pennsylvania, Virginia, California, and Alabama, the superintendents were the decisive figures in securing a sterilization law's adoption; in all states with such laws, the superintendents had lobbied for them; and in every state, superintendents determined how and how often sterilizations would occur. These seemingly minor state figures anchored almost a century of coerced sterilization and gave materiality to the phrase "sterilized by the state." They illustrated once again the "potentially lethal combination," as Paul Weindling puts it, "of medical power and biological knowledge."[5]

Until the 1930s, the odds were heavily stacked in favor of sterilization: it had the support of eugenic thought, it had the backing of middle- and upper-middle-class women's groups, it promised substantial savings for taxpayers, and the feebleminded did not vote.

But, at the same time, it is important both to a sound research design (for a theoretical account to be more than a tautology, it must be falsifiable) and a thorough historical account to explain situations in which coercive sterilization occurred, as well as those in which it did not. The most important factor in this account was the Roman Catholic Church, which formed, from the late 1920s, the only national organization that opposed coerced eugenic sterilization. As the discussion of Ohio shows, it could be a singularly powerful point of opposition. The Church was able to combine high rhetoric and populist politics particularly effectively in the public debate. In the former, theologians – a highly educated, thoughtful, and articulate lot – could draw up anti-sterilization literature in powerful, persuasive language. In regard to populist politics, Roman Catholic priests, never inclined to an overly strict interpretation of the distinction between politics and religion, could thunder from the pulpit against eugenic sterilization with even more force than they did (and do, at least in the last century) against contraception and abortion. Sensitive to the Roman Catholic vote, legislators in states with large Catholic populations had ample reason to steer clear of eugenic sterilization. This was particularly true of Eastern and Midwestern states in which Roman Catholic populations were longstanding and in which the archdioceses

[5] Paul Weindling, *Nazi Medicine and the Nuremburg Trials: From Medical War Crimes to Informed Consent* (New York: Palgrave Macmillan, 2004), 5. Weindling draws here on Michel Foucault and Ivan Illich.

were established and well-organized. In Massachusetts, Pennsylvania, and Ohio, there was no sterilization law. Those states with large Roman Catholic populations that did secure a law – New Jersey, New York, and Connecticut – either soon saw the legislation struck down (New Jersey) or made relatively rare recourse to it (Connecticut and, above all, New York). Infrequent use, however, may have had more to do with the vagaries of institutions and superintendents than with persistent Roman Catholic opposition.

The one state that stands as a notable exception to this broad trend is California, which had both a relatively large Roman Catholic population (around 14.5 percent in 1916 and around 13.5 percent in 1926) and also one of the most ambitious sterilization programs in the United States. The California example partly reflects the dynamic nature of growth in the state's Catholic population. Increasing largely through immigration, the state's dioceses were unable to organize that population against coerced sterilization. "We in California have a very large Catholic constituency," observed Frank C. Reid of the Human Betterment Foundation, "but our law was enacted before there was any decided Catholic opposition."[6] Importantly, the majority of the state's Catholic population was made up of recent Latino immigrants who lacked the vote. As immigrants, they were the target of fears of racial degeneration and thus acted as an argument in favor of rather than against eugenic sterilization. Even in California, however, Roman Catholic opposition was eventually felt: once the Church finally organized, pro-sterilization advocates found it impossible to replace the state's "outdated" law with "up-to-date provisions."[7]

Perhaps unremarkably, the factors that explained the institution of coerced sterilization before the Second World War in part – but only in part – explain its continuation afterward. As noted in Chapter 9, superintendents, like pro-sterilization advocates generally, thought the experience of National Socialism said nothing to their arguments in favor of coerced sterilization, and many superintendents – Butler at Sonoma, Le Vann in Red Deer – carried on sterilizing more patients right into the 1960s. The survival of their position and their commitment to sterilization past the end of the war enabled the practice to persist well into the second half of the twentieth century.

At the same time, new considerations argued in favor of continued, and often increased, sterilization. The institutional development of greatest significance was indirect as well as direct federal government involvement in sterilization policy. In the former, federal insistence that African Americans enjoy full access to welfare programs established under the New Deal swelled the number of black Southerners on welfare rolls. In North Carolina, which uniquely allowed sterilization by social workers outside institutions, the result was a large uptick in sterilizations under the state's eugenic sterilization law. The subsequent

---

[6] Frank C. Reid to Lewis Williams, Director Charitable Institutions, Boise, Idaho, January 16, 1942, 7.11, Gosney Papers.
[7] Ibid.

federal decision in 1970 to provide federal funds for sterilization broke the dam: supported by most likely racist doctors and nurses, thousands of coerced sterilizations, mainly of African Americans, became tens and then hundreds of thousands.

While all of this was occurring, the original ideational frame justifying coerced sterilization – eugenics – faltered in the face of increasing skepticism from scientific and, to some degree, political observers.[8] The practice continued largely unabated at the state level (although with great variation among states), and activists and public intellectuals required a new paradigm to justify continued coerced sterilization. Unreconstructed, prewar supporters of eugenics – Paul Popenoe, Marian S. Olden, Robert Dickinson, and Fred O. Butler – were central to this shift. The first effort, led by Olden's Sterilization League of New Jersey, involved recasting sterilization's purpose: it was protective rather than eugenic, and it protected unborn children. The League defined this protection as a right – indeed, as a birthright. This new paradigm meant that, as incredible as it might seem, the move was prescient of events that would occur fifteen years later. By the 1960s, major social movements cast their claims – for women, African Americans, and gays and lesbians – in terms of rights. These groups do the same to this day, and wisely so, for it is a very hard argument to resist. Individuals and groups opposing these claims do not, unless they wish to lose the argument, reject the importance of rights. They instead claim that the issue in question – for instance, gay marriage – is not a question of rights but rather of some other principle – such as the "protection" of the family. It is for this reason that it is common to hear minority rights arguments dismissed as being claims for "special treatment," which can then be rejected on the assumption that preferential treatment is never a right in a way that equal treatment is.

In the case of coerced sterilization, the claim that every child has a right to good parents was powerful enough, but it conceded a major principle: that unborn children have rights. This tenet underpinned the Roman Catholic Church's arguments against abortion and contraception. Because supporters of eugenics, coerced sterilization, birth control, and abortion were part of a single movement, it was not a principle that they wished to concede.[9]

The linking of sterilization with world population growth was much more successful. It focused on real, visible problems – apparently infinite population growth in the context of a finite world, poverty, and hunger – rather than abstract, negatively stated rights. The fight against world population growth also allowed supporters of coercive sterilization to achieve the old eugenic goal of bringing sterilization out of the institutions. The target became whole countries with high population growth: the United States itself to a degree but, above all, the developing countries of China and India. The apostles of population

---

[8] Kevles, *In the Name of Eugenics*, chapter 11.

[9] Dowbiggin, "'Rational Coalition,'" 223–52; Dowbiggin, *A Merciful End: The Euthanasia Movement in Modern America* (New York: Oxford University Press, 2003).

growth arguments – Paul Ehrlich, author of *The Population Bomb*, Sir Charles Galton, and John Maynard Smith – kept their public interventions soundly eugenic by repackaging old arguments about differential fertility and genetic decline. The result was that their calls for population control by means of sterilization, among other things, were wrapped in the same apocalyptic language that characterized prewar eugenic debates. The stakes in population reduction were, as they had been in eugenic sterilization, nothing less than the survival of the human race.

The results of this commitment and moral fervor were immense: birth control clinics were bankrolled and operated more by the choices and preferences of their funders than according to the needs of their clients throughout the Southern United States. This was followed by the partial underwriting of a quasi-colonial campaign (although one that enjoyed significant support among Southern elites) to impose the views of the wealthy, Northern elite on the poor populations of the lesser-developed South.

In India, population lobbies formed by individuals with varyingly deep commitments to eugenic principles provided extensive rhetorical and material support for population control policies that resulted in immense illness, suffering, and death.

The research supporting the population lobbies' activism had several common characteristics: the extensive use of graphs showing the population curve making a sharp vertical uptick; somewhat amateurish cut-and-pasted photos showing every square inch of Earth covered in people; and confident predictions that mass starvation, institutional breakdown, and war would result from unrestrained population growth.[10] The population projections themselves were based on calculations that were highly sensitive to assumptions that were often little more than guesswork.[11]

These comments lead to the difficult issue of the relationship between the movements for coerced sterilization and the birth control movement. As we argued in Chapter 9, the membership rolls of both the Association for Voluntary Sterilization (AVS entirely) and Planned Parenthood (to a large degree) contained avowed eugenicists. AVS in particular was motivated well into the 1960s not by the concerns of individual choice or a woman's autonomy over her body but, rather, by the social and economic effects of differential class fertility and high overall population growth. But the picture here is complex, and it is difficult to find true heroes, as the pro-choice movement would have it, or true villains, as the pro-life movement would have it. Marian Olden's hatred for people with mental disabilities was nearly genocidal in the intensity of its vision. But she was at the same time genuinely moved to political action by the terrible, unnecessary deaths of women during childbirth and by the burdens placed on poor families by too many children. Moreover, Margaret Sanger both subscribed to

[10] Greenhalgh, *Just One Child*, 159–60.
[11] Ibid., 159–60.

some core eugenic ideas and threw her by-then-considerable influence behind campaigns to limit developing world populations for reasons that had precious little to do with the rights of women from the Southern United States, India, and elsewhere. At the same time, she was deeply committed to the right of women, and particularly poor women, to control their own reproduction.

Like many movements, the early birth control movements drew on complex and contradictory motives and had both liberal and illiberal strains. Rather than sweeping the latter under the historical rug, supporters of choice should acknowledge and openly address them. Until they do so, the eugenic under-current in its history will be fodder for pro-life activists.

OBSERVABLE IMPLICATIONS: EUGENICS
AND MODERN POLITICS

The arguments advanced in this book have implications for studies outside the fields of eugenics and coerced sterilization policy. Two issues are central: first, the relationship between institutions and abuse and, second, the relationship between North American and National Socialist eugenics.

Institutions and Abuse

Within the media and scholarly literature, there has been extensive discussion of the abuse of children by Roman Catholic priests worldwide and by their care-takers (some Roman Catholic, many not) in residential schools established for aboriginal Canadians. Commentators have explained these abuses with refer-ence to, respectively, Roman Catholicism and racism: the homophobia and repressed sexuality of the former encourages sexually maladjusted men to enter the priesthood, and extensive anti-aboriginal racism led to the physical and sexual abuse of native children.[12] We do not provide a definitive ruling on the doubtlessly complex causes lying behind these scandals, but the similarities between the experiences among Roman Catholic and residential school pupils, on the one hand, and the institutionalized mentally handicapped, on the other, are striking from a historical-comparative point of view. They draw attention to power hierarchies in institutions rather than to mere individual wickedness. All of these victims were housed in institutions cut off from regular contact with the

---

[12] On Roman Catholics, see Barney Zwartz, "Child Sex Abuse Link to Celibacy," *The Age* (Melbourne, Australia), January 24, 2013; Breda O'Brien, "Distorted View of Abuse Hides the Real Picture," *The Irish Times*, November 5, 2011; Hadley Freeman, "Church Culture of Shame 'Feeds Abuse,'" *The Guardian*, December 4, 2002; Daniel J. Wakin, "Ideas & Trends: Facing a Sin of the Fathers," *New York Times*, February 17, 2002. On residential schools, see Don Marks, "Canadians Can't Be Smug about Racism," *Winnipeg Free Press*, January 23, 2013; "A Century of RCMP Ignorance; Report Says Police Were Seldom Aware of Abuse at Residential Schools," *Toronto Star*, October 30, 2011; Michael Oliveira, "Details Demanded on Children Missing from Residential Schools," *The Globe and Mail* (Canada), February 9, 2008.

outside world. All of these institutions accorded largely unregulated power to those who operated them. And all were residential, placing empowered employees and powerless children together through the night, when sexual desires are least encumbered. Residential institutions of this sort, in short, encourage abuse.

A related way of theorizing sterilization is as a form of abuse. The often arbitrary way in which sterilization was applied meant that it became a tool within the institutions for, to put it in Foucaultian terms, disciplining and punishing patients. This dimension features above all in the stories told by patients describing their experiences in such institutions. We heard in Chapter 11 the voices of those subjected to sterilization, some of the close to two dozen survivors of coerced sterilization in Oregon and Alberta with whom we spoke. Their stories provide information on the nature of the commitment process, on the daily routine of the institutions in which they were housed, and on the nature of the sterilization operation itself. They also convey a deeper sense of what it was to be a person with mental disabilities in the middle decades of the last century. We later include testimonies of those who supported and performed coerced sterilization and who, in some cases, still believe in it. This second set of voices is important both as part of the historical record and as a corrective. There is an understandable tendency in discussions of eugenics to moralize, presenting those who supported eugenics as the next thing to evil. Doing so may be viscerally satisfying, but it does little to advance an understanding of why reasonable, educated people who thought of themselves as progressive could so wholeheartedly support a project that today seems to be the embodiment of illiberal and abusive state power. To this end, we presented in Chapter 14 the results of interviews conducted with two doctors who performed sterilizations and another who sat on a Eugenics Board recommending them.

Matters become more complicated when we consider the grisly case of medical experimentation. There is a large literature on this topic, but that literature largely succeeds more at saying what happened rather than why it happened. When these practices occurred within institutions – the oatmeal radiation experiments at in Massachusetts, the cosmetics experiments in Pennsylvania – then we could speculate that the factors encouraging abuse also encouraged experimentation. Institutions in which power was centralized in the superintendent and wielded by his attendants, and over which courts and legislatures exercised little oversight, made this a sort of revocation of rights. We remarked earlier that there was a sense in which patients' rights were checked at the door when they entered homes for the insane and feebleminded. On entering it, they became less than human. There is no other way to explain researchers' willingness to expose young boys to radiation disguised as food or to test scarring chemicals on prisoners' skin unless it is conceded that they viewed these boys and men as far less worthy of respect than most American citizens. But this line of argument only carries us so far, as such experimentation also occurred outside the institutions. Neither the Tuskegee syphilis sufferers nor the Cincinnati cancer patients were institutionalized. They were, however, mostly

African American and mostly poor. In these sad histories, those subject to experimentation were, like many who ended up in institutions, poor in resources, education, and connections.

The Nazi Connection

A final issue is one that excites the press and popular writers: the relationship between National Socialist and American eugenics. As some would have it, American eugenics "inspired" the National Socialists. It is true, as we are often reminded, that Hitler approvingly cited existing American sterilization policy. It is also true that German eugenicists, who were generally not National Socialists, shared a number of key propositions with American eugenicists. That is, however, the extent of the connection. Those who believe that American policy determined Nazi policy lack even a basic understanding of Germany's National Socialism. The sterilization and murder of the mentally handicapped was only one feature of a coherent, horrible plan for the demographic reordering of Central and Eastern Europe. The larger plan involved the forcible transfer of tens of millions, the deliberate starvation of millions of Slavs, mass sterilization, the murder of the physically handicapped (including war veterans), and, of course, the extermination of all European Jews.[13] The sterilization and murder of the mentally handicapped was the first part of this program, and it would have been implemented with or without the American experience. Hitler was glad to cite precedents – American eugenic laws, the expulsion or genocide of North American Indians – but he certainly did not need them. Those writers who excitedly suggest a tight causal connection between 1930s Germany and 1930s–1950s America should reflect on the distinction, clear to most high school students, between a reason and an excuse.

ESCHEWING JUDGMENT

This book, particularly in its later chapters, raises at least two sorts of issues deserving serious debate. The first concerns history and moral judgment. As noted, in books on the history of eugenics, there is an almost irresistible tendency to moralize about and often to condemn with righteous indignation the history of eugenic sterilization (a tendency to which this book may not be free). Leaving the complex case of Germany aside, this academic hand-wringing does the job of establishing the authors' progressive credentials (which, perhaps inadvertently, suggests that, as good progressives, they might have supported precisely the policies that they are now confidently condemning), but it adds little to an understanding of why these policies were enacted and sustained. Particularly when talking about the decades before 1945, these arguments can come across

---

[13] On this, see Gerhard L. Weinberg, *A World at Arms: A Global History of World War II* (Cambridge: Cambridge University Press, 1994), and Snyder, *Bloodlands*.

as an ineffectual howl of frustration at the way the world was. In the 1910s and 1920s, eugenics was an accepted truth, and it is for this reason that it attracted so much support from so many people from all points of the ideological spectrum. When there was opposition, such as that articulated by Roman Catholic intellectual G. K. Chesterton, it was grounded in reasons – namely, theological ones – that we would today regard as idiosyncratic and not terribly convincing.

The case becomes more complicated in the interwar and postwar years, as some members of the scientific community turned against eugenics. But, at the same time, other developments – the effect of the war on the population of young, healthy men in many countries, the sharp drop-off in the birth rate, and the spiraling public costs associated with unemployment and poverty – seemed to confirm eugenicists' warnings. A large number of educated, seemingly reasonable men and women therefore continued to support coerced sterilization, even if only in particular circumstances. This fact creates a sort of cognitive crisis for us today, particularly for the politically liberal and broadminded, who are in the same literate, affluent, and often "progressive" categories in which eugenicists were once found in such large numbers.

One way to resolve this crisis is to make the case that eugenicists were, if not monsters, then fatally flawed people. Such a reaction could be seen in the campaign by the University of Alberta against its one-time favorite son, John MacEachran (see Chapter 14). The Department of Psychology in the university had, after his death in 1971, named a seminar room and a lecture series after MacEachran, and it awarded scholarships in his name. Following a critical newspaper article about his legacy in the *Edmonton Journal*, the psychology department appointed an internal review committee. The committee recommended removing his name from the seminar room, the lecture series, and the awards. Those visiting the department today would not know that MacEachran ever existed. In defending the removal, Douglas Wahlstein quoted the judge's ruling in Leilani Muir's suit against the Alberta government:

> The circumstances of Ms. Muir's sterilization were so high-handed and so contemptuous of the statutory authority to effect sterilization, and were undertaken in an atmosphere that so little respected Ms. Muir's human dignity that the community's, and the court's, sense of decency is offended.[14]

The court and Wahlstein confuse two claims: that what the Alberta Eugenics Board did was illegal and that it was offensive. If the grounds for revocation were the former, they do not need to add the latter. In doing so, they miss the important point that coerced eugenic sterilization in the 1950s and 1960s offended neither the court's nor the community's sense of decency; this practice was fully consistent with the prevailing views of both. In chiseling over

---

[14] Douglas Wahlsten, "Leilani Muir versus the Philosopher King: Eugenics on Trial in Alberta." *Genetica* 99, nos. 2–3 (1997): 185–98.

MacEachran's name, the psychology department, whatever its protests to the contrary, simply buried history.[15]

Like many retrospective reevaluations, the episode is difficult to interpret. Another way of coping – or, even better, understanding – the history of coerced sterilization is to cast one's mind back to an age before the rights revolution. Although the language of rights goes back some five hundred years, arguments grounded in rights have won political salience only in the second half of the twentieth century. In the prewar years, an individual's sense of duty – to his or her class, race, country, or God – was as powerful, probably more powerful, than his or her sense of personal entitlement. One example illustrates the point. In 1940, the American reform president, Franklin D. Roosevelt, appointed Harry L. Stimson as his Secretary of War. Stimson was a Harvard-educated lawyer, the son of a Union veteran of the Civil War, and a member of upper-class New York. His upbringing emphasized the puritan virtues of work and abstention (although he lived very comfortably) and the not-uncommon early-twentieth-century view of war as a cleansing antidote to the soft materialism associated with American affluence.[16] "Every man," Stimson wrote in 1915, "owes to his country not only to die for her if necessary, but also to spend a little of his life in learning how to die for her effectively." Stimson was in no way an extremist or hostile to a rights culture. After the Second World War, it was he who argued against Churchill and Henry Morgenthau's suggestion that the major German war criminals be apprehended and shot. Stimson persuaded influential opinion and, most importantly Roosevelt, that they had to be dealt with in a manner consistent with the U.S. Bill of Rights and American democratic values.[17]

In such a cultural context, child rearing could hardly be conceived of as a personal right equivalent to the freedom of belief or association but, rather, as a responsibility. And, viewed as such, it becomes much more plausible to see limitations on responsibility: if people are incapable of raising children, and if children are a responsibility rather than a right, then preventing people from having children is less shocking to moral senses. Another example helps to illustrate his point. It is standard practice across North America to remove children from their parents' care when those parents cannot care for them. Abstractly, one might ask what the significant difference is between taking away one's children and denying one the right to have them in the first place. It is not obvious that one or the other involves less intrusive state power, creates less personal suffering, or is less of a restriction on the freedoms of the person

---

[15] One of their fiercest local critics argued that they should, instead, have left the name and added a plaque on eugenics and sterilization in Alberta instead.

[16] Ronald Schaffer, *Wings of Judgment: American Bombing in World War II* (New York: Oxford University Press, 1985), 5–6.

[17] On this, see Richard Overy, *Interrogations: The Nazi Elite in Allied Hands, 1945* (New York: Viking, 2001), introduction, and Michael Bess, *Choices under Fire: Moral Dimensions of World War II* (New York: Knopf, 2006), chapter 11.

most affected – the person sterilized or "robbed" of potential children. Only the notion that procreation is a right immutable and beyond challenge makes eugenic sterilization so unthinkable. Such a view has, ironically, more than a little of the religious in it.

In making these points, it is important to emphasize that we are not talking here about the examples discussed in Chapter 11. They cover the easy cases, people who could clearly function in society and thus were unjustifiably targeted for sterilization. They suffered serious injustices. That they are given so much attention reflects both methodological and strategic considerations. Methodologically, they are the only people to whom researchers can talk. The severely mentally handicapped or severely mentally ill do not make capable or reliable interview partners. This means that, in all accounts – including this one – the non- or only slightly mentally handicapped make the perfect victim. They should not have been defined as feebleminded or mentally retarded, they should not have been institutionalized, and they should not have been sterilized. Yet they make up only part of the total population of sterilized people. We hear less, if anything, of the severely mentally handicapped who could not take care of themselves much less their children. We also hear less, although Johanna Schoen's work is a powerful corrective to this neglect, about those who requested sterilization and who saw it as a means to enhancing their freedom.

That such varying categories exist in no way makes the sterilization of Leilani Muir, Velma Hayes, and thousands of others any less of a personal and social injustice. It means instead that reaching conclusions, much less drafting policy, for people with mental disabilities is exceedingly difficult. Journalists are keen to draw out of North America's experience of coerced sterilization the "lessons of history," but there are, in fact, few of them. Indeed, there is perhaps only one: any claim that ambitious and personally intrusive policies will solve the challenges of dealing with mental disabilities or mental health should be treated with the utmost caution.

Bibliography

Archival Sources

American Philosophical Society Archive (APSA), Philadelphia, Pennsylvania, USA
Archiv zur Geschichte der Max-Planck-Institut (MPA), Berlin-Dahlem, Germany
Association for Voluntary Sterilization Records (AVS), Social Welfare History Archives, University of Minnesota Libraries Minneapolis, USA
British Columbia Archives (BCA), Victoria, British Columbia, Canada
California State Archives (CSA), Sacramento, California, USA
Clarence James Gamble Papers, Robert L. Dickinson Papers, Center for the History of Medicine Francis A. Countway Library of Medicine, Harvard University, Cambridge, Massachusetts, USA
E. S. Gosney Papers and Human Betterment Foundation Records, California Institute of Technology (Caltech) Archives, Pasadena, California, USA
Irving Fisher Papers (IFP), Yale University Library Manuscripts and Archives, New Haven, Connecticut, USA
John A. Ryan Papers, American Catholic Research Center and University Archives, Catholic University of America Archives (CUA), Washington, DC, USA
Oregon State Archives (OSA), Salem, Oregon, USA
Provincial Archives of Alberta (PAAB), Edmonton, Alberta, Canada
Rockefeller Foundation Archives, Rockefeller Archive Center (RAC), Sleepy Hollow, New York, USA
United Kingdom National Archives (UKNA), Kew, London, UK
Washington State Archives (WSA), Olympia, Washington, DC, USA

Court Decisions

Buck v. Bell, 274 U.S. 200 (1927).
Griswold v. Connecticut, 381 U.S. 479 (1965).
People v. Harley Blankenship, 16 Cal. App. 2d 606 (1936).

Interviews

Anonymous interviewees (four)

Dr. Ernest Brown, September 5, 2003, Lumberton, NC

Gordon Bullivant (Executive Director, Foothills Academy, Calgary), October 16, 2003, Calgary, AB

Jon Faulds (attorney for plaintiffs in class action suit), October 17, 2003, Edmonton, AB

Velma Hayes, July 21, 2003, Portland, OR

Dennis Heath, July 19, 2003, Portland, OR

Dr. Warren Hern, June 2003, Boulder, CO

Dr. Robert Lampard, October 17, 2003, Red Deer, AB

Dr. Everett W. Lovrien, July 22, 2003, Fairview, OR

Bill Lynch (Council on Developmental Disabilities), July 16, 2003, Portland, OR

Judy Lytton, October 19, 2003, Edmonton, AB

Ted McNeil, May 16, 2003, Portland, OR

Ruth Morris, July 22, 2003, Eugene, OR

Leilani Muir, October 18, 2003, Edmonton, AB

John Murphy (President, Portland Habilitation Center), July 21, 2003, Portland, OR

Ken Nelson, October 14, 2003, Edmonton, AB

Ken Newman, July 18, 2003, Portland, OR

Laverne Throckhorn, July 17, 2003, Portland, OR

Bruce Uditsky (CEO, Alberta Association for Community Living), Edmonton, AB, October 18, 2003

William West (Adult and Family Case Coordinator, Association for Retarded Citizens [ARC]), Portland, OR, July 16–18, 2003

Secondary Sources

Acemoglu, Daron, and James A. Robinson. *Why Nations Fail: The Origins of Power, Prosperity, and Poverty*. New York: Crown Publishers, 2012.

Allen, Robert Loring. *Irving Fisher: A Biography*. Cambridge, MA: Blackwell, 1993.

Allen, William R. Review of *Irving Fisher: A Biography*, by Robert Loring Allen. *Journal of the History of Economic Thought* 17, no. 1 (1995): 153–5.

Applebaum, Nancy, Anne S. MacPherson, and Karin Alejandra Rosemblatt, eds. *Race and Nation in Modern Latin America*. Chapel Hill: University of North Carolina Press, 2003.

Baker, Jean H. *Margaret Sanger: A Life of Passion*. New York: Hill and Wang, 2011.

Baker-Benfield, G. J. *The Horrors of the Half-Known Life: Male Attitudes Toward Women and Sexuality in Nineteenth Century America*. New York: Harper & Row, 1976.

Baragar, C. A., G. A. Davidson, W. J. McAlister, and D. L. McCullough. "Sexual Sterilization: Four Years Experience in Alberta." *American Journal of Psychiatry* 91, no. 5 (1935): 897–923.

Barber, William J. "Irving Fisher (1867–1947): Career Highlights and Formative Influences." In *The Economics of Irving Fisher: Reviewing the Scientific Work of a Great Economist*, edited by Hans-E. Loef and Hans G. Monissen, 3–21. Cheltenham: Edward Elgar Publishing, 1999.

Barkan, Elazar. *The Retreat of Scientific Racism: Changing Concepts of Race in Britain and the United States between the Two World Wars.* Cambridge: Cambridge University Press, 1992.

Barker, David. "The Biology of Stupidity: Genetics, Eugenics and Mental Deficiency in the Inter-War Years." *British Journal of the History of Science* 22, no. 3 (1989): 347–75.

Bashford, Alison and Philippa Levine, eds. *The Oxford Handbook of the History of Eugenics.* Oxford: Oxford University Press, 2010.

Baumgartner, Frank R. *Conflict and Rhetoric in French Policymaking.* Pittsburgh, PA: University of Pittsburgh Press, 1989.

and Bryan D. Jones. *Agendas and Instability in American Politics.* Chicago: University of Chicago Press, 1993.

Baur, Erwin, Eugen Fischer, and Fritz Lenz. *Grundriss der menschlichen Erblichkeitslehre und Rassenhygiene.* 2 vols. Munich: J. F. Lehmanns Verlag, 1923.

Bearman, Peter, ed. "Exploring Genetics and Social Structure." Special issue, *American Journal of Sociology* 114, no. S1 (2008).

Béland, Daniel and Robert H. Cox, eds. *Ideas and Politics in Social Science Research.* New York: Oxford University Press, 2011.

Berman, Sheri. "Ideas, Norms and Culture in Political Analysis." *Comparative Politics* 33, no. 2 (2001): 231–50.

Bess, Michael. *Choices under Fire: Moral Dimensions of World War II.* New York: Knopf, 2006.

Bessel, Richard. *Germany 1945: From War to Peace.* New York: Harper, 2009.

Binder, Sarah A., R. A. W. Rhodes, and Bert A. Rockman, eds. *The Oxford Handbook of Political Institutions.* Oxford: Oxford University Press, 2006.

Black, Edwin. *War against the Weak: Eugenics and America's Campaign to Create a Master Race.* New York: Four Walls Eight Windows, 2003.

Bliven, Bruce. "A Gloomy Prophecy." *New Republic* 131, no. 7 (August 16, 1954): 20.

Bloom, Mark. "Sterilization Guidelines: 22 Months on the Shelf." *Medical World News,* November 9, 1973.

Blyth, Mark. "Any More Bright Ideas? The Ideational Turn in Comparative Political Economy." *Comparative Politics* 29, no. 2 (1997): 229–50.

Bock, Gisela. "Racism and Sexism in Nazi Germany: Motherhood, Compulsory Sterilization and the State." *Signs: Journal of Women in Culture and Society* 8, no. 3 (1983): 400–21.

Zwangssterilisation im Nationalsozialismus. Opladen: Westdeutscher Verlag, 1986.

Braslow, Joel T. "In the Name of Therapeutics: The Practice of Sterilization in a California State Hospital." *Journal of the History of Medicine & Allied Sciences* 51, no. 1 (1996): 29–51.

Mental Ills and Bodily Cures: Psychiatric Treatment in the First Half of the Twentieth Century. Berkeley: University of California Press, 1997.

Broberg, Gunnar, and Nils Roll-Hansen, eds. *Eugenics and the Welfare State: Sterilization Policy in Denmark, Sweden, Norway and Finland.* East Lansing: Michigan State University Press, 1996.

Bruinius, Harry. *Better for All the World: The Secret History of Forced Sterilization and America's Quest for Racial Purity.* New York: Knopf, 2006.

Burleigh, Michael. *Death and Deliverance: "Euthanasia" in Germany, c. 1900–1945.* Cambridge: Cambridge University Press, 1994.

"Eugenic Utopias and the Genetic Present." *Totalitarian Movements and Political Religions* 1, no. 1 (2000): 56–77.

The Third Reich: A New History. London: Macmillan, 2000.

Cahn, Susan K. *Sexual Reckonings: Southern Girls in a Troubling Age*. Cambridge: Harvard University Press, 2007.

Carlson, Elof Axel. *The Unfit: A History of a Bad Idea*. Cold Spring Harbor, NY: Cold Spring Harbor Laboratory Press, 2001.

Caron, Simone M. *Who Chooses? American Reproductive History since 1830*. Gainesville: University of Florida Press, 2008.

Castles, Katherine. "Quiet Eugenics: Sterilization in North Carolina's Institutions for the Mentally Retarded, 1945–1965." *Journal of Southern History* 68, no. 4 (2002): 849–78.

Cattel, Raymond B. *The Fight for Our National Intelligence*. London: P. S. King & Son, 1937.

Chesterton, G. K. *Eugenics and Other Evils: An Argument against the Scientifically Organized Society*. Seattle: Inkling Books, 2000. First published 1922 by Cassel, London.

Collins, Ronald K. L., ed. *The Fundamental Holmes*. Cambridge: Cambridge University Press, 2010.

Connelly, Mark T. *The Response to Prostitution in the Progressive Era*. Chapel Hill: University of North Carolina Press, 1980.

Connelly, Matthew. *Fatal Misconception: The Struggle to Control World Population*. Cambridge: Harvard University Press, 2008.

Cook, Robert C. *Human Fertility: The Modern Dilemma*. New York: W. Sloane, 1951.

Crackenthorpe, Montague H. *Population and Progress*. London: Chapman and Hall, 1907.

Crane, R. Newton. "Recent Eugenic and Social Legislation in America." *Eugenics Review* 10, no. 1 (April 1918): 24–9.

Critchlow, Donald T. *Intended Consequences: Birth Control, Abortion, and the Federal Government in Modern America*. New York: Oxford University Press, 1999.

D'Antonio, Michael. *The State Boys Rebellion*. New York: Simon & Schuster, 2004.

Darwin, Francis. ed., *The Life and Letters of Charles Darwin*. New York: D. Appleton and Co., 1887.

Darwin, Charles Galton. *The Next Million Years*. London: R. Hart-Davis, 1952.

Decker, Julio. "The Immigration Restriction League and the Political Regulation of Immigration, 1894–1924." Ph.D. diss., University of Leeds, 2012.

Degler, Carl N. *In Search of Human Nature: The Decline and Revival of Darwinism in American Social Thought*. New York, Oxford University Press, 1991.

DeWolfe Howe, Mark, ed. *Holmes-Laski Letters: The Correspondence of Mr. Justice Holmes and Harold J. Laski, 1916–1935*. 2 vols. Cambridge: Harvard University Press, 1953.

ed. *Holmes-Pollock Letters: The Correspondence of Mr. Justice Holmes and Sir Frederick Pollock, 1874–1932*. 2 vols. Cambridge: Harvard University Press, 1941.

Dickinson, Robert L. "Sterilization without Unsexing." *Journal of the American Medical Association* 92, no. 5 (1929): 373–9.

Dorr, Gregory M. *Segregation's Science: Eugenics and Society in Virginia*. Charlottesville: University of Virginia Press, 2008.

Dowbiggin, Ian R. *Keeping America Sane: Psychiatry and Eugenics in the United States and Canada, 1880–1940*. Ithaca, NY: Cornell University Press, 1997.

A Merciful End: The Euthanasia Movement in Modern America. New York: Oxford University Press, 2003.

" 'A Rational Coalition': Euthanasia, Eugenics, and Birth Control in America, 1940–1970." *Journal of Social Policy History* 14, no. 3 (2002): 223–60.

The Sterilization Movement and Global Fertility in the Twentieth Century. Oxford: Oxford University Press, 2008.

Ehrlich, Paul R. *The Population Bomb*. New York: Ballantine, 1968.

Evans, Richard J. *The Coming of the Third Reich*. London: Allen Lane, 2003.

The Third Reich in Power. London: Allen Lane, 2005.

The Third Reich at War. New York: Penguin, 2010.

Fangerau, Heiner. *Etablierung eines Rassenhygienischen Standardwerkes 1924–1941: der Baur-Fischer-Lenz im Spiegel der zeitgenössischen Rezensionsliteratur*. Frankfurt: Peter Lang, 2001.

Fogarty, Gerald P. *Commonwealth Catholicism: A History of the Catholic Church in Virginia*. Notre Dame: University of Notre Dame Press, 2001.

Franks, Angela. *Margaret Sanger's Eugenic Legacy: The Control of Female Fertility*. Jefferson, NC: McFarland & Co., 2005.

Freeden, Michael. "Eugenics and Progressive Thought: A Study in Ideological Affinity." *Historical Journal* 22, no. 3 (1979): 645–71.

Gallagher, Nancy L. *Breeding Better Vermonters: The Eugenics Project in the Green Mountain State*. Hanover, NH: University Press of New England, 1999.

Gamble, Clarence, and William P. Richardson. "The Sterilization of the Insane and Mentally Deficient in North Carolina." *North Carolina Medical Journal* 8, no. 1 (January 1947): 19–21.

Gerstle, Gary. *American Crucible: Race and Nation in the Twentieth Century*. Princeton, NJ: Princeton University Press, 2001.

"The Resilient Power of the States across the Long Nineteenth Century: An Inquiry into a Pattern of American Governance." In Jacobs and King, *The Unsustainable American State*, 61–87. New York: Oxford University Press, 2009.

Gessler, Bernhard. *Eugen Fischer (1874–1967): Leben und Werk des Freiburger Anatomen, Anthropologen und Rassenhygienikers bis 1927*. Frankfurt: Peter Lang, 2000.

Gillham, Nicholas Wright. *A Life of Sir Francis Galton: From African Exploration to the Birth of Eugenics*. New York: Oxford University Press, 2001.

Goddard, Henry H. *The Kallikak Family: A Study of the Heredity of Feeblemindedness*. New York: Macmillian, 1912.

Gordon, Linda. *Woman's Body, Woman's Right: A Social History of Birth Control in America*. New York: Grossman Publishers, 1976.

Gould, Stephen Jay. *The Mismeasure of Man*. New York: Norton, 1981.

Graham, Loren R. "Science and Values: The Eugenics Movement in Germany and Russia in the 1920s." *American Historical Review* 82, no. 5 (1977): 1133–64.

Greenhalgh, Susan. *Just One Child: Science and Policy in Deng's China*. Berkeley: University of California Press, 2008.

Grekul, Jana. "Sterilization in Alberta, 1928 to 1972: Gender Matters." *Canadian Review of Sociology* 45, no. 3 (2008): 247–66.

"A Well-Oiled Machine: Alberta's Eugenics Program, 1928–1972." *Alberta History* 59, no. 3 (2011): 16–23.

Harvey Krahn, and Dave Odynak. "Sterilizing the 'Feeble-minded': Eugenics in Alberta, Canada 1929–1972." *Journal of Historical Sociology* 17, no. 4 (2004): 358–84.

Grob, Gerald N. *The Inner World of American Psychiatry, 1890–1940: Selected Correspondence*. New Brunswick, NJ: Rutgers University Press, 1985.

The Mad Among Us: A History of the Care of America's Mentally Ill. Cambridge: Harvard University Press, 1994.

Mental Illness and American Society, 1875–1940. Princeton, NJ: Princeton University Press, 1983.

Grotjahn, Alfred. *Die Hygiene der menschlichen Fortpflanzung: Versuch einer praktischen Eugenik*. Berlin: Urban & Schwarzenberg, 1926.

Haffner, Sebastian. *Geschichte eines Deutschen: die Erinnerungen 1914–1933*. Stuttgart: Deutsche Verlags-Anstalt, 2000.

Hall, Peter. *Governing the Economy: The Politics of State Intervention in Britain and France*. New York: Oxford University Press, 1986.

Haller, Mark. *Eugenics: Hereditarian Attitudes in American Thought*. New Brunswick, NJ: Rutgers University Press, 1963.

Hansen, Bent Sigurd. "Something Rotten in the State of Denmark: Eugenics and the Ascent of the Welfare State." In Broberg and Roll-Hansen, *Eugenics and the Welfare State*, 9–76. East Lansing: Michigan State University Press, 1996.

Hansen, Randall, and Desmond King. "Eugenic Ideas, Political Interests, and Policy Variance: Immigration and Sterilization Policy in Britain and the US." *World Politics* 53, no. 2 (2001): 237–63.

Harkavy, Oscar, and Krishna Roy. "Emergence of the Indian National Family Planning Program." In *The Global Family Planning Revolution: Three Decades of Population Policies and Programs*, edited by Warren C. Robinson and John A. Ross, 301–24. Washington, DC: The World Bank, 2007.

Hendricks, Melissa. "Raymond Pearl's 'Mingled Mess.'" *Johns Hopkins Magazine* 58, no. 2 (April 2006): 50–6.

Herlitzius, Anette. *Frauenbefreiung und Rassenideologie. Rassenhygiene und Eugenik im politischen Programm der "Radikalen Frauenbewegung" (1900–1933)*. Wiesbaden: Deutscher Universitäts-Verlag, 1995.

Hern, Warren. "Biological Tyranny." *New Republic* (February 27, 1971): 15–17.

Hodges, Jeffrey Alan. "Dealing with Degeneracy: Michigan Eugenics in Context." Ph.D. diss., Michigan State University, 2001.

Hodgson, Dennis. "The Ideological Origins of the Population Association of America." *Population and Development Review* 17, no. 1 (1991): 1–34.

"Notestein, Frank W." In *The Encyclopedia of Population*, edited by Paul Demeny and Geoffrey McNicoll, 1: 696. New York: Macmillan Reference USA, 2003.

Hornblum, Allen M. *Acres of Skin: Human Experiments at Holmesburg Prison; A True Story of Abuse and Exploitation in the Name of Medical Science*. New York: Routledge, 1998.

Immergut, Ellen. *Health Politics: Interests and Institutions in Western Europe*. New York: Cambridge University Press, 1992.

Jacobs, Lawrence, and Desmond King, eds. *The Unsustainable American State*. New York: Oxford University Press, 2009.

Jacobson, Matthew Frye. *Barbarian Virtues: The United States Encounters Peoples at Home and Abroad, 1876–1917*. New York: Hill and Wang, 2000.

Johnson, Kimberley. *Governing the States: Congress and the New Federalism, 1877–1929*. Princeton, NJ: Princeton University Press, 2007.

Reforming Jim Crow: Southern Politics and State in the Age before Brown. New York: Oxford University Press, 2010.

Jones, James H. *Bad Blood: The Tuskegee Syphilis Experiment*. New York: The Free Press, 1981.

Jones, Kathleen. *Asylums and After: A Revised History of the Mental Health Services; From the Early 18th Century to the 1990s*. London: Athlone Press, 1993.

Judt, Tony. *Thinking the Twentieth Century*. With Timothy Snyder. New York: Penguin Press, 2012.

Kaelber, Lutz. "Eugenics: Compulsory Sterilization in 50 American States." University of Vermont. Available at http://www.uvm.edu/~lkaelber/eugenics/.

Kantor, William M. "Beginnings of Sterilization in America." *Journal of Heredity* 28, no. 11 (1937): 374–6.

Kerlin, Isaac Newton. *The Mind Unveiled; or, A Brief History of Twenty-Two Imbecile Children*. Philadelphia, PA: U. Hunt and Son, 1858.

Kevles, Daniel J. *In the Name of Eugenics: Genetics and the Uses of Human Heredity*. Cambridge: Harvard University Press, 1995.

Kincheloe, Marsha R., and Herbert G. Hunt, Jr. *Empty Beds: A History of Vermont State Hospital*. Barre, VT: Northlight Studio Press, 1989.

King, Desmond. *In the Name of Liberalism: Illiberal Social Policy in the United States and Britain*. Oxford: Oxford University Press, 1999.

Making Americans: Immigration, Race, and the Origin of the Diverse Democracy. Cambridge: Harvard University Press, 2000.

and Randall Hansen. "Experts at Work: State Autonomy, Social Learning and Eugenic Sterilisation in 1930s Britain." *British Journal of Political Science* 29, no. 1 (1999): 77–107.

and Rogers M. Smith. "Racial Orders in American Political Development." *American Political Science Review* 99, no. 1 (2005): 75–92.

Kline, Wendy. *Building a Better Race: Gender, Sexuality, and Eugenics from the Turn of the Century to the Baby Boom*. Berkeley: University of California Press, 2001.

"Eugenics in the United States." In Bashford and Levine, *Oxford Handbook of the History of Eugenics*, 511–22. Oxford: Oxford University Press, 2010.

Klinkner, Philip A., and Rogers M. Smith. *The Unsteady March: The Rise and Decline of Racial Equality in America*. Chicago: University of Chicago Press, 1999.

Kluchin, Rebecca M. *Fit to Be Tied: Sterilization and Reproductive Rights in America, 1950–1980*. Brunswick, NJ: Rutgers University Press, 2009.

Knodel, John E. *The Decline of Fertility in Germany, 1871–1939*. Princeton, NJ: Princeton University Press, 1974.

Kröner, Hans-Peter. *Von der Rassenhygiene zur Humangenetik: Das Kaiser-Wilhelm-Institut für Anthropologie, menschliche Erbe und Eugenik nach dem Kriege*. Stuttgart: Gustav Fischer Verlag, 1998.

Kühl, Stefan. *The Nazi Connection: Eugenics, American Racism and German National Socialism*. New York: Oxford University Press, 1994.

"The Relationship between Eugenics and the So-called 'Euthanasia Action' in Nazi Germany." In Szöllösi-Janze, *Science in the Third Reich*, 185–211.

Lael, Richard L., Barbara Brazos, and Margot Ford McMillen. *Evolution of a Missouri Asylum: Fulton State Hospital 1851–2006.* Columbia: University of Missouri Press, 2007.

Lagemann, Ellen Condliffe. *The Politics of Knowledge: The Carnegie Corporation, Philanthropy, and Public Policy.* Middletown, CT: Wesleyan University Press, 1989.

Lampard, Robert. *Alberta's Medical History: "Young and Lusty, and Full of Life."* Red Deer, AB: privately printed, 2008.

Lantzer, Jason S., and Alexandra M. Stern. "Building a Fit Society: Indiana's Eugenics Crusaders." *Traces of Indiana and Midwestern History* 19, no. 1 (2007): 4–11.

Largent, Mark. "'The Greatest Curse of the Race': Eugenic Sterilization in Oregon, 1909–1983." *Oregon Historical Quarterly* 103, no. 2 (2002): 188–209.

Breeding Contempt: The History of Coerced Sterilization in the United States. New Brunswick, NJ: Rutgers University Press, 2008.

Larson, Edward J. *Sex, Race, and Science: Eugenics in the Deep South.* Baltimore, MD: Johns Hopkins University Press, 1995.

and Leonard J. Nelson. "Involuntary Sexual Sterilization of Incompetent in Alabama: Past, Present and Future." *Alabama Law Review* 43, no. 2 (1992): 399–444.

Latham, Michael E. *The Right Kind of Revolution: Modernization, Development, and U.S. Foreign Policy from the Cold War to the Present.* Ithaca, NY: Cornell University Press, 2011.

Lawrence, Jane. "The Indian Health Service and the Sterilization of Native American Women." *American Indian Quarterly* 24, no. 3 (2000): 400–19.

Lenz, Fritz. *Die Rasse als Wertprinzip. Zur Erneuerung der Ethik.* Munich: J. F. Lehmanns Verlag, 1933.

Leon, Sharon M. "'A Human Being, and Not a Mere Social Factor': Catholic Strategies for Dealing with Sterilization Statutes in the 1920s." *Church History* 73, no. 2 (2004): 383–411.

Lieberman, Robert C. *Shifting the Color Line: Race and the American Welfare State.* Cambridge: Harvard University Press, 1998.

Lidbetter, E. J. *Heredity and the Social Problem Group.* London: E. Arnold, 1933.

Littlewood, Thomas B. *The Politics of Population Control.* Notre Dame, IN: University of Notre Dame Press, 1977.

Lombardo, Paul A., ed. *A Century of Eugenics in America: From the Indiana Experiment to the Human Genome Era.* Bloomington: Indiana University Press, 2011.

Three Generations, No Imbeciles: Eugenics, the Supreme Court, and Buck v. Bell. Baltimore: The Johns Hopkins University Press, 2008.

London, Gary D. "Family Planning Programs of the Office of Economic Opportunity: Scope, Operation, and Impact." *Demography* 5, no. 2 (1968): 924–30.

Mahoney, James, and Kathleen Thelen, eds. *Explaining Institutional Change: Ambiguity, Agency, and Power.* New York: Cambridge University Press, 2010.

Marshall, Dominique. "Children's Rights and Children's Action in International Relief and Domestic Welfare: The Work of Herbert Hoover between 1914 and 1950." *Journal of the History of Childhood and Youth* 1, no. 3 (2008): 351–88.

Martin, Matthew D., III. "The Dysfunctional Progeny of Eugenics: Autonomy Gone AWOL." *Cardozo Journal of International and Comparative Law* 15, no. 2 (2007): 371–421.

Mastin, Joseph T. *Mental Defectives in Virginia: A Special Report of the State Board of Charities and Corrections to the General Assembly 1916, on Weak Mindedness in the*

State of Virginia; together with a Plan for the Training, Segregation and Prevention of the Procreation of the Feebleminded. Buck v Bell Documents. Paper 2. Available at http://digitalarchive.gsu.edu/col_facpub/2.

Maynard Smith, John. "Eugenics and Utopia," *Daedalus* 94, no. 2 (1965): 487–505.

McLaren, Angus. "The Creation of a Haven for 'Human Thoroughbreds': The Sterilization of the Feeble-Minded and the Mentally Ill in British Columbia." *Canadian Historical Review* 67, no. 2 (1986): 127–50.

Our Own Master Race: Eugenics in Canada, 1885–1945. Toronto: McClelland & Stewart, 1990.

Mehta, Jal. "The Varied Roles of Ideas in Politics: From 'Whether' to 'How.'" In Béland and Cox, *Ideas and Politics in Social Science Research*, 23–46. New York: Oxford University Press, 2011.

Mennel, Robert M., and Christine L. Compston, eds. *Holmes and Frankfurter: Their Correspondence, 1912–1934.* Hanover, NH: University Press of New England, 1996.

Moreno, Jonathan D. *Undue Risks: Secret State Experiments on Humans.* New York: Routledge, 2001.

Morison, Elting E., John Morton Blum, and John J. Buckley, eds. *The Letters of Theodore Roosevelt.* 8 vols. Cambridge: Harvard University Press, 1951–54.

Morning, Ann. "Reconstructing Race in Science and Society: Biology Textbooks, 1952–2002." In "Exploring Genetics and Social Structure," edited by Peter Bearman: 106–37. Special issue, *American Journal of Sociology* 114, no. S1 (2008).

Mottier, Véronique. "Eugenics and the State: Policy-Making in Comparative Perspective." In Bashford and Levine, *Oxford Handbook of the History of Eugenics*, 134–53. Oxford: Oxford University Press, 2010.

Müller-Hill, Benno. *Murderous Science: Elimination by Scientific Selection of Jews, Gypsies, and others, Germany, 1933–1945.* Translated by George R. Fraser. Oxford: Oxford University Press, 1998. Originally published as *Tödliche Wissenschaft: die Aussonderung von Juden, Zigeunern und Geisteskranken 1933–1945* (Reinbek bei Hamburg: Rowohlt Taschenbuch Verlag, 1984).

Ngai, Mae M. *Impossible Subjects: Illegal Aliens and the Making of Modern America.* Princeton, NJ: Princeton University Press, 2004.

Noll, Steven. *Feeble-Minded in Our Midst: Institutions for the Mentally Retarded in the South, 1900–1940.* Chapel Hill: University of North Carolina Press, 1995.

"The Public Face of Southern Institutions for the 'Feeble-Minded.'" *The Public Historian* 27, no. 2 (2005): 25–41.

"Southern Strategies for Handling the Black Feeble-Minded: From Social Control to Profound Indifference." *Journal of Policy History* 3, no. 2 (1991): 130–51.

Notestein, Frank W. "Class Differences in Fertility." *Annals of the American Academy of Political and Social Sciences* 188, no. 1 (1936): 26–36.

Nowak, Kurt. *Geschichte des Christentums in Deutschland.* Munich: C.H. Beck, 1995.

Nyiszli, Miklos. *Auschwitz: A Doctor's Eyewitness Account.* New York: Fell, 1960.

Ochsner, A. J. "Surgical Treatment of Habitual Criminals." *Journal of the American Medical Association* 32, no. 16 (April 22, 1899): 867–8.

Office of Economic Opportunity, Executive Office of the President. "OEO Instruction 6130–1, May 18, 1971." [Copy provided to authors by Dr. Warren Hern.]

Office of Policy Planning and Research. *The Negro Family: The Case for National Action.* U.S. Department of Labor, March 1965. Available at http://www.dol.gov/oasam/programs/history/webid-meynihan.htm.

Ordover, Nancy. *American Eugenics: Race, Queer Anatomy, and the Science of Nationalism*. Minneapolis: University of Minnesota Press, 2003.

Osborn, Frederick. *The Future of Human Heredity; An Introduction to Eugenics in Modern Society*. New York: Weybright and Talley, 1968.

Overy, Richard. *Interrogations: The Nazi Elite in Allied Hands, 1945*. New York: Viking, 2001.

Paul, Diane B. "Eugenics and the Left," in *Politics of Heredity: Essays on Eugenics, Biometrics, and the Nature-Nature Debate*. Albany: State University of New York Press, 1998.

Paul, Julius. "The Return of Punitive Sterilization Proposals: Current Attacks on Illegitimacy and the AFDC Program." *Law & Society Review* 3, no. 1 (1968): 77–106.

"State Eugenic Sterilization History: A Brief Overview." In *Eugenic Sterilization*, edited by Jonas Robitscher, 25–40. Springfield, IL: Charles C. Thomas, 1973.

"Three Generations of Imbeciles Are Enough": State Eugenic Sterilization Laws in American Thought, and Practice. Washington, DC: Walter Reed Army Institute of Research, 1965.

Pearl, Raymond. *The Biology of Population Growth*, Rev. ed. New York: Knopf, 1930.

"The Biology of Superiority." *The American Mercury* 12, no. 47 (1927): 257–66.

Peart, Hartley F. "Vasectomy and Salpingectomy under California Law," *California and Western Medicine* 6 (May–June 1941): 1–8, Box 73, AVS.

Pernick, Martin S. *The Black Stork: Eugenics and the Death of "Defective" Babies in American Medicine, and Motion Pictures Since 1915*. Oxford: Oxford University Press, 1996.

Perske, Robert. *Hope for the Families: New Directions for Parents of Persons with Retardation of Other Disabilities*. Nashville, TN: Abingdon Press, 1981.

Pierson, Paul. "Increasing Returns, Path Dependence, and the Study of Politics." *American Political Science Review* 94, no. 2 (2000): 251–67.

Ploetz, Alfred. *Die Tüchtigkeit unserer Rasse und der Schutz der Schwachen*. Berlin: S. Fischer, 1895.

Popenoe, Paul. "The Progress of Eugenic Sterilization." *Journal of Heredity* 25, no. 1 (1934): 19–26.

Porter, Roy. *Madness: A Brief History*. Oxford: Oxford University Press, 2002.

Proctor, Robert. "From *Anthropologie* to *Rassenkunde* in the German Anthropological Tradition." In *Bones, Bodies, and Behavior: Essays on Biological Anthropology*, edited by George W. Stocking, Jr., 138–79. Madison: University of Wisconsin Press, 1988.

Racial Hygiene: Medicine under the Nazis. Cambridge: Harvard University Press, 1988.

Rafter, Nicole Hahn. *White Trash: The Eugenic Family Studies, 1877–1919*. Boston: Northeastern University Press, 1988.

Reilly, Philip R. "Involuntary Sterilization in the United States: A Surgical Solution." *Quarterly Review of Biology* 62, no. 2 (1987): 153–70.

The Surgical Solution: A History of Involuntary Sterilization in the United States. Baltimore, MD: Johns Hopkins University Press, 1991.

"The Surgical Solution: The Writings of Activist Physicians in the Early Days of Eugenical Sterilization." *Perspectives in Biology and Medicine* 26, no. 4 (1983): 637–56.

Rein, Martin, and Donald A. Schön. "Problem-Setting in Policy Research." In *Using Social Research in Public Policy*, edited by Carol H. Weiss, 235–51. Lexington, MA: D. C. Heath and Co., 1977.

ReVelle, Penelope, and Charles ReVelle. *The Global Environment: Securing a Sustainable Future*. Boston: Jones and Barlett, 1992.

Richardson, William P., and Clarence J. Gamble. "The Sterilization of the Insane and Mentally Deficient in North Carolina." *North Carolina Medical Journal* 8, no. 1 (1947): 19–21.

Rissom, Renate. *Fritz Lenz und die Rassenhygiene*. Husum: Matthiesen, 1983.

Rodgers, Daniel T. *Atlantic Crossings: Social Politics in a Progressive Age*. Cambridge, MA: Belknap Press, 1998.

Rosen, Christine. *Preaching Eugenics: Religious Leaders and the American Eugenics Movement*. New York: Oxford University Press, 2004.

Rosenberg, Charles E. "Charles Benedict Davenport and the Beginning of Human Genetics." *Bulletin of the History of Medicine* 35 (1961): 266–76.

Rosenfeld, Bernard, Sidney M. Wolfe, and Robert E. McGarrah, Jr. *A Health Research Group Study on Surgical Sterilization: Present Abuses and Proposed Regulations*. Washington, DC: Health Research Group, 1973.

Ryan, Patrick J. "'Six Blacks from Home': Childhood, Motherhood, and Eugenics in America." *Journal of Policy History* 19, no. 3 (2007): 253–81.

"Unnatural Selection: Intelligence Testing, Eugenics and American Political Cultures." *Journal of Social History* 30, no. 3 (1997): 669–85.

Rydell, Robert W. *World of Fairs: The Century-of-Progress Expositions*. Chicago: University of Chicago Press, 1993.

Sabbatini, Renato M. E. "The History of Shock Therapy in Psychiatry." Available at http://www.cerebromente.org.br/n04/historia/shock_i.htm.

Sanger, Margaret. "A Plan for Peace." *Birth Control Review* 16, no. 4 (April 1932): 107–8.

Sanger, Margaret. *An Autobiography*. New York: Dover Publications, 1971.

Schaffer, Ronald. *Wings of Judgment: American Bombing in World War II*. New York: Oxford University Press, 1985.

Schmidt, Vivien A. "Does Discourse Matter in the Politics of Welfare State Adjustment?" *Comparative Political Studies* 35, no. 2 (2002): 168–93.

Schneider, William H. *Quality and Quantity: The Biological Request for Regeneration in Twentieth-Century France*. Cambridge: Cambridge University Press, 1990.

Schön, Donald A., and Martin Rein. *Frame Reflection: Toward the Resolution of Intractable Policy Controversies*. New York: Basic Books, 1994.

Schoen, Johanna. *Choice and Coercion: Birth Control, Sterilization, and Abortion in Public Health and Welfare*. Chapel Hill: University of North Carolina Press, 2005.

Searle, G. R. *Eugenics and Politics in Britain, 1900–1914*. Leyden: Noordhof International Publishers, 1976.

Sen, Amartya Kumar. *Poverty and Famines: An Essay on Entitlement and Deprivation*. Oxford: Oxford University Press, 1981.

Shapiro, Joseph P. *No Pity: People with Disabilities Forging a New Civil Rights Movement*. New York: Times Books, 1993.

Shepsle, Kenneth. "Rational Choice Institutionalism." In Binder, Rhodes, and Rockman, *Oxford Handbook of Political Institutions*, 23–38. Oxford: Oxford University Press, 2006.

Siemens, Hermann Werner. *Die biologischen Grundlagen der Rassenhygiene und der Bevölkerungspolitik. Für Gebildete aller Berufe.* Munich: J. F. Lehmann, 1917.
Grundzüge der Vererbungslehre, Rassenhygiene, und Bevölkerungspolitik. 8th ed. Munich: J. F. Lehmann, 1937.
Sieveking, G. Herman. *"Sterilisierungsgesetze des Auslandes."* Deutsches Ärzteblatt 35, no. 64 (August 25, 1934): 829–31.
Simmons, Harvey G. "Explaining Social Policy: The English Mental Deficiency Act of 1913." *Journal of Social History* 11, no. 3 (1978): 387–403.
Snyder, Timothy. *Bloodlands: Europe between Hitler and Stalin.* New York: Basic Books, 2010.
Soloway, Richard A. *Demography and Degeneration: Eugenics and the Declining Birthrate in Twentieth Century Britain.* Chapel Hill: University of North Carolina Press, 1995.
Spektorowski, Alberto. "The Eugenic Temptation in Socialism: Sweden, Germany and the Soviet Union." *Comparative Studies in Society and History* 46, no. 1 (2004): 84–106.
Steinmo, Sven, Kathleen Thelen, and Frank Longstreth, eds. *Structuring Politics: Historical Institutionalism in Comparative Analysis.* New York: Cambridge University Press, 1992.
Stepan, Alfred, and Juan J. Linz. "Comparative Perspectives on Inequality and the Quality of Democracy in the United States." *Perspectives on Politics* 9, no. 4 (2011): 841–56.
Stepan, Nancy Leys. *The Hour of Eugenics.* Ithaca, NY: Cornell University Press, 1991.
Stephens, Martha. *The Treatment: The Story of Those Who Died in the Cincinnati Radiation Tests.* Durham: Duke University Press, 2002.
Stern, Alexandra M. *Eugenic Nation: Faults and Frontiers of Better Breeding in Modern America.* Berkeley: University of California Press, 2005.
"From Legislation to Lived Experience: Eugenic Sterilization in California and Indiana, 1907–79." In Lombardo, *A Century of Eugenics,* 95–116. Bloomington: Indiana University Press, 2011.
"From Mestizophilia to Biotypology: Racialization and Science in Mexico, 1920–1960." In Applebaum, MacPherson, and Rosemblatt, *Race and Nation in Modern Latin America,* 187–210. Chapel Hill: University of North Carolina Press, 2003.
"Gender and Sexuality: A Global Tour and Compass." In Bashford and Levine, *Oxford Handbook of the History of Eugenics,* 173–91. Oxford: Oxford University Press, 2010.
"'The Hour of Eugenics' in Veracruz, Mexico: Radical Politics, Public Health, and Latin America's only Sterilization Law." *Hispanic American Historical Review* 91, no. 3 (2011): 431–43.
"Sterilized in the Name of Public Health: Race, Immigration, and Reproductive Control in Modern California." *American Journal of Public Health* 95, no. 7 (2005): 1128–38.
"'We Cannot Make a Silk Purse Out of a Sow's Ear': Eugenics in the Hoosier Heartland." *Indiana Magazine of History* 103, no. 1 (2007): 3–38.
Stoddard, Lothrop. *The Rising Tide of Color Against White-Supremacy.* New York: Scribner, 1920.
Stopes, Marie C. *Married Love.* Edited by Ross McKibbin. Oxford: Oxford University Press, 2004.

Radiant Motherhood: A Book for Those who Are Creating the Future. London: G. P. Putnam's Sons, 1920.

Strange, Carolyn, and Jennifer A. Stephen. "Eugenics in Canada: A Checkered History, 1850s–1990s." In Bashford and Levine, *Oxford Handbook of the History of Eugenics*, 523–38. Oxford: Oxford University Press, 2010.

Swift, Katherine. "Sinister Science: Eugenics, Nazism, and the Technocratic Rhetoric of the Human Betterment Foundation." *Lore* 6, no. 2 (May 2008): 1–11.

Szöllösi-Janze, Margit, ed. *Science in the Third Reich*. Oxford: Berg Publishers, 2001.

Thelen, Kathleen. "Historical Institutionalism in Comparative Politics." *Annual Review of Political Science* 2, no. 1 (1999): 369–404.

How Institutions Evolve: The Political Economy of Skills in Germany, Britain, the United States, and Japan. Cambridge: Cambridge University Press, 2004.

Thompson, Warren S. "Eugenics and the Social Good." *The Journal of Social Forces* 3, no. 3 (March 1925): 414–9.

Thomson, Mathew. *The Problem of Mental Deficiency: Eugenics, Democracy, and Social Policy in Britain, c. 1870–1959*. Oxford: Oxford University Press, 1998.

Timm, Annette F. *The Politics of Fertility in Twentieth-Century Berlin*. Cambridge: Cambridge University Press, 2010.

Tomes, Nancy. *A Generous Confidence: Thomas Story Kirkbride and the Art of Asylum-Keeping, 1840–1883*. Cambridge: Cambridge University Press, 1984.

Tone, Andrea. *Devices and Desires: A History of Contraception in America*. New York: Hill and Wang, 2001, 271.

Trent, James W., Jr. *Inventing the Feeble Mind: A History of Mental Retardation in the United States*. Berkeley: University of California Press, 1994.

"'Who shall say who is a useful person?': Abraham Myerson's Opposition to the Eugenics Movement." *History of Psychiatry* 12, no. 45 (2001): 33–57.

Trombley, Stephen. *The Right to Reproduce: A History of Coercive Sterilization*. London: Weidenfeld and Nicolson, 1988.

Trudeau, Chad. F., ed. *Public School Laws of Louisiana and Sanitary Regulations of the State Board of Health, 11th compilation*. Baton Rouge, LA: Ramires-Jones Printing Company, 1919.

Tsebelis, George. *Veto Players: How Political Institutions Work*. Princeton, NJ: Princeton University Press, 2002.

Turda, Marius. "Race, Science, and Eugenics in the Twentieth Century." In Bashford and Levine, *Oxford Handbook of the History of Eugenics*, 62–79. Oxford: Oxford University Press, 2010.

Tydén, Mattias "The Scandinavian States: Reformed Eugenics Applied," in *The Oxford Handbook of the History of Eugenics*, ed. Alison Bashford and Philippa Levine. Oxford: Oxford University Press, 2010.

Valles, Sean A. "Lionel Penrose and the Concept of Normal Variation in Human Intelligence." *Studies in History and Philosophy of Biological and Biomedical Sciences* 43, no. 1 (2012): 281–9.

Valone, David A. "*Eugenic Science in California: Guide to E. S. Gosney Papers and Records of the Human Betterment Foundation*." Pasadena: California Institute of Technology, 1996.

Verschuer, Otmar von. "Eugen Fischer zum 70. Geburtstag am 5. Juni 1944." *Der Erbarzt* 12, nos. 5–6 (May/June 1944): 57–9.

Wahlsten, Douglas. "Leilani Muir versus the Philosopher King: Eugenics on Trial in Alberta." *Genetica* 99, nos. 2–3 (1997): 185–98.

Waylen, Georgina, Karen Celis, Johanna Kantola, and Laurel Weldon, eds. *The Oxford Handbook of Gender and Politics*. Oxford: Oxford University Press, 2013.

Weinberg, Gerhard L. *A World at Arms: A Global History of World War II*. Cambridge: Cambridge University Press, 1994.

Weindling, Paul. *Health, Race and German Politics between National Unification and Nazism*. Cambridge: Cambridge University Press, 1989.

Nazi Medicine and the Nuremburg Trials: From Medical War Crimes to Informed Consent. New York: Palgrave Macmillan, 2004.

"The Survival of Eugenics in 20th Century Germany." *American Journal of Human Genetics* 52, no. 3 (1993): 643–9.

"Weimar Eugenics: The Kaiser Wilhelm Institute for Anthropology, Human Heredity and Eugenics in Social Context." *Annals of Science* 42, no. 3 (1985): 303–18.

Weingart, Peter. "German Eugenics between Science and Politics." *Osiris* 2nd series, vol. 5 (1989): 260–82.

Weiss, Sheila F. "Race and Class in Fritz Lenz's Eugenics." *Medizinhistorisches Journal: Internationale Vierteljahresschrift für Wissenschaftsgeschichte* 27, nos. 1–2 (1992): 5–25.

Race Hygiene and National Efficiency: The Eugenics of Wilhelm Schallmayer. Berkeley: University of California Press, 1987.

"The Race Hygiene Movement in Germany, 1904–1945." In *The Wellborn Science: Eugenics in Germany, France, Brazil, and Russia*, edited by Mark B. Adams, 8–68. New York, Oxford University Press, 1990.

Whitney, E. A. "Presenting Mental Deficiency to Students." *American Journal of Mental Deficiency* 50, no. 1 (July 1945): 54–8.

Wood, H. C. "A Prescription for the Alleviation of Welfare Abuses and Illegitimacy." *Journal of the Kentucky State Medical Association* 61, no. 4 (1963): 319–23.

Zenderland, Leila. *Measuring Minds: Henry Herbert Goddard and the Origins of American Intelligence Testing*. Cambridge: Cambridge University Press, 1998.

Ziegler, Mary. "The Framing of a Right to Choose: *Roe v. Wade* and the Changing Debate on Abortion Law." *Law and History Review* 27, no. 2 (summer 2009): 281–330.

"Reinventing Eugenics: Reproductive Choice and Law Reform after World War II." *Cardozo Journal of Law & Gender* 14, no. 2 (April 2008): 319–50.

Index